REDUCING FAT IN MEAT ANIMALS

REDUCING FAT IN MEAT ANIMALS

Edited by

J. D. WOOD and A. V. FISHER
University of Bristol, Langford, Bristol, UK

ELSEVIER APPLIED SCIENCE
LONDON and NEW YORK

ELSEVIER SCIENCE PUBLISHERS LTD
Crown House, Linton Road, Barking, Essex IG11 8JU, England

Sole Distributor in the USA and Canada
ELSEVIER SCIENCE PUBLISHING CO INC
655 Avenue of the Americas, New York, NY 10010, USA

WITH 110 TABLES AND 28 ILLUSTRATIONS

British Library Cataloguing in Publication Data

Reducing fat in meat animals.
1. Livestock. Meat animals. Fats reduction
I. Wood, J. D. II. Fisher, A. V.
636.0883

ISBN 1-85166-455-6

Library of Congress Cataloging-in-Publication Data

Reducing fat in meat animals/edited by J. D. Wood and A. V. Fisher.
p. cm.
Includes bibliographical references.
ISBN 1-85166-455-6
1. Meat. 2. Animal food—Fat content. I. Wood, J. D.
II. Fisher, A. V.
TX556.M4R43 1990
664'.907—dc20

90-2734
CIP

Photoset by Enset (Photosetting), Midsomer Norton, Bath, Avon
Printed in Great Britain by Page Bros (Norwich) Ltd

Preface

It has been clear for several years that consumers in many countries are expressing a wish to eat less fat in meat, mainly for reasons of the perceived benefits to health. This has motivated animal scientists to search for ways of producing carcasses with lower levels of fat than formerly. These extend from simple changes in production factors (breed, sex, feeding system, etc.) to more sophisticated techniques of metabolic control.

However, progress towards meaningful reductions in fat content has often been slow and has differed between the meat producing species. Most progress has been achieved for pigs, but for beef in particular there are strong views held by some in the meat industry and the catering sector that fat is an important contributor to eating satisfaction, and that to reduce fat is to reduce the quality of meat. The subjective nature of fatness assessment in live animals and carcasses has also been a barrier to progress. Marketing schemes based on objective measures related to fatness/leanness in the carcass are beginning to emerge, but there has often been a lack of incentive to change.

The subject of fat deposition in meat animals has to be considered from several points of view if successful strategies for change are to be developed and if the full implications of a move to leaner carcasses are to be understood. In this book we have attempted to bring together recent evidence from all the relevant areas: the medical significance of reducing fat in the diet, consumer attitudes, methods of reducing fat in growing animals, techniques for measuring fat in animals and carcasses, the role of fat in meat quality and, finally, marketing schemes to change carcass composition.

The contributors, who are all experts in their particular fields, are drawn from six countries, and together they present a comprehensive review of the subject of fat deposition. This book will, we hope, be a valuable reference source for animal and food scientists, advanced students in agriculture and animal production, food manufacturers and those involved in planning strategies within the meat and food industries.

J D WOOD
A V FISHER
Langford, Bristol

Contents

Chapter 2 *Dietary Fat and Human Health*
G. ROSE

Chapter 3 *Consumer Attitudes to Fat in Meat*
JUDITH WOODWARD & VERNER WHEELOCK

Chapter 10 *Marketing Procedures to Change Carcass Composition*
A. J. KEMPSTER

List of Contributors

PAUL ALLEN
National Food Centre, Dunsinea, Castleknock, Dublin 15, Ireland

J. J. BASS
MAFTech, Ruakura Agricultural Centre, Private Bag, Hamilton, New Zealand

B. W. BUTLER-HOGG
MAFTech, Ruakura Agricultural Centre, Private Bag, Hamilton, New Zealand

J. P. DUMONT
Laboratoire d'Etude des Interactions des Molécules Alimentaires, INRA, BP 527, 44026 Nantes Cedex 03, France

A. V. FISHER
Department of Meat Animal Science, University of Bristol, Langford, Bristol, BS18 7DY, UK

R. GOUTEFONGEA
Laboratoire d'Etude des Interactions des Molécules Alimentaires, INRA, BP 527, 44026 Nantes Cedex 03, France

A. J. KEMPSTER
Meat and Livestock Commission, PO Box 44, Winterhill House, Snowdon Drive, Milton Keynes MK6 1AX, UK

A. H. KIRTON
MAFTech, Ruakura Agricultural Centre, Private Bag, Hamilton, New Zealand

H. J. MERSMANN
US Department of Agriculture, Roman L. Hruska, US Meat Animal Research Center, ARS, Clay Center, Nebraska 68933, USA
Present address: USDA/ARS Children's Nutrition Research Center, Baylor College of Medicine, Houston, Texas 77030, USA

KERIN O'DEA
Department of Human Nutrition, Deakin University, Victoria 3217, Australia

G. ROSE
London School of Hygiene and Tropical Medicine, University of London, Keppel Street, London WC1, UK

ANDREW J. SINCLAIR
Department of Applied Biology, Royal Melbourne Institute of Technology, GPO Box 2476V, Melbourne, Victoria 3001, Australia

VERNER WHEELOCK
Food Policy Research Unit, School of Biomedical Sciences, University of Bradford, Bradford, Yorkshire BD7 1DP, UK

JUDITH WOODWARD
Food Policy Research Unit, School of Biomedical Sciences, University of Bradford, Bradford, Yorkshire BD7 1DP, UK

J. D. WOOD
Department of Meat Animal Science, University of Bristol, Langford, Bristol, BS18 7DY, UK
Formerly Carcass and Abattoir Department of the AFRC Institute of Food Research—Bristol Laboratory, Langford, Bristol, UK.

Chapter 1

Fats in Human Diets Through History: Is the Western Diet out of Step?

ANDREW J. SINCLAIR
Department of Applied Biology,
Royal Melbourne Institute of Technology,
Victoria, Australia
&
KERIN O'DEA
Department of Human Nutrition,
Deakin University,
Victoria, Australia

INTRODUCTION

Chronic degenerative diseases are the major cause of premature mortality and morbidity in affluent Western societies. These are: occlusive vascular disease (coronary heart disease, stroke, atherosclerosis and thrombosis), hypertension, obesity, type 2 diabetes, gallstones, bowel diseases and certain cancers. In general, these conditions become obvious in the post-reproductive years and are therefore unlikely to be subject to the pressures of natural selection. They are also a relatively recent phenomenon, having become highly prevalent only in the 20th century. Such diseases are still rare or unknown in the less developed 'Third world' and in traditionally-living populations.

Among the lifestyle factors responsible for these recent increases in the prevalence of chronic degenerative diseases, diet is arguably the most important. Major changes in dietary composition have occurred in Western societies during the 20th century. Despite total energy and protein remaining fairly constant since early this century, the consumption of fibre-rich carbohydrate has fallen by half and that of refined carbohydrate and fat has risen proportionately (Gortner, 1975). The type of fat responsible for this increase was predominantly of animal origin (Rizek *et al.*, 1983; English, 1987). Thus refined carbohydrate and saturated fat have replaced the fibre-rich starchy staples of our forebears. Such foods are energy-dense: for example, a 100 g chocolate bar contains the same amount of energy as 700 g of steamed potatoes or 2 kg of mixed vegetables. It can also be argued that energy-dense foods are easier to over-consume due to their lack of bulk. Furthermore, we are now less physically active than were our grandparents and great-grandparents, and what used to be a weight-maintaining diet earlier this century is now providing excess energy. The increasing prevalence of obesity in Western societies is the consequence of this relative over-consumption. Many of the other chronic degenerative diseases of Western lifestyles can be linked directly to over-consumption of particular nutrients, the most well-documented of which is the association of excessive saturated animal fat consumption with a range of diseases: coronary heart disease, atherosclerosis, thrombosis, hypertension, bowel diseases, and cancers of the colon, endometrium and breast (Doll & Peto, 1981; Bristol *et al.*, 1985; Renaud, 1985; Keys *et al.*, 1986; Pietinen & Huttunen, 1987).

FAT IN THE WESTERN DIET

The Western diet is characterised by a high level of fat (~ 40% energy), about 40–50% of which is saturated (Table 1). The saturated fats are largely derived either directly from animal fats (visible meat fat, dairy products), indirectly from manufactured foods containing these fats, and from industrially hardened (hydrogenated) vegetable oils.

Foods rich in saturated fat together with cholesterol-rich foods such as eggs, organ meats and certain shellfish have been targets for criticism by health professionals. This has been based on epidemiological evidence, animal experimentation and observations in humans with hereditary hyperlipoproteinaemia providing strong evidence that excessive levels of dietary saturated fat are associated with increased plasma cholesterol levels. Hypercholesterolaemia in turn, is associated with an increased risk of coronary heart disease (Keys *et al.*, 1986).

Table 1
Fatty acid intake in the Western diet[a]

Diet	*Saturated*	*Monounsaturated*	*Polyunsaturated*
United Kingdom[b]	45	36	13
Australia[c]	42	40	17
Australia[d]	31	30	12

[a]g fatty acid type/person/day.
[b]National average household food supply (Bull *et al.*, 1983).
[c]1983 National Dietary Survey (English, 1987).
[d]Weighed food intake over a 2-week period for 13 adults (Sinclair *et al.*, 1987).

The main saturated fatty acids in the diet are palmitic, stearic and myristic acids (Bull *et al.*, 1983). Although saturated fat adversely affects both atherosclerosis and thrombosis, it is now clear that individual fatty acids can affect the conditions quite differently. For example, the most hypercholesterolaemic fatty acids are palmitic (16:0) myristic (14:0) and lauric (12:0) acids, with stearic acid (18:0) considered to be either neutral or only slightly hypercholesterolaemic (Hegsted *et al.*, 1965). In contrast, stearic acid is the most thrombogenic dietary fatty acid, with myristic acid being less so and palmitic acid being neutral (Renaud, 1985). Thus, it cannot be assumed that all saturated fat will be similarly hypercholesterolaemic and thrombogenic.

Vegetable fats such as palm oil and cocoa butter are also rich in saturated fatty acids (Table 2), however, their hypercholesterolaemic effects are less than the saturated fatty acids found in milk fat. Baudet *et al.* (1984) fed Benedictine nuns diets which differed only in the type of fat: 20% energy derived from sunflower oil, peanut oil, palm oil or milk fat over a period of five months. The only diet which raised cholesterol levels was that enriched with milk fat.

Although dietary saturated fatty acids and cholesterol have independent effects on plasma cholesterol levels, there is an interaction between their effects: the plasma cholesterol response to dietary cholesterol becoming attenuated on diets with increased proportions of polyunsaturated fatty acids relative to saturated fatty acids (increasing P:S ratio) or on diets with a low fat content (Schonfeld *et al.*, 1982).

In the Australian diet over the last 45 years there has been a rise in the proportion of vegetable oils and fats (including margarines and processed

Table 2
Fatty acid content[a] of edible fats and oils

Fat/oil	*Saturated*	*Mono-unsaturated*	*Poly-unsaturated*	*Ratio of 18:2n-6 to 18:3n-3*
Cow's milk fat	62	29	4	1·6
Beef tallow (dripping)	48	42	4	6·5
Sheep fat	43	39	8	1·8
Pig fat (lard)	40	44	11	6·9
Chicken fat	33	45	18	15
Cocoa butter	66	30	1	—
Palm oil	48	38	9	27
Soyabean oil	15	23	58	7·5
Olive oil	14	73	9	11
Corn oil	13	25	58	72
Safflower oil	9	13	74	147

[a]g fatty acid type/100 g fat or oil (Posatti *et al.*, 1975; Brignoli *et al.*, 1976).

foods) relative to animal fats (English, 1987). This has meant a replacement of some of the saturated fat by polyunsaturated fat. The main polyunsaturated fatty acid (PUFA) in vegetable oils and in the Western diet is linoleic acid (Table 2). Increasing the proportion of dietary PUFA and decreasing saturated fat is a well-accepted effective way of reducing plasma cholesterol levels (Horrobin & Huang, 1987). The dietary goals of most Western countries advocate reducing the consumption of total fat and saturated fat and in some cases increasing the consumption of PUFA at the expense of saturated fatty acids (English, 1987).

The Western diet is also characterised by relatively high levels of *trans* unsaturated fatty acids. These fatty acids are derived from two main sources: industrial hardening (hydrogenation) of naturally occurring *cis* polyunsaturated fatty acids found in vegetable oils, non-ruminant fats and marine oils, and by the biohydrogenation by rumen micro-organisms of dietary PUFA found in plant material ingested by ruminants. The major type of *trans* fatty acid produced by these processes is *trans* octadecenoic acid (*t*-18:1). The position of the double bond can vary from C-5 through to C-15 (Sommerfeld, 1983). The concentration of *trans* fatty acids in various foods, processed vegetable oils and margarines varies with the extent and type of processing. Margarines generally contain between 10–40% *trans* fatty acids (Adlof & Emken, 1986). The *trans*-

octadecenoic acid content of the diet in West Germany has been estimated to be between 4·5–6·4 g/person/day (Beare-Rogers, 1983) and in the US between 4·5–12·1 g/person/day (Senti, 1985).

Trans fatty acids have a higher melting point than the corresponding *cis* form which allows the product to acquire a harder consistency. In this respect their structure resembles more closely that of a saturated fatty acid as do some aspects of their metabolism in the body (Beare-Rogers, 1983). There has therefore been some concern about their presence in the diet with particular reference to their effect on plasma cholesterol levels. There have been divergent results and a lack of consensus as to their effects on plasma cholesterol levels. These inconsistencies may be explained by the different preparations of *trans* fatty acids used between experiments. In general however, their effect, if any, is to raise the cholesterol level (Beare-Rogers, 1983).

The monounsaturated fatty acids are the other major group of fatty acids in the Western diet (Bull *et al.*, 1983). Octadecenoic acid (18:1) accounts for about 83% of the total monounsaturates with oleic acid (*cis* 9–18:1) being the predominant component (Bull *et al.*, 1983). Until recently this group of fatty acids was considered to be 'neutral' in terms of cholesterol-lowering properties (Mattson & Grundy, 1985). However recent epidemiological findings have forced a re-evaluation of this view. The most powerful evidence has come from the 15-year follow-up of the Seven Countries Study (Keys *et al.*, 1986) which demonstrated a particularly low frequency of coronary heart disease in the men from Crete despite their relatively high fat intake (36% energy). Olive oil was the major component of the dietary fat and oleic acid the predominant fatty acid. Analysis of the dietary composition and coronary heart disease death rates in the 15 participating populations from the Seven Countries indicated that the ratio of monounsaturated to saturated fat in the diet was strongly and inversely predictive of the death rate due to coronary heart disease. Thus, the higher the proportion of olive oil and the lower the proportion of saturated fat, the lower the death rate. It was also significant that total death rate from all causes was also inversely related to this same ratio (Keys *et al.*, 1986). Metabolic studies by Grundy and coworkers (Mattson & Grundy, 1985; Grundy, 1986) have shed light on the mechanism by which oleic acid may confer this protection. They have shown that oleic acid is just as effective as linoleic acid in lowering plasma low density lipoprotein (LDL) cholesterol, but has the advantage of not simultaneously lowering high density lipoprotein (HDL) cholesterol. Linoleic acid lowers both LDL and HDL cholesterol by decreasing their production. Oleic acid selectively lowers LDL cholesterol by increasing its clearance (Grundy, 1987) but leaves the 'protective' HDL unchanged.

THE NEED FOR FAT IN THE DIET

Although the Western diet is rich in fat (40% energy) there are communities in which the fat intake is considerably lower and in which there is a lower prevalence of coronary heart disease (Keys *et al.*, 1986). The public in Western societies have been advised to reduce their fat consumption to 25–30% of energy and ensure that the saturated fat component is no more than 10% of energy. One could ask why not reduce it even further or eliminate fat altogether. This latter option would be impossible to accomplish given that almost all foods contain fat or structural lipid and that there is a definite need for fat in the human diet, since dietary fat is the source of the essential fatty acids and fat soluble vitamins.

Burr and Burr (1929) first demonstrated that fat was an essential component of the diet of rats. Their work led to the discovery of the essential fatty acids (EFA) linolenic (18:2n-6) and linoleic acids (18:3n-3). Since that time the effect of fat deficiency has been studied in many species and EFA deficiency symptoms have been observed in birds, fish, insects and many mammalian species including man (Holman, 1968; Yamanaka *et al.*, 1981). Common deficiency symptoms in the rat include reduced growth, poor food conversion, increased metabolic rate, scaly skin, hair loss, dandruff, infertility in both sexes, increased water loss through the skin and fatty infiltration of the liver (Aaes-Jorgensen, 1961; Holman, 1968, 1970). Skin changes appear to be a characteristic of EFA deficiency in most species.

Biochemical studies revealed that the EFA were metabolised in the body to longer chain, more unsaturated fatty acids. Klenk and Mead established that there were three main families of polyunsaturated fatty acids (Mead, 1968). These are known as the n-9, n-6 and n-3 families and the parent fatty acids in each family are oleic acid (18:1n-9), linoleic acid (18:2n-6) and linolenic acid (18:3n-3), respectively. From each parent acid, a series of desaturated and chain elongated PUFA are derived by the sequential action of $\Delta 6$ desaturation, chain elongation, $\Delta 5$ desaturation, chain elongation and $\Delta 4$ desaturation enzymes (Table 3). There is no interconversion between the PUFA in the different families and it is only when linoleic and linolenic acids are both lacking in the diet that oleic acid is converted to eicosatrienoic acid (20:3n-9). The appearance of this

Table 3
The essential fatty acids and their long chain PUFA metabolites

n-6 series	*Enzyme function*	*n-3 series*	*n-9 series*
Linolenic acid (18:2n-6)		α-Linolenic acid (18:3n-3)	Oleic acid (18:1n-9)
↓	Δ6 desaturase (rate limiting)	↓	↓
γ-Linolenic acid (18:3n-6)		Octadecatetraenoic acid (18:4n-3)	Octadecadienoic acid (18:2n-9)
↓	Chain elongation	↓	↓
Eicosatrienoic acid (20:3n-6)		Eicosatetraenoic acid (20:4n-3)	Eicosadienoic acid (20:2n-9)
↓	Δ5 desaturase	↓	↓
Arachidonic acid (20:4n-6)		Eicosapentaenoic acid (20:5n-3)	Eicosatrienoic acid (20:3n-9)
↓ ⇡	Chain elongation	↓ ⇡	↓
Docosatetraenoic acid (22:4n-6)		Docosapentaenoic acid (22:5n-3)	Docosatrienoic acid (22:3n-9)
↓	Δ4 desaturase	↓ ⇡	
Docosapentaenoic acid (22:5n-6)		Docosahexaenoic acid (22:6n-3)	

fatty acid in tissue lipids is therefore a unique and sensitive biochemical marker of EFA deficiency (Mead, 1968). The Δ6 desaturase is the rate-limiting step in the conversion of C_{18} to C_{20} and C_{22} PUFA (Sprecher, 1977). There is evidence that although these conversions do occur in man, the process is extremely inefficient (Emken *et al.*, 1987). Consequently, fatty acids such as arachidonic, eicosapentaenoic (EPA) and docosahexaenoic (DHA) acids may be derived most efficiently by the ingestion of dietary sources of preformed PUFA (eggs, liver and meat for arachidonic acid and fish and fish oils for EPA and DHA).

The functions of the EFA can be broadly grouped into two categories. Firstly, the EFA and their longer chain, more unsaturated derivatives are constituents of cell membrane phospholipids which play an important structural role in all mammalian cell membranes. Extensive circumstantial evidence suggests that EFA are necessary for normal membranes. This may be related to the fact that the physical properties of the structural phospholipids are in large part determined by the chain length and degree of unsaturation of their component fatty acids. For example, linoleic acid is a major fatty acid of an epidermal sphingolipid and the replacement of linoleic acid by oleic acid in this sphingolipid in EFA deficiency is associated with increased water loss through the skin and development of skin lesions (Hansen, 1986).

The second main function of the EFA is that three of their derivatives (20:3n-6, 20:4n-6 and 20:5n-3) serve as precursors for a group of potent hormone-like substances with short half-lives known as eicosanoids and leukotrienes. These include prostaglandins, thromboxanes, prostacyclins, leukotrienes and lipoxins (Willis, 1981; Higgs, 1985; Samuelsson *et al.*, 1987). Dihomogammalinolenic acid (20:3n-6) and arachidonic acid (20:4n-6) are precursors of the 1- and 2-series of prostaglandins respectively, and 20:4n-6 of the 2-series of thromboxane and prostacyclin and of the 4-series of leukotrienes. The physiological role of the metabolite of 20:3n-6 (prostaglandin E_1) has often been overlooked. PGE_1 has some of the desirable actions of PGE_2 and PGI_2 and some unique actions of its own (Horrobin, 1988). EPA (20:5n-3) is the precursor of the 3-series of thromboxane and prostacyclin and of the 5-series of leukotrienes (Fig. 1). These compounds are produced in a number of tissues in response to a variety of stimuli following the release of the PUFA substrate from the membrane phospholipids and act near their site of synthesis to produce varied effects such as the regulation of platelet aggregation, thrombosis and inflammation. The eicosanoids and leukotrienes derived from arachidonic acid and EPA differ in biological activity: thromboxane A_3 is

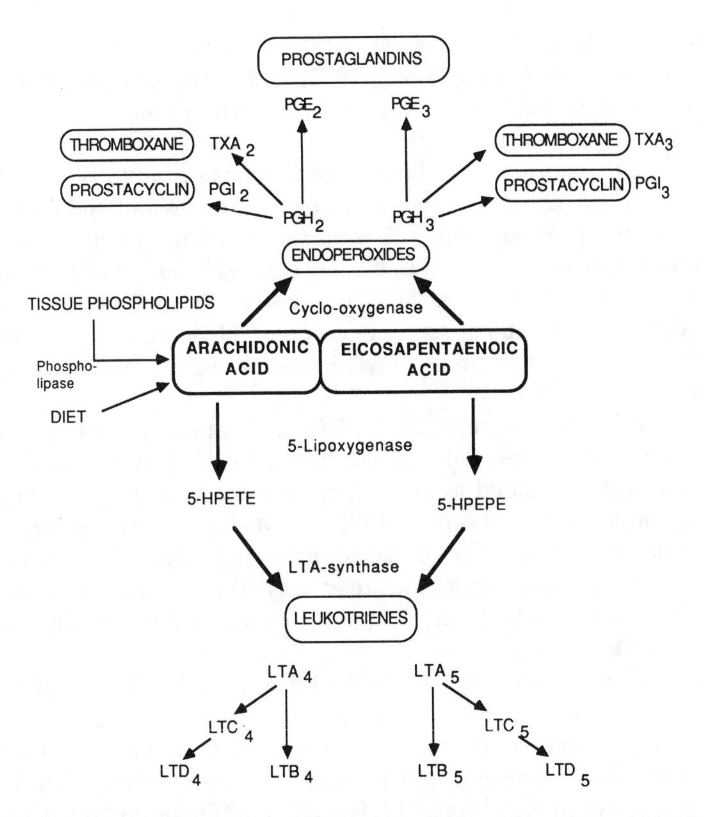

Fig. 1. Scheme for the major metabolic transformations of arachidonic and eicosapentaenoic acids to eicosanoids and leukotrienes.

significantly less pro-aggregatory than thromboxane A_2 and leukotriene B_5 is less active as a chemotaxic compound than leukotriene B_4, but prostacyclin I_3 is as potent as prostacyclin I_2 as an anti-aggregatory compound (Mehta *et al.*, 1987). DHA is also metabolised to oxygenated compounds in mammalian tissues via the lipoxygenase pathway (Aveldano & Sprecher, 1983) and these compounds can inhibit platelet aggregation (Croset *et al.*, 1988).

There is a substantial body of evidence which suggests that linoleic acid is the main EFA for most mammalian species including man with estimates of the minimum requirements being between 1–2% of the dietary energy (Holman, 1970). The recommended dietary requirements for linoleate in humans have ranged from 1–2% of dietary energy (Holman, 1968) up to 15% (Vergroessen, 1977) and between these two limits various recommendations can be found (Crawford *et al.*, 1978; Lasserre *et al.*, 1985). Hansen and co-workers (1963) fed infants on five milk formula diets with levels of linoleic acid ranging from 0·04 to 7·3% of dietary energy. Dermal lesions were evident on two of the diets and, based on disappearance of these lesions and the triene–tetraene ratio, the requirements were estimated to be between 1–2% of the dietary energy (Holman *et al.*, 1964). EFA deficiency has also been induced in infants and adults maintained entirely with parenteral feeding and one estimate of the requirements for an adult was in excess of 2·2% of dietary energy (Collins *et al.*, 1971).

It is likely that the dietary linoleic acid intake is well above the minimum requirement. For example the PUFA level in British diets has been recently calculated to be 9·8 g/person/day, based on the national average household food supply (Bull *et al.*, 1983). This represents about 3% of dietary energy. The intake of breast-fed infants is also likely to exceed requirements since Crawford *et al.* (1978) have shown that human milk from mothers in several countries contains about 8–10% of the energy as EFA.

Although it has been well established that the n-3 PUFA are the main EFA for fish and other marine species (Tinoco, 1982) the existence of a dietary requirement for the n-3 fatty acids has been an unresolved issue in human and mammalian nutrition for several decades (Simopoulos, 1989). The occurrence of high levels of DHA (22:6n-3) in brain grey matter of mammals (Crawford & Sinclair, 1972) and retinal photoreceptors (Neuringer & Connor, 1986) of a variety of different mammals despite wide variations in type and proportion of dietary PUFA suggest an important role for this fatty acid in brain and retinal function. Although it has been suggested that increasing levels of DHA in membrane phospholipids lead to an increase in membrane fluidity (Salem *et al.*, 1986), our understanding of the roles of highly unsaturated fatty acids in membranes is still in a very preliminary state. Specific changes in the brain and retina of rats and monkeys fed diets deficient in n-3 PUFA provide strong evidence for the essential role of n-3 PUFA although the molecular basis remains to be elucidated (Salem *et al.*, 1986; Neuringer *et al.*, 1989). In addition, one group has reported alopecia, scaly dermatitis and fatty livers in Capuchin monkeys on n-3 deficient diets (Fiennes *et al.*, 1973). In the last eight years there have been several reports which have suggested the occurrence of n-3 deficiency in humans (Holman *et al.*, 1982; Bjerve *et al.*, 1987; Simopoulos, 1989). In each case the patients were fed either intravenously or by gastric tube on preparations which were linoleic acid-rich and containing only very low levels of linolenic acid. The symptoms observed included neurological and visual changes, scleroderma and scaly dermatitis. The addition of linolenic acid to the diet of these patients was reported to correlate with the disappearance of the clinical symptoms. The estimated minimal requirements for linolenic acid were 0·2–0·3% of dietary energy in one report (Bjerve *et al.*, 1987) and 0·5–0·6% in another (Holman *et al.*, 1982).

There has been considerable interest over recent years in the n-3 PUFA, particularly those derived from marine sources (EPA and DHA), due to the low incidence of coronary heart disease in Eskimos despite the high level of fat in their traditional diet (Dyerberg & Bang, 1979). There is considerable evidence now that dietary EPA can modulate the production of eicosanoids derived from arachidonic acid, in addition to being metabolised to its own series of eicosanoids and leukotrienes. It has been generally assumed that the beneficial effects of fish oils are due to EPA or its oxygenated metabolites; most fish oils also contain DHA and this fatty acid can also influence the production of eicosanoids directly or indirectly by the retroconversion of DHA to EPA. Much effort is being directed towards establishing whether dietary EPA or marine PUFA can protect against coronary heart disease by reducing plasma triglyceride levels and/or decreasing thrombosis tendency and whether these PUFA can reduce pain and stiffness in rheumatoid arthritis (Gibson, 1988; Harris, 1989; Simopoulos, 1989).

THE WESTERN DIET AS THE EXCEPTION IN HUMAN HISTORY

The modern Western diet is out of step with other diets in human history. Humans have existed as a genus for about two million years (Pilbeam, 1984), and for the vast majority of that time they were hunter-gatherers. Agriculture is a relatively recent development, becoming established only in the last 10 000 years of our two million year history. The industrial revolution and subsequent developments affecting our food supply (e.g. agribusiness and food processing) culminating in the Western diet of developed, affluent societies (less than 200 years) represent less than the last 10 s in the 24-h clock of human existence spanning two million years. Using this same clock, agriculture began only 8 min before midnight. In evolutionary terms, the human genetic constitution has changed little in the past 40 000 years since the appearance of modern man. Accordingly, as suggested by Eaton and Konner (1985) in their provocative discussion of paleolithic nutrition and its implications for present day human health, the diets available to pre-agricultural humans provided nutrition for which we, as modern Western humans, are genetically programmed.

Archeological evidence indicates that once stone tools were developed by *Homo habilis* about two million years ago, animal foods became an important component of the diet of paleolithic man. It has been estimated that animal foods contributed about 50% of the energy intake from this period until the introduction of agriculture 8000–10 000 years ago (Eaton & Konner, 1985). With the emergence of modern man 40 000–50 000 years ago, big game hunting increased and animal foods may have provided as much as 80% of the energy intake. The remainder of the diet was derived from gathered, uncultivated vegetable foods: tuberous roots, fruits, nuts, beans and seeds. Cereal grains did not become a major component of the diet until the development of agriculture relatively recently. The proportions of animal and vegetable foods in the diet of paleolithic man appeared to vary with the availability of big game (large herbivores in particular). Despite the high proportion of meat in such diets, they were not high in fat since the wild animals eaten had much lower carcass fat contents than modern domesticated meat animals such as beef cattle and sheep (Ledger, 1968; Crawford *et al.*, 1970).

RECENT HUNTER-GATHERER DIETS: AUSTRALIAN ABORIGINES

The study of the diets, eating patterns and lifestyles of those few hunter-gatherer societies which have survived into the 20th century has provided not only a window on our pre-agricultural past, but also a reference standard with which to compare our current Western diet and lifestyle. A series of studies have been conducted into the risk factors for type 2 diabetes and cardiovascular disease in Aborigines from north-western Australia who had temporarily reverted to a hunter-gatherer lifestyle. These have given the opportunity to make detailed observations on their traditional diets and eating patterns. Before European settlement of Australia 200 years ago, Aborigines lived as hunter-gatherers all over the continent, from the tropical coastal region of the north (latitude 11°S) through the vast arid regions of the centre (latitude 20–30°S) to the cooler temperate regions of the south (30–43°S). Due to their higher rainfall the coastal areas of the northern, eastern and southern parts of Australia could sustain higher populations than the drier inland areas. Despite these major differences in climate and geographical location, available data indicate that Aborigines from all over Australia had a rich and varied diet in which animal foods were an important component (Meehan, 1977; O'Dea, 1984; Woodward *et al.*, 1987).

In a study examining the effects of temporarily reverting to a traditional lifestyle on glucose tolerance and lipid levels in a group of diabetic Aborigines, we had the opportunity to estimate energy intake and dietary composition of such a hunter-gatherer diet (O'Dea, 1984). All food hunted and gathered over a two-week period was weighed and samples taken for subsequent nutrient analysis. Average energy intake over this period was only 1200 kcal/person/day, which is much lower than the usual estimated urban intake. Under the circumstances of this study animal foods contributed almost two-thirds of the energy intake. This may have been higher than in the truly traditional setting since guns, bullets and fish hooks were provided and would have increased the chances of a successful hunt or fishing trip. However, despite the high proportion of animal foods the fat content of the diet was estimated to be 13% of dietary energy with approximately equal contributions from saturated, monounsaturated and polyunsaturated fatty acids (O'Dea & Sinclair, 1985). The diet was very high in protein (54% energy) and low in carbohydrate (33% energy).

Animal Foods in the Aboriginal Diet

Everything edible on an animal carcass is eaten by Aborigines living traditionally: muscle, fat depots, and internal organs (sometimes even intestinal contents). Fat depots are highly prized, as are brain, liver and other cholesterol-rich organ meats. However, fat depots tend to be small in wild animals throughout most of the year, and have to be shared out among many people. Muscle provides the largest contribution in volume and energy from an animal carcass. Large animals (such as kangaroo) are baked in a pit after having the intestinal contents removed. If the liver is not cooked immediately by the hunter and eaten at the site of the kill, it is baked inside the carcass with the other internal organs. When the lightly-cooked carcass is opened, all present share in eating the organs and 'soup' (peritoneal fluid and blood). Retroperitoneal fat is eaten by the hunter or senior men. The hunter also takes the mesenteric fat and shares it at his discretion. However, fat depots on wild animals are usually small. The hunter takes the head (including the highly-prized brain), with the remainder of the carcass (legs, shoulders and other joints) being distributed according to formal rules depending upon the relationship of each person to the hunter. This pattern of meat distribution ensures that the more successful a hunter, the higher his cholesterol intake (in the form of liver and brains). Such high cholesterol intakes were not, however, associated with high plasma cholesterol levels (O'Dea *et al.*, 1980; O'Dea, 1984). This is probably explained by the low saturated fat content of the diet overall (Nichaman & Hamon, 1987).

We have analysed animal foods eaten traditionally by Aborigines from different parts of the continent, ranging from the tropical north-west, through the central desert regions, to the cool south-east (Naughton *et al.,* 1986). There were striking similarities in their lipid content and fatty acid compositions independent of geographical location. The muscle samples were uniformly low in fat ($\leq$ 2·6% wet weight), with a high proportion of PUFA which is consistent with the data of Crawford *et al.* (1970) for large, wild African herbivores. All muscle samples analysed (84 samples from 30 species of mammal, bird, reptile, fish, crustacea) contained more than 20% PUFA, most contained more than 30%, and some more than 50% PUFA. These meats were relatively rich in long chain (C_{20} and C_{22}) PUFA. In terms of their actual PUFA composition, these meats could be grouped loosely into three main categories:

(i) those rich in n-3 PUFA (EPA and DHA) which were sea and estuarine in origin (fish, shellfish);
(ii) those rich in n-6 PUFA (linoleic and arachidonic acids) which were land-based, including marsupial mammals, reptiles and birds; and
(iii) those rich in both n-3 (linolenic, EPA, docosapentaenoic and DHA) and n-6 PUFA (linoleic and arachidonic acid) which included land-based, freshwater and coastal animals (marsupial mammals, reptiles, tropical fish).

However, these differences notwithstanding, the land-based muscle meats were often relatively good sources of linolenic acid.

Since muscle is the most important component of the animal carcass in terms of its contribution to dietary energy, its low fat content and high proportion of structural lipid (polyunsaturated) has a major influence on the fat content and composition of the hunter-gatherer diet overall (Fig. 2).

The other major sources of dietary fat in the Aboriginal hunter-gatherer diet are organ meats and depot fat. We have measured the lipid

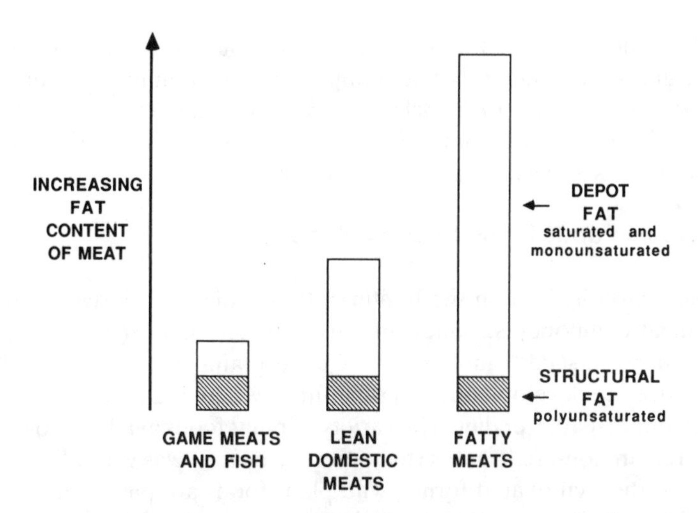

Fig. 2. The relationship between total fat content in meat and the relative proportion of structural lipid. The higher the total fat content, the lower the proportion of structural lipid.

content and fatty acid composition of liver and depot fat samples from a variety of animals eaten by Aborigines in different regions of Australia. The lipid content of the livers was almost always between 3–10% wet weight, with a high proportion of PUFA (28–58% total fatty acids). There was one exception to this—the liver of the mangrove ray (a stingray from northern Australian coastal waters) which was extremely high in lipid (59% wet weight) with a lower proportion of PUFA (15%). However, mangrove ray was not a major dietary component of northern Australian Aborigines and when its fatty liver was eaten it was mixed together with its extremely lean muscle and thereby greatly diluted. The PUFA of most of the livers were predominantly n-6 (linoleic and arachidonic acids) although they also contained a range of n-3 PUFA. Platypus liver was particularly rich in arachidonic acid (26%) while most others were rich in both arachidonic and linoleic acids. Depot fat samples varied widely in their PUFA content, ranging from around 5% total fatty acids in the fat from the commonly eaten kangaroo and small goanna to 30–40% (perente, duck). The type of PUFA in the depot fat samples was predominantly shorter chain (linoleic and linolenic acid) with little of the longer chain PUFA of either the n-3 or n-6 series.

The major sources of fat in the diet of a group of northern Australian Aborigines that had reverted temporarily to a hunter-gatherer diet (O'Dea, 1984; O'Dea & Sinclair, 1985) were kangaroo and freshwater bream. The fatty acid composition of the meat, liver and fat consumed from these two species is presented in Table 4.

Vegetable Foods in the Aboriginal Diet

Dietary carbohydrate in the hunter-gatherer diet was derived from the non-meat components: honey and uncultivated plants (tuberous roots, fruits, berries, seeds, nuts, beans). Cereal grains, the dietary staples of man since the development of agriculture, were not a major component of the hunter-gatherer diet. The variety of plant foods available to hunter-gatherers in non-arctic areas (temperate, tropical) was wide. Relative to many of their cultivated forms, wild plant foods are particularly rich in protein and vitamins. For example, a species of wild plum eaten by Aborigines in northern Australia has the highest vitamin C content of any known food (2–3% wet weight!) (Brand *et al.*, 1982). A commonly eaten yam in northern Australia is considerably more nutrient-dense than its modern equivalent, the potato: being higher in carbohydrate, protein, fibre, zinc and iron (Brand *et al.*, 1983). The wild vegetable foods are also rich in dietary fibre (Brand *et al.*, 1983, 1985). The seeds which made significant contributions to the diet of Aborigines in certain seasons were not only rich in protein, but also contained significant quantities of linoleic acid. For example the oil content of twenty varieties of indigenous acacia seeds ranged from 3–22%, with linoleic acid as the predominant PUFA (12–71% of total fatty acids). There was little linolenic acid in any of the seeds or nuts analysed (Brown *et al.*, 1987). Rivett *et al.* (1983) have also analysed the seeds from 8 different indigenous Australian plants which were reported to have been eaten by Aborigines. The lipid content of the 18 species ranged from 2–25% with linoleic acid also being the major PUFA. In 5 of the 18 species, there were also appreciable quantities of linolenic acid (linoleic/linolenic $< 10:1$). Although many of the vegetable foods are not rich fat sources (tuberous roots, leafy vegetables, fruits and berries), they nevertheless contain both n-6 and n-3 PUFA and the ratio of n-6/n-3 was often much less than that seen in most seeds (Hepburn *et al.*, 1986).

The carbohydrate in many of these traditional foods has been shown to be more slowly digested and absorbed than the carbohydrate in equivalent domesticated plant foods (Thorburn *et al.*, 1987*a*). These observa-

Table 4

Lipid content and fatty acid composition of important sources of dietary fat of Aborigines living traditionally in north-western Australia[a,b]

Food	*Lipid content (% wet weight)*	*Saturated*	*Monounsaturated*	*Total PUFA*	*n-6 PUFA*		*n-3 PUFA*		
					18:2	*20:4*	*18:3*	*20:5*	*22:6*
Kangaroo									
Muscle	1·2	30·0	26·9	43·1	22·1	7·7	6·1	2·0	0·8
Liver	4·3	32·7	34·2	33·1	15·0	7·3	3·6	0·8	1·3
Fat	8·6	50·9	45·0	4·1	2·7	—	1·4	—	—
Freshwater bream									
Muscle	2·4	40·2	30·4	29·4	7·7	5·9	3·7	1·8	4·2
Fat	78	39·0	43·5	17·5	8·7	1·5	2·8	0·5	0·9

[a]O'Dea (1984).
[b]Fatty acids expressed as percentage of total fatty acids (Naughton *et al.*, 1986).

tions, originally made on the bushfoods of the Australian Aborigines, have now been extended to wild plant foods eaten traditionally by other populations, e.g. the Pima Indians (Snow *et al.*, 1987). The lower post-prandial glucose and insulin levels elicited by the ingestion of these slowly digested wild plant foods may have been a factor in helping protect these populations from developing type 2 diabetes (Thorburn *et al.*, 1987*b*)—a condition to which they are particularly vulnerable when they make the transition from a traditional to a Western diet and lifestyle (Wise *et al.*, 1976; Bastian, 1979).

Relative to their modern domesticated equivalents, the vegetable foods in traditional hunter-gatherer diets were higher in protein, fibre and vitamins, contributed PUFA from both the n-6 and n-3 series and contained carbohydrate which is slowly digested and absorbed. In general, the wild vegetable foods were bulky, with high nutrient density but low energy density. The only carbohydrate source with high energy density was wild honey. This was a favourite food of hunter-gatherers, but usually available in limited quantity with an associated high work component. The great popularity of honey raises the possibility that the 'sweet tooth' of present day Western societies may have its origin in our hunter-gatherer past: that it may even have been a taste preference critical to survival of hunter-gatherers providing, like fat, a rare source of energy-rich food.

THE FIRST POST HUNTER-GATHERER DIET: THE ADVENT OF AGRICULTURE

The first profound changes to the human diet accompanied the advent of agriculture. Cereal grains were not a major component of the hunter-gatherer diet. Australian Aborigines, for example, harvested various seeds and nuts (from wild grasses and trees such as acacia) in certain seasons and ground them into a paste after initial processing (winnowing and/or baking). However these foods were apparently not major dietary components for most of the year. Wild cereal grains are extremely small and usually difficult to harvest, because once they ripen, the brittle rachis readily breaks and facilitates dispersal of the seed in the wind. Despite these problems, reasonable harvests were sometimes achieved using quite novel approaches. For example, in the south-east Kimberley region of Western Australia the Aborigines utilised their intimate knowledge of animal behaviour to harvest one particular type of grass seed early in the dry season. The seeds in question were assiduously gathered by ants and placed in convenient piles at the mouths of their holes (presumably in readiness for transfer underground for storage as a food supply for the dry season). Aborigines were able to gather these seeds readily and use them to prepare unleavened patties or damper (Olive Bieundurry, personal communication). There is no evidence that the Aborigines accumulated sufficient quantities of grass seeds in this way to provide a continuing food supply, or if the possibility of storing the seed for future use was ever considered.

Cereal seeds came into their own as a potential dietary staple only when two major changes occurred to facilitate their harvest in large quantity: increased rachis toughness and increased seed size. One of the earliest and most marked differences between domesticated cereals and their wild progenitors was increased rachis toughness which resulted in a reduced tendency for seed dispersal (less effective dissemination) and also, most importantly, facilitated efficient harvesting by ensuring that the seeds remained conveniently clustered on the head of the cereal stem after ripening. The survival of such domesticated species depended on human intervention for seed dissemination (sowing). Once man began actively cropping such cereals, active selection pressures would also have operated for increased seed size (to enable speedier and more efficient harvesting) and indehiscent seeds (which did not split open when ripe and were therefore more stable for long-term storage).

There is evidence that cereals were domesticated in several areas (the Middle East, Asia, Africa and South America) over long periods of time. Throughout the 'Fertile Crescent' of the Middle East extensive, dense stands of the probable wild progenitors of wheat, barley and oats can still be found, and archeological research has revealed evidence of sequential changes from gathered plant to cultivated cereals between 7000 and 5000 BC. Recent analysis of wild einkorn wheat in south-eastern Turkey indicated higher protein content but reduced starch content relative to modern wheat. Such wild cereals would have provided a balanced if monotonous diet: starch, protein and essential fatty acids in a form which could be stored stable over long periods of time (Evans, 1975). Such a cereal-based food supply would have enabled larger groups to live in the same area over extended time periods in a more secure setting than that offered by hunting and gathering. However, it is likely that hunting and gathering continued to provide important variety to the newly-acquired cereal-based diet in the early agricultural period.

From the time of their domestication in the very early days of agriculture 9000 years ago, until very recent times, cereals have been man's dietary staples in diverse populations all over the world. Although the consumption of cereals has fallen greatly in the affluent societies of the Western world during the 20th century, on a global scale, four cereals (wheat, rice, maize and barley) provide more than half of the protein and most of the energy consumed by mankind (Evans, 1975).

The cereal-based diet of agriculturalists was low in fat (10–15% energy) with linoleic acid being the major PUFA (5–10 g/person/day). The content of n-3 PUFA was usually low, but did vary with the contribution of meat and fish to the diet (seasonally variable and seldom more than 20% energy). The meat consumed by early agriculturalists (and socio-economically deprived populations in the Third World in the present day) would not have been from animals subjected to intensive management or genetic selection and is likely to have been low in fat, particularly saturated fat. The marine food chain is also relatively low in saturated fat and is based on n-3 PUFA rather than n-6 PUFA (the major PUFA in most marine species are EPA, docosapentaenoic acid and DHA). Although most fish, molluscs and crustacea are low in fat (< 5%) with a high proportion of n-3 PUFA, even those with high levels of lipid in their flesh are rich in n-3 PUFA since these fatty acids are found in both storage fat and structural phospholipids (Fogerty *et al.*, 1986).

TYPES OF PUFA IN DIETS THROUGH HUMAN HISTORY

The major PUFA in the diet of present day humans is linoleic acid, irrespective of whether they live in an advanced Western society or as a peasant farmer in the Third World. Although these two extremes have vastly different diets and lifestyles, both rely upon cereal grains either directly or indirectly as their predominant source of PUFA, and linoleic acid is the major PUFA in these seeds. We have estimated the linoleic acid intake in a group of adult Australians to be 10–15 g/person/day (Sinclair *et al.*, 1987). This is consistent with data gathered in UK and Australia (Table 1). The data of Bull *et al.* (1983) show that about half the linoleic acid is derived from fats and oils and one-fifth from cereal products and one-sixth from each of vegetables, fruits and nuts and from meat and meat products. In our study (Sinclair *et al.*, 1987a,b), the consumption of n-3 PUFA was low (about 1 g/person/day) and the content of long chain PUFA (C_{20} and C_{22}) of either the n-6 or n-3 series was low (about

Table 5
Polyunsaturated fatty acid content[a] in typical Western and hunter-gatherer diets

Diet	*Linoleic and linolenic acids*	*Long chain n-6 and n-3 PUFA*	*n-6/n-3 ratio*
Western diet[b]	12·3	0·2	12
Hunter-gatherer diet[c]	3·3	2·3	2·4
Experimental low-fat diets			
Southern fish[b]	2·9	1·5	1·8
Tropical fish[b]	2·7	1·2	3·4
Kangaroo meat[b]	4·1	0·7	5·2

[a]g fatty acid type/person/day.
[b]Sinclair *et al.* (1987).
[c]O'Dea and Sinclair (1985).

0·2 g/person/day). Although agriculturalists relying on a cereal staple consume much higher quantities of cereal each day than do people in Western societies, because they are not consuming oils additionally, their total consumption of linoleic acid is lower (5–10 g/person/day). As in Western societies, the consumption of n-3 PUFA by agriculturalists is generally very low unless fish is an important component of their diet. In our group of Australian adults we found that the ratio of n-6/n-3 PUFA was about 12:1 with the predominant fatty acids being linoleic and linolenic acids (Table 5). Non-fish eating agriculturalists would also have had a relatively high ratio of n-6/n-3 PUFA in their diet. However, it is important to note that the diet of early agriculturalists would usually have been supplemented with locally caught fish since early agriculture developed around sources of freshwater e.g. river valleys. Relative to the Western diet, the hunter-gatherer diet had a quite different PUFA composition (Table 5). Our calculations on the diet of northern Australian Aborigines suggest that the total PUFA content may have been lower than that in our current diet (3–6 vs 10–15 g/person/day), but with marked differences in the type of PUFA: much less linoleic acid, much more arachidonic acid and n-3 PUFA including the highly unsaturated elongation products of linolenic acid (mainly EPA and DHA). So not only was the ratio of n-6/n-3 PUFA much lower in the hunter-gatherer diet, but it was much richer in preformed long chain (C_{20} and C_{22}) PUFA of both the n-6 and n-3 series (Table 5).

We believe that both of these factors (proportionately more n-3 PUFA and much more preformed long chain PUFA) could have important physiological implications. The long chain n-6 PUFA arachidonic acid is a component of cell membrane phospholipids throughout the body and a precursor of numerous prostaglandins including those involved in modulation of haemostatic function, prostacyclin and thromboxane A_2. The long chain n-3 PUFA DHA has an important role in the development of retina and nervous tissue (including the brain). The n-3 equivalent of arachidonic acid, EPA, may also have a role in modulating haemostatic function by competing with, or substituting for, arachidonic acid. It is generally assumed that mammals can synthesise these long chain PUFA from their vegetable precursors (linoleic and linolenic acids) in sufficient quantity to fulfil these important physiological requirements. However, there is evidence that although these conversions do occur in man, the process is extremely inefficient. Whenever the higher polyunsaturated fatty acids such as arachidonic acid are present in the diet they are rapidly and preferentially incorporated into membrane lipids of different tissues (Kulmacz *et al.*, 1986; Sinclair *et al.*, 1987). We have shown that when people consume a low-fat diet rich in very lean meat (rich in arachidonic acid) or fish (rich in DHA, EPA), the proportion of linoleic acid in their plasma phospholipids falls while those of arachidonic acid and/or DHA and EPA rise, despite linoleic acid still being the most abundant dietary PUFA (Sinclair *et al.*, 1987). Given the major physiological roles of these long chain PUFA and their inefficient conversion from their plant precursors, it would seem important to ensure their supply in the diet. The hunter-gatherer dict provided them, whereas the Western diet does not (Table 5).

Furthermore, in the Western diet the ratio of linoleic/linolenic acids is much higher than in the hunter-gatherer diet due to the consumption of vegetable seed oils which contain only low levels of linolenic acid. This can also have a bearing on the synthesis of the long chain PUFA. Although the n-3 PUFA linolenic acid has a higher affinity for the desaturase enzymes than does the n-6 PUFA linoleic acid, if the ratio linoleic/linolenic acids is too high then there is very little synthesis of the linolenic elongation products, EPA and DHA (Fiennes *et al.*, 1973; Bjerve *et al.*, 1987; Martinez & Ballabriga, 1987).

In the studies in which we have attempted to mimic the composition of the Australian Aboriginal hunter-gatherer diet (low-fat, rich in lean meat or fish and rich in fruits and vegetables), the plasma fatty acid composition was substantially modified (Table 6), reflecting the PUFA composi-

Table 6
Plasma n-6 and n-3 PUFA profiles in total lipids (TL) or phospholipids (PL) from people consuming different diets (percent total fatty acids)

Group	*Plasma fraction*	*PUFA type*				*n-6/n-3*	*18:2n-6/C_{20}+C_{22} PUFA*
		18:2n-6	*20:4n-6*	*20:5n-3*	*22:6n-3*		
1. Typical Western diet[a]	PL	26	11	1	4	6·7	1·3
Low-fat diet—Tropical fish	PL	16	16	1	7	3·8	0·6
—Southern fish	PL	16	12	3	9	2·2	0·6
—Kangaroo	PL	16	15	1	4	5·1	0·6
—Vegetarian	PL	20	10	1	4	6·4	1·1
2. Australian Aborigines[b]							
Baseline (urban)	TL	15	4	2	1	6·5	1·8
Post 2 weeks low-fat fish	TL	9	11	3	2	3·6	0·5
3. Australian Aborigines[c]							
Baseline (urban)	TL	16	3	0·6	1	8·3	2·2
Post 7 weeks on hunter-gatherer diet	TL	16	12	1	3	5·4	0·9
4. Australian Aborigines[d]							
Living traditionally-oriented lifestyle	PL	13	16	1	4	4·4	0·5
5. Danes[e]	PL	21	8	0·2	3	8·4	1·6
Greenland Eskimos[e]	PL	7	1	7	7	0·7	0·4

[a]Sinclair *et al.* (1987).
[b]O'Dea & Sinclair (1982).
[c]O'Dea & Sinclair (1985).
[d]O'Dea *et al.* (1988*b*).
[e]Dyerberg *et al.* (1975).

tion of the foods eaten (Sinclair *et al.*, 1987). The provision of a dietary supply of a range of PUFA of different chain lengths and with a 'balanced' ratio of n-6 to n-3 PUFA may allow the tissues to incorporate fatty acids from the blood into their structural and metabolically active phospholipid pools on the basis of reaction rates (affinity) of esterifying enzymes rather than being limited by the availability of particular PUFA (e.g. 18:3n-3, 20:5n-3, 22:6n-3, or 20:4n-6) in the diet. The effect of different dietary habits on the plasma lipid PUFA profiles is well illustrated in Table 6: very low-fat diets are characterised by low levels of linoleic acid, lower n-6/n-3 ratios and an increased proportion of long chain PUFA (C_{20} and C_{22}). Other diets which are not particularly low in fat but which are rich in fish such as the Eskimo's diet (Dyerberg *et al.*, 1975), also show similar characteristics.

An important implication of these differences in dietary PUFA between the Western and hunter-gatherer diets relates to the risk of occlusive vascular disease. A widely accepted scheme for the regulation of thrombosis involves a balance between the anti-aggregatory vasodilator prostacyclin I_2 (synthesised from arachidonic acid derived from endothelial phospholipids in the blood vessel wall) and the pro-aggregatory vasoconstrictor thromboxane A_2 (synthesised from arachidonic acid derived from platelet phospholipids). It has been proposed that EPA is the factor responsible for the decreased tendency for thrombosis in fish-eating populations by promoting the formation of its own series 3 prostanoids, PGI_3 (anti-aggregatory, vasodilatory) and TXA_3 (inactive) (Dyerberg *et al.*, 1979). Fischer *et al.* (1986) have measured the urinary metabolites of arachidonic acid and EPA in Eskimos and age- and sex-matched Danish controls. They showed that there was a reduction in thromboxane A_2 and A_3 and an increased level of both PGI_2 and PGI_3 in the Eskimos compared with the controls. EPA has also been shown to inhibit cyclo-oxygenase, the first step in prostaglandin synthesis (Needleman *et al.*, 1979) (Fig. 1) and to displace arachidonic acid from cell membrane phospholipids (Morita *et al.*, 1983). The essence of these schemes is that EPA may act to reduce thrombosis tendency by reducing the formation of thromboxane (A_2 and A_3) and thereby altering the prostanoid balance in favour of prostacyclin (PGI_2 and PGI_3).

Budowski and Crawford (1985) have suggested that the low level of n-3 PUFA in the Western diet has led to an unrestrained metabolism of arachidonic acid to eicosanoids and leukotrienes and they have speculated that this may be implicated in the pathogenesis of some of the diseases of affluence of Western societies (e.g. occlusive vascular disease, arthritis). In fact much of the current interest in the use of fish oils rich in EPA and DHA in Western countries is aimed at raising tissue levels of EPA and DHA in anticipation that there will be a reduction in the metabolism of arachidonic acid to its series of eicosanoids and leukotrienes accompanied by an increased formation of eicosanoids and leukotrienes derived from EPA (Kinsella, 1987; Mehta *et al.*, 1987). We believe that this approach is unbalanced since it is making the assumption that all that is required to 'normalise' the Western diet is the addition of long chain n-3 PUFA. We have argued above that the Western diet is already unbalanced (high linoleic acid levels, low n-3 PUFA and low long chain PUFA) and that our evolutionary diet would have contained a range of PUFA (C_{18}–C_{22}) including long chain PUFA of both n-6 and n-3 series; that is the diet would have contained more arachidonic acid, and EPA and DHA than at present and with an n-6/n-3 ratio of between 1:1 up to 4:1.

Although much effort is being directed at replacing tissue arachidonic acid by the fish oil fatty acids EPA and DHA, we believe that dietary arachidonic acid can be viewed in a more positive manner. Whilst arachidonic acid is the precursor of the platelet aggregating thromboxane A_2, it is also the precursor of the most powerful inhibitor of platelet aggregation, prostacyclin I_2. PGI_2 also produces vasodilation and could therefore directly increase blood flow. Thus, it is possible that under conditions where plasma phospholipid arachidonic acid is elevated, the thrombosis tendency could be controlled through increased PGI_2 production relative to TXA_2 production. Indeed, we have recently shown that dietary saturated fat in the form of butter resulted in a dose-dependent reduction in arterial PGI_2 production in rats which was paralleled by a reduction in arachidonic acid and an increase in EPA in plasma phospholipids (O'Dea *et al.*, 1988*a*). Supplementation with oral arachidonic acid (80 μl/rat/day) fully reversed all of these changes (Steel *et al.*, 1987).

Although the hunter-gatherer diet contained significant quantities of arachidonic acid it would also have contained long chain n-3 PUFA. This relationship may have ensured adequate tissue levels of arachidonic acid for metabolic and structural functions, however it would also have ensured a role for EPA in modulating the arachidonic acid metabolism by competing with or substituting for arachidonic acid, in addition to being metabolised to a physiologically active prostacyclin PGI_3 (a potent inhibitor of platelet aggregation). EPA can modulate the production of leukotriene B_4 (synthesised from arachidonic acid), as well as being metabolised to leukotriene B_5 which has attenuated chemotactic and aggregatory activities for human neutrophils and thereby can modulate the inflammatory response (Lee *et al.*, 1985). In addition to its role in influencing metabolism of eicosanoids and leukotrienes, EPA together with DHA have a dramatic effect in lowering plasma triglyceride levels mainly through the inhibition of the hepatic synthesis of very low density lipoproteins (VLDL) (Nestel *et al.*, 1984).

IMPACT OF DIETARY CHANGES ON HEALTH

Australian Aborigines are particularly prone to obesity and type 2 diabetes when they make the transition from a traditional hunter-gatherer to a Western lifestyle. The prevalence of diabetes among adult Australians of European origin is 3–4% (Glatthaar *et al.*, 1985), while that of Aborigines is three to six times higher (Wise *et al.*, 1976; Bastian, 1979). Hypertension and coronary heart disease are also widespread among westernised Aborigines (Wise *et al.*, 1976; Bastian, 1979). We have shown that healthy, lean, young Aborigines from a community in which obesity and diabetes were highly prevalent in those over 35 years of age, exhibited mild impairment of glucose tolerance, hyperinsulinaemia and hypertriglyceridaemia (O'Dea *et al.*, 1982). These results are consistent with underlying insulin resistance in these people even though they were young and lean. In both short-term (2 weeks) and longer-term (3 months) studies, we have shown that temporary reversion to a traditionally-oriented diet (no Western foods) and lifestyle in non-diabetic Aborigines was associated with improved glucose tolerance, reduction in the hyperinsulinaemia and a reduction in plasma triglyceride concentrations (O'Dea *et al.*, 1980, 1982; O'Dea & Spargo, 1982). The change from a Western to a traditional lifestyle involved three factors which directly improve insulin sensitivity: increased physical activity; reduced energy intake and weight loss; and a low-fat, high-fibre diet. Thus, the change back to a traditional hunter-gatherer lifestyle should improve all abnormal aspects of carbohydrate and lipid metabolism which are linked to insulin resistance in Aboriginal diabetics.

This hypothesis was tested in a study in which a group of overweight, middle-aged Aboriginal diabetics from northern Australia was examined before and after living for seven weeks as hunter-gatherers in an isolated location in the north-west Kimberley region of Western Australia (O'Dea, 1984). The only food eaten by the subjects (10 diabetic, 4 non-diabetic) was that which they hunted and gathered. The

Table 7
Design of the study and composition of the diet during the seven-week lifestyle change period

Phase of Study	*1. Travelling (1·5 weeks)*		*2. Coast (2 weeks)*		*3. Inland (3·5 weeks)*	
Main foods (as % total calories)	Beef	75	Fish	80	Kangaroo	36
					Freshwater fish (bream)	19
					Yams	28
	Kangaroo, Turtle, Bream, Yams, Honey	25	Birds, Kangaroo, Crocodile	20	Honey, figs, birds, crocodiles, turtle, yabbies	17
Composition of diet	*Estimate only*		*Estimate only*		*Measured over two-week period*	
Carbohydrate (%)	10		< 5		33	
Protein (%)	50		80		54	
Fat (%)	40		20		13	
Saturated fat (% total energy)	22		8		4	
Monounsaturated	15		4		5	
Polyunsaturated	3		7		4	

7-week period was spent as follows: en route to the coastal location, 1·5 weeks; at the coastal location, 2 weeks; at the inland river location, 3·5 weeks. The diets during these three phases are summarised in Table 7. The diet during the period at the inland river location was considered to be most 'typical' of the usual traditional one in that it contained a range of animal foods, fish, vegetables and honey. The diet during the 2-week period on the coast was derived almost exclusively from seafood. During the final phase (inland river) all food hunted and collected was weighed before it was eaten and samples were taken and stored in liquid nitrogen before being flown back to Melbourne for analysis. Energy intake over this period averaged 1200 kcal/person/day. Three foods accounted for more than 80% of the total energy consumed over this 2-week period (kangaroo 36%, yams 28%, freshwater bream 19%). Despite the high content of animal foods (64% total energy) the diet was not high in fat (13% total energy, with approximately equal contributions from saturated, monounsaturated and polyunsaturated fatty acids).

All subjects lost weight during the study (3–11 kg, mean 8 kg) and there was a marked improvement in all of the metabolic abnormalities of diabetes: fasting glucose fell from 11·6±1·2 mM to 6.6±0·5 mM; glucose tolerance improved greatly but was not completely normalised; fasting insulin concentrations fell and the insulin secretory response to glucose improved but was not normalised; hypertriglyceridaemia (4·0±0·5 mM) was abolished (1·2±0·03 mM). In addition there was a reduction in a number of risk factors for cardiovascular disease: blood pressure fell (9 mmHg); hypertriglyceridaemia was abolished; bleeding times increased significantly after two weeks on the seafood diet (4·1±0·4 min to 5·3±0·4 min), and continued to rise (5·9±0·4 min) on the mixed traditional diet, the total increase being 44% over five weeks.

This latter result was particularly significant as it occurred in association with a doubling in the concentration of arachidonic acid in the plasma lipids (O'Dea & Sinclair, 1985). In a previous study we reported a 3-fold increase in the proportion of arachidonic acid in plasma lipids following two weeks on a diet derived almost exclusively from tropical seafood (O'Dea & Sinclair, 1982). In this previous study the proportion of the n-3 PUFA usually associated with seafood, EPA and DHA, also rose, although not as strikingly as arachidonic acid. Analysis of the tropical seafoods eaten (Sinclair *et al.*, 1983) indicated that they were low in fat (0·6–3·3% wet weight) and rich in arachidonic acid (5–14% total fatty acids) and DHA (6–28% total fatty acids), but with lower than expected levels of EPA (3–10% total fatty acids). In both studies (O'Dea & Sinclair,

1982, 1985) the proportion of arachidonic acid rose while that of linoleic acid fell, despite the fact that the omnivorous diet of the latter part of the second study contained more linoleic than arachidonic acid (Table 5). That there appeared to be preferential incorporation of arachidonic acid into the plasma lipids was further supported by the observation that the rise in arachidonic acid in the cholesterol ester and phospholipid fractions was almost exactly counter-balanced by the fall in linoleic acid. The association of a longer bleeding time with a marked rise in arachidonic acid concentration in the plasma lipids (in contrast to the observations made with the EPA-rich fish from the cold northern hemisphere waters) (Dyerberg & Bang, 1979) suggests that the mechanism by which dietary PUFA modulate haemostatic function may be more complex than generally believed. We are currently investigating the possibility that diets which raise arachidonic acid in plasma lipids may reduce the tendency to thrombosis by favouring the production of the anti-thrombotic vasodilatory prostanoid, PGI_2, at the expense of the pro-aggregatory, vasoconstrictor prostanoid, TXA_2 (O'Dea *et al.*, 1988*a*). We are also investigating the effect (if any) of increased DHA levels in plasma phospholipids on the balance between prostacyclin and thromboxane. DHA is the major PUFA in the flesh of most marine species (Hepburn *et al.*, 1986; Dunstan *et al.*, 1988) and appears to be readily incorporated from the diet into tissues (Sinclair & Crawford 1975; Sinclair *et al.*, 1987). In addition to its presumed role in the membranes of brain and retina, DHA may have a direct role in modulating platelet responsiveness and thrombosis tendency (von Schacky & Weber, 1985). DHA has been demonstrated to be a competitive inhibitor in the conversion of arachidonate to prostanoids (Corey *et al.*, 1983) and to suppress the synthesis of leukotrienes (Lokesh *et al.*, 1988). Furthermore DHA is a precursor of EPA (via a retroconversion process, Table 3) and it can act as a precursor of prostacyclin I_3 via retroconversion to EPA (Fischer *et al.*, 1987).

Effect of the Hunter-Gatherer Diet on Westerners

In order to investigate the physiological implications of these findings in Aborigines in more detail we have attempted to mimic the hunter-gatherer diet as closely as possible under controlled conditions in the urban setting (Sinclair *et al.*, 1987). The subjects in these studies were healthy weight-stable Caucasian volunteers. Each study ran for four weeks. During the first and fourth weeks the subjects ate their usual diets, weighing and recording food intake. In the second and third weeks they consumed one of four experimental diets which were similarly low in fat (about 7% energy) but contained quite different PUFA: the diets contained either 500 g/day of tropical fish (rich in arachidonic acid and DHA), fish from Australia's temperate southern waters (rich in DHA, with some EPA) or kangaroo meat (rich in linoleic and arachidonic acids). The fourth diet was vegetarian with a similarly low fat content and high P:S ratio but containing no long chain ($> C_{20}$) PUFA. All diets were based on fruit, vegetables, non-fat dairy products and cereal foods. Fats, oils, nuts, meat and fish (other than that supplied) and processed foods with added fat were excluded. The experimental diets containing fish or kangaroo meat (extremely lean) were qualitatively similar to the diets consumed by the Aborigines in our previous studies in that they contained no added fat or processed foods. However, they differed quantitatively in that they contained much less lean meat or fish and more vegetables, fruits and cereals and were therefore higher in carbohydrate and lower in protein. This compromise was forced upon us due to the unacceptability of extremely high meat and fish intakes for the participants. Since the effects of the diets on plasma fatty acid profiles were similar to those observed in the Aborigines when they reverted temporarily to traditional diets based on tropical fish or kangaroo meat, the dietary model was considered valid.

Some of the changes in response to the diets appeared to be due to the low fat content. Fasting cholesterol fell 19–24% over the two-week period on all four diets with the most pronounced fall being in the first week. The fall in cholesterol appeared to be equally rapidly reversible on all diets, rising within one week of resuming the baseline diet. Falls in both HDL and LDL cholesterol contributed to the fall in total cholesterol, with effects on HDL being more consistent (Sinclair *et al.*, 1987).

The fatty acid composition of plasma lipids (phospholipids and cholesterol esters) changed significantly on all diets (Sinclair *et al.*, 1987). The proportion of linoleic acid fell on all four diets, probably in response to the low linoleic acid content of the experimental diets relative to the usual Western diet (Table 5). The changes in the proportions of long chain ($\geq C_{20}$) PUFA were quite diet-specific and reflected the composition of the major food eaten (Table 6). The proportion of arachidonic acid increased on the two diets which were good sources of this PUFA (tropical fish and kangaroo meat); the proportion of DHA rose in the two diets which were good sources of it (tropical and southern fish); and EPA increased most markedly on the southern fish diet. The vegetarian diet did not affect the proportion of the long chain PUFA. These changes in

PUFA composition of plasma lipids were reversed within one week of the subjects resuming their usual diets. In terms of effects on haemostatic function these diets also had variable effects. Platelet aggregation was not affected significantly by any of the diets (Butcher *et al.*, 1989). Both fish diets resulted in decreased plasma and whole blood viscosity and the southern fish diet also reduced platelet adhesion (Butcher, unpublished observations). The tropical fish and kangaroo meat diets reduced the cold pressor response (i.e. there was a smaller reduction in blood flow in response to a cold stimulus) (Butcher *et al.*, 1990). We believe that the effect of tropical fish and kangaroo meat on the cold pressor response is particularly significant as there is good evidence that this response is inversely related to the prostacyclin-like activity of the vascular system: infusion of PGI_2 has been shown to reduce the cold pressor response (Cowley *et al.*, 1985). These results are being cautiously interpreted as indicating that the diets rich in tropical fish or kangaroo meat stimulated PGI_2 production, which would be consistent with the increased concentration of arachidonic acid in the plasma phospholipids. Although it has been assumed that the production of eicosanoids depends on the stimulated release of arachidonic acid from membrane phospholipids, it has been shown that oral doses of arachidonic acid lead to a very rapid generation of eicosanoids (Ramesha *et al.*, 1985), thus questioning the relative roles of dietary PUFA and tissue phospholipids as eicosanoid precursors. The southern Australian fish diet was associated with an enhanced cold pressor response. It would be of great interest to see whether the changes in cold pressor response were correlated with changes in the output of prostacyclin metabolites in the urine (Fischer & Weber, 1984), and to measure the cold pressor response following a diet rich in the EPA-rich northern hemisphere fish which lowers plasma arachidonic acid and raises the concentration of EPA in plasma lipids (Dyerberg & Bang, 1979).

Independent Effects of Lean Beef and Beef Fat on Plasma Cholesterol Levels

Over the last 30 years there has been much debate about saturated fat and cholesterol in the Western diet. Foods perceived as rich sources of these have been under attack by health professionals. In particular red meat, dairy products and eggs have been labelled as foods which should be avoided by those wishing to reduce their risk of coronary heart disease. Beef has been seen as being rich in fat even if all external visible fat has been removed. In the USA, where marbled (intramuscular fat) beef is common the minimum fat content of beef cuts trimmed of all external visible fat is stated to be about 9% (Greenberg, 1984). In Australia, recent analyses of beef destined for the domestic retail market have shown that the minimum fat content of lean beef, completely trimmed of external visible fat, is between 2–5% (Sinclair & O'Dea, 1987*a*). If the normal retail fat-trim (4–6 mm) is included the fat content is between 7–18% depending on the cut selected (Table 8) and from 15–27% if the meat cuts contain a 12 mm thick fat-trim (Sinclair & O'Dea, unpublished observations). Thus, for this type of Australian meat, the consumer now has the choice, albeit an expensive one, of trimming all visible fat or buying fat trimmed cuts enabling the consumption of meat with a relatively low fat content provided appropriate low-fat cooking methods are adopted.

Table 8
Minimum and maximum lipid content[a] of selected cuts of beef[b]

Cut	*Maximum lipid*[c] *(lean and visible fat)*	*Minimum lipid (lean only)*	*% yield*[d]
Topside[e]	6·7	2·1	82
Blade	9·8	2·9	70
Rump	13·3	3·0	70
Scotch fillet	14·6	5·0	66
Sirloin	17·7	2·7	59

[a]g lipid/100 g wet weight of cut.
[b]Carcass weights ranged from 175 to 220 kg.
[c]The fat-trim on each cut was between 4 and 6 mm.
[d]Weight of fat trimmed lean meat as percent of original sample weight (lean and all external visible fat).
[e]Four samples of each cut were analysed.

Having shown that a diet rich in a very lean red meat (kangaroo), but low in fat was just as effective in lowering plasma cholesterol levels as similarly low-fat vegetarian or fish-supplemented diets (Sinclair *et al.*, 1987) it was important to establish whether low-fat diets rich in lean beef had similar effects, and to determine the effect of adding beef fat (dripping) back to such a diet. This study design made it possible to differentiate between *lean* beef and beef *fat* (dripping) as risk factors for cardiovascular disease. When healthy, weight-stable, normocholesterolaemic subjects consumed a low-fat (9% energy) diet rich in lean beef (500 g/2000 kcal/day), there was a significant reduction (20%) in plasma cholesterol levels within one week. This fall was due entirely to reductions in LDL cholesterol. Both total and LDL cholesterol levels increased significantly when dripping was added back to the diet in a step-wise fashion over two weeks (10% energy, 20% energy). HDL cholesterol was not affected by any of these dietary changes. These results indicate that lean beef can be included in a cholesterol-lowering diet provided the overall fat content of the diet is kept low. However, the reversal of these beneficial effects by addition of dripping highlights the importance of selecting extremely lean beef and trimming it of all visible fat before cooking. Beef currently available to the consumer in Australia was not suitable for inclusion into the diet study without considerable trimming of visible fat, however, the results should be seen as being very positive for the meat industry in that, under the conditions of this study, only LDL was lowered, with HDL being unaffected (Traianedes *et al.*, 1987; O'Dea *et al.*, 1990). In terms of effects on the plasma lipid PUFA composition, the low-fat lean beef diet resulted in a fall in linoleic acid, a rise in arachidonic acid and very small rises in the n-3 PUFA (DHA and EPA). These changes did not appear to be reversed by the addition of dripping to the diet. In contrast to kangaroo meat there were no consistent effects of the lean beef diet on forearm blood flow. This may have been related to the smaller rise in arachidonic acid in plasma lipids with the lean beef. There were no effects on platelet aggregation.

These results indicate that it is possible to modulate plasma lipid fatty acid composition in man using Western foods, although the changes are less pronounced than those seen in Aborigines who temporarily revert to a hunter-gatherer diet. The physiological implications of these different degrees of change are not yet understood and deserve further study. It is important to establish how lean domestic meats need to become in order to produce the health benefits associated with wild meats in the hunter-gatherer diet. It may not be necessary to reverse the domestication process fully but rather to concentrate on management practices which minimise fat deposition in the carcass and/or by the use of leaner breeds of cattle such as Charolais or Limousin.

Dietary Cholesterol and Plasma Cholesterol Levels

Although it is now generally accepted by health professionals and researchers world-wide that high plasma cholesterol levels increase the risk of coronary heart disease (Keys *et al.*, 1986; Stamler *et al.*, 1986), the relationship between the level of cholesterol in the diet and the level of cholesterol in the plasma has not been fully clarified and remains a source of controversy (Fisher *et al.*, 1983; Connor *et al.*, 1986). In a number of studies examining the effect of temporary reversion to a traditional hunter-gatherer lifestyle in Australian Aborigines (O'Dea *et al.*, 1980; O'Dea & Spargo, 1982; O'Dea, 1984), we have observed either no change or a fall in the plasma cholesterol levels despite high intakes of dietary cholesterol in the form of brain, liver and other organ meats. The failure of such diets to increase plasma cholesterol levels was attributed to their low fat content and particularly low saturated fat content (O'Dea, 1984). A number of more quantitative dietary studies have addressed this question directly. Faber and co-workers (1986) showed that there was no relationship between plasma cholesterol level and cholesterol intake in a group of body builders consuming between 0 and 81 eggs per week. Similarly, high intakes of lobster, crab and shrimp have been shown to be associated with only mild effects on cholesterol levels in normal subjects while clams, oysters and scallops were without effect (Connor & Lin, 1982). Doubling the intake of cholesterol from eggs had no sustained effect on plasma cholesterol levels in subjects being instructed to follow a reduced-fat, high-fibre diet (Edington *et al.*, 1987). Katan and Beynen (1987) have shown in numerous studies that there is wide individual variability in the response to dietary cholesterol under conditions of the 'normal' Western diet.

The cholesterol-elevating effect of dietary cholesterol appears to be a function of the saturated fat content of the diet. Diets rich in saturated fat are associated with increased plasma cholesterol levels even if the diet itself is low in cholesterol, and if the diet is enriched with cholesterol this hypercholesterolaemic effect of saturated fat is enhanced (Schonfeld *et al.*, 1982; Fisher *et al.*, 1983). However, if the polyunsaturated : saturated fatty acid (P : S) ratio is high (Schonfeld *et al.*, 1982; Fisher *et al.*, 1983) or the total fat content of the diet low (Whyte *et al.*, 1977) enrichment of the diet with cholesterol does not affect plasma cholesterol levels. Under these circumstances, cholesterol synthesis and/or excretion are reduced to balance the increased cholesterol intake in the diet (Whyte *et al.*, 1977). By a process which is not yet fully understood, high levels of saturated fat in the diet appear to interfere with this finely-tuned system (Spady & Dietschy, 1982) and allow the cholesterol concentrations in blood to rise to unacceptably high levels.

COMPARISON OF WILD AND DOMESTICATED MEATS

Animal foods provided essentially all of the dietary fat consumed by the Aborigines in north-west Australia when we examined their traditional diet (O'Dea, 1984; O'Dea & Sinclair, 1985). Despite the high proportion of animal foods in the diet, and despite the clear preference for depot fat and high-fat organ meats, the overall fat content of the diet was low (< 20% energy). This was due to the particularly low fat content of the muscle meats eaten and the relatively small fat depots on most animals for most of the year. Moreover, the fat of wild animals is proportionately much less saturated than that of domesticated animals which provide most of the meat and animal fats in Western societies (Table 9).

A combination of improved breeding and management practices and a readily available food supply has resulted in overfat cattle, sheep and pigs. Carcasses and meat cuts from these animals have large amounts of associated visible fat; as a visit to a butcher's shop shows. There have been positive and encouraging trends within the Australian meat industry recently to promote lean cuts of meat which are substantially fat trimmed. In contrast, there is very little visible fat on the carcasses of wild animals. As Crawford has pointed out (Crawford, 1968), modern intensive cattle rearing produces carcasses with about 30% adipose fat and about 50% lean. However, the lean contains 3–10% of its fresh weight as infiltrated triglyceride (intramuscular fat). Thus, the animal is at best producing only about 45% of its carcass as actual muscle cell. Since muscle cells are 80% water, less than 10% of the intensively reared beef carcass is structural nutrient (mainly protein). This means that the intensively reared carcass provides three times more fat than structural material. The same calculations on wild species or domestic species not subjected to selective breeding or intensive feeding (such as rabbit, horse, fish) give 15% structural nutrient (mainly protein, structural fat) and 2–5% fat, i.e. three to five times as much protein as fat (Crawford *et al.*, 1970). The high fat carcasses of intensively produced meat have a high proportion of saturated fat. The higher the total fat content of meat, the lower the relative proportion of structural fat, and the lower the proportion of PUFA (Fig. 2). Meats from wild animals including wild ruminants, although containing very low levels of total lipid (< 2%) (Crawford *et al.*, 1976; Sinclair *et al.*, 1982; Naughton *et al.*, 1986) have a high proportion of PUFA relative to saturated fatty acids (P:S > 1·0) (Table 10). Meat

Table 9

Lipid content (% wet weight) and fatty acid composition (% total fatty acids) of the lean portion of a range of meats and fish

Meat type	*Lipid content*	*Saturated FA*	*Monounsaturated FA*	*Total PUFA*	*n−6 PUFA*		*n-3 PUFA*		*n-6/n-3*
					18:2	*Long chain*	*18:3*	*Long chain*	
Lamb[a] (mid-loin chop)	4·6	43	42	9	4	1	2	2	1·3
Beef[a] (T-bone) pasture	4·4	45	43	7	4	1	1	1	2·5
(topside) fed	2·4	42	41	11	5	2	2	2	1·8
(topside) grain-fed	2·1	43	34	14	10	2	1	1	6·0
Kimberley beef[b]	2·6	37	42	20	10	5	2	3	3·0
Pork[c] leg	3·5	37	44	14	11	2	0·5	1	8·7
Chicken: breast[c]	1·4	38	37	20	14	4	0.5	2	7·2
thigh	2·6	38	40	22	16	4	0·5	2	8·0
Rabbit[c]	1·5	37	16	40	18	5	11	8	1·2
Horse[d]	1·0	28	13	43	27	8	4	3	5·0
Buffalo[d]	1·1	31	27	29	15	6	3	4	3·0
Sambar deer[d]	0·8	28	14	31	15	9	2	4	4·0
Kangaroo[d]	1·1	28	20	38	20	10	4	4	7·5
Wild pig[e] (leg)	2·4	35	36	28	18	4	4	2	3·7
Barramundi[f]	0·6	37	17	44	2	17	1	24	0·8
Whiting[g] (King George)	1·0	29	26	45	2	11	—	28	0·5
Flake[g]	1·1	34	19	47	1	21	—	25	0·9
Flathead[g]	0·8	28	11	61	1	10	—	50	0·2
King Prawn[g]	0·5	31	25	44	2	9	—	33	0·3

[a] Sinclair & O'Dea (1987*a*).
[b] O'Dea & Sinclair (1985).
[c] Sinclair & O'Dea (1987*b*).
[d] Sinclair *et al.* (1982).
[e] Naughton *et al.* (unpublished observations).
[f] Sinclair *et al.* (1983).
[g] Dunstan *et al.* (1988).

from domesticated ruminants even when trimmed of all visible fat usually contains up to 5% total lipid (Sinclair & O'Dea, 1987*a*) and has a low P:S ratio (< 0·4). This is due to the fact that the PUFA in ruminant meat lipids are found in the phospholipid (structural lipid) fraction of the meat (P:S ratio > 1·2) which is a relatively constant concentration (0·5–0·8%), whereas the triglycerides (fat) in ruminant meat are rich in saturated and monounsaturated fatty acids with a P:S ratio of about 0·1. Thus very low-fat meats have a relatively high proportion of phospholipid to triglyceride and have a higher P:S ratio than fattier meats (Table 10). The P:S ratio in most ruminant meat cuts with even a modest amount of visible fat (say 4 mm) would approach the P:S ratio of the fat (0·1). Even in wild meats with their relatively high proportion of PUFA, the concentration of long chain PUFA is low (e.g. the arachidonic acid level in kangaroo is about 50 mg/100 g wet weight). Nevertheless, despite the low absolute amounts of these PUFA, they appear to be incorporated very efficiently into human plasma lipids (Sinclair *et al.*, 1987).

Table 10
Lipid content and P:S ratios in meat from domesticated and non-domesticated species

Meat	*Lipid content*[a]	*Total lipid P:S*[b]	*Triglyceride P:S*	*Phospholipid P:S*
Beef	2·5	0·22	0·06	1·10
Sheep	3·1	0·26	0·09	1·28
Goat	2·3	0·36	0·09	1·23
Buffalo	1·1	0·91	0·07	1·52
Sambar deer	0·8	1·12	0·11	1·54
Horse	1·0	1·52	0·46	1·82
Kangaroo	1·1	1·37	0·33	1·57
Pig	1·4	0·75	0·24	1·38

Data from Sinclair *et al.* (1982).
[a]g lipid/100 g wet weight.
[b]Ratio of polyunsaturated to saturated fatty acids.

WHY HAS THE WESTERN DIET DEVELOPED AS IT HAS?

These differences in fat content and fatty acid compositon between wild and domestic meats are particularly significant in terms of dietary fat intake and its relation to the risk of chronic degenerative diseases. It is interesting to speculate why we have changed our food supply in the way we have. Why have we produced a surfeit of energy-dense foods high in fat and sugar? It may well be directly explicable in terms of our precarious survival as hunter-gatherers in the past: the two most highly-prized food items for the Aboriginal hunter-gatherers of northern Australia were depot fat and honey which due to their energy density would have been important to survival. However, they were available in limited quantities and their acquisition was usually associated with hard work and high energy expenditure. Perhaps the biblical quest of our ancestors for a secure life in 'the land of milk and honey' was a direct response to the insecure existence of their forebears as they relied less on hunting and gathering and settled into an agricultural-based system of food supply. It is also possible that such distant 'race memories' of our hunter-gatherer past underlie the changes we have wrought in our own food supply. Our 'promised land' has become the land of saturated fat and sugar! By removing the natural constraints (limited availability of energy-dense foods combined with hard work in obtaining them) from what was an important survival concept in the hunter-gatherer lifestyle, we have produced conditions favouring unrestrained over-consumption which now threaten the health of Western societies. By being energy-dense our diet facilitates excessive energy intake and obesity, a problem compounded by our low physical activity and energy output. Excessive consumption of particular nutrients, such as saturated fat, also directly increases the risk of many of the chronic degenerative diseases.

CONCLUSIONS

The hunter-gatherer, or paleolithic diet, has important therapeutic implications for the treatment and prevention of many of the diseases of developed affluent societies (obesity, type 2 diabetes, cardiovascular diseases, gallstones, bowel diseases, certain cancers). Its principles are simple: a combination of very lean meat, organ meats, fish and shellfish and a range of vegetables and fruits in whatever proportions are preferred. Our research with Aborigines has shown that even when such a diet is very rich in animal foods it can significantly reduce the major risk factors for type 2 diabetes and cardiovascular disease in a relatively short period. The major characteristics of the hunter-gatherer lifestyle are summarised in Table 11. The advantages of such a diet are:

(1) low in total fat;
(2) particularly low in saturated fat;
(3) a wide range of different polyunsaturated fatty acids (PUFA) including the physiologically important long chain PUFA of both the n-3 and n-6 series;
(4) nutrient-rich (vitamins, minerals);
(5) carbohydrate which is slowly digested and absorbed;
(6) high in different types of dietary fibre;
(7) low in sodium, high in potassium;
(8) *evolutionary precedent.*

Table 11
Characteristics of hunter-gatherer and Western lifestyles

Characteristic	*Hunter-gatherer lifestyle*	*Western lifestyle*
Physical activity level	High	Low
Diet		
Energy density	Low	High
Energy intake	Moderate	High
Protein	High	Low–moderate
Animal	High	Low–moderate
Vegetable	Very low	Low–moderate
Carbohydrate	Low–moderate (slowly absorbed)	Moderate (rapidly absorbed)
Fibre	High	Low
Fat	Low	High
Vegetable	Very low	Moderate
Animal	Low, polyunsaturated	High, saturated
n-6/n-3 ratio	Low	High
Long chain PUFA	High	Low

This type of diet is also unusually rich in animal protein and high in cholesterol—characteristics not generally favoured by nutritionists in making recommendations for better health. However, it is important to appreciate that the most positive feature of the hunter-gatherer diet overall is its evolutionary precedent. Because it is the diet which humans have consumed for the greater part of their history, it seems reasonable to assume that it is the diet to which we are best suited.

REFERENCES

Aaes-Jorgensen, E. (1961). Essential fatty acids. *Physiol. Rev.*, **41**, 1–51.

Adlof, R. O. & Emken, E. A. (1986). Distribution of hexadecanoic, octadecaenoic and octadecadienoic acid isomers in human tissue lipids. *Lipids*, **21**, 543–7.

Aveldano, M.I. & Sprecher, H. (1983). Synthesis of hydroxy fatty acids from 4,7,10,13,16,19-(1-^{14}C) docosahexaenoic acid by human platelets. *J. Biol. Chem.*, **258**, 9339–43.

Bastian, P. (1979). Coronary heart disease in tribal Aborigines—the West Kimberley Survey. *Aust. NZ J. Med.*, **9**, 284–92.

Baudet, M. F., Dachet, C., Lasserre, M., Esteva, O. & Jacotot, B. (1984). Modification in the composition and metabolic properties of human low density and high density lipoproteins by different dietary fats. *J. Lipid Res.*, **25**, 456–68.

Beare-Rogers, J. L. (1983). Trans- and positional isomers of common fatty acids. In *Advances in Nutrition Research*, ed. H. H. Draper. Plenum Publishing Corporation, New York, pp. 171–200.

Bjerve, K. S., Mostad, I. L. & Thoresen, L. (1987). Alpha-linolenic acid deficiency in patients on long-term gastric-tube feeding: estimation of linolenic acid and long-chain unsaturated n-3 fatty acid requirement in man. *Am. J. Clin. Nutr.*, **45**, 66–77.

Brand, J. C., Cherikoff, V., Lee, A., Truswell, A. S. (1982). An outstanding food source of vitamin C. *Lancet*, **2**, 873.

Brand, J. C., Rae, C., McDonnell, J., Lee, A., Cherikoff, V. & Truswell, A. S. (1983). The nutritional composition of Australian Aboriginal bushfoods, 1. *Food Technol. Aust.*, **35**, 293–8.

Brand, J. C., Cherikoff, V. & Truswell, A. S. (1985). The nutritional composition of Australian Aboriginal bushfoods, 3. Seeds and Nuts. *Food Technol. Aust.*, **37**, 275–9.

Brignoli, C. A., Kinsella, J. E. & Weihrauch, J. L. (1976). Comprehensive evaluation of fatty acids in food V. Unhydrogenated fats and oils. *J. Am. Dietet. Assoc.*, **68**, 224–9.

Bristol, J. B., Emmett, P. M., Heaton, K. W. & Williamson, R. C. N. (1985). Sugar, fat and the risk of colorectal cancer. *Brit. Med. J.*, **291**, 1467–70.

Brown, A. J., Cherikoff, V. & Roberts, D. C. K. (1987). Fatty acid composition of seeds from some Australian Acacia species. *Lipids*, **22**, 490–4.

Budowski, P. & Crawford, M. A. (1985). α-Linolenic acid as a regulator of the metabolism of arachidonic acid: dietary implications of the ratio, n-6 : n-3 fatty acids. *Proc. Nutr. Soc.*, **44**, 221–9.

Bull, N. L., Day, M. J. L., Burt, R. & Buss, D. H. (1983). Individual fatty acids in the British household food supply. *Human Nutr: Appl. Nutr.*, **37A**, 373–7.

Burr, G. O. & Burr, M. M. (1929). New deficiency disease produced by rigid exclusion of fat from the diet. *J. Biol. Chem.*, **82**, 345–67.

Butcher, L. A., O'Dea, K., Sinclair, A. J., Parkin, J. D., Smith, I. L. & Blombery, P. (1990). The effects of very low fat diets enriched with fish or kangaroo meat on cold-induced vasoconstriction and platelet function. *Prost. Leukotr. EFA.*, **39**, 221–6.

Collins, F. D., Sinclair, A. J., Royle, J. P., Coats, D. A., Maynard, A. J. & Leonard, R.F. (1971). Plasma lipids in human linoleic acid deficiency. *Nutr. Metabol.*, **13,** 150–67.

Connor, S. L., Gustafson, J. R., Artaud-Wild, S. M., Flavell, D. P., Classick-Kohn, C. J., Hatcher, L. F. & Connor, W. E. (1986). The cholesterol/saturated fat index: an indication of the hypercholesterolemic and atherogenic potential of food. *Lancet,* **1,** 1229–32.

Connor, W. E. & Lin, D. S. (1982). The effect of shellfish in the diet upon the plasma lipid levels in humans. *Metabolism,* **31,** 1046–51.

Corey, E. J., Shih, C. & Cashman, J. R. (1983). Docosahexaenoic acid is a strong inhibitor of prostaglandin but not leukotriene biosynthesis. *Proc. Natl. Acad. Sci.*, **80,** 3581–4.

Cowley, A. J., Heptinstall, S. & Hampton, J. R. (1985). Effects of prostacyclin and the stable prostacyclin analogue ZK 36374 on forearm blood flow and blood platelet behaviour in man. *Thrombos. Haemost.*, **53,** 90–4.

Crawford, M. A. (1968). Fatty-acid ratios in free-living and domestic animals: Possible implications for atheroma. *Lancet,* **1,** 1329–33.

Crawford, M. A. & Sinclair, A. J. (1972). Nutritional influences in the evolution of the mammalian brain. In *Lipids, Malnutrition and the Developing Brain,* Ciba Foundation Symposium. ASP, Amsterdam, pp. 267–92.

Crawford, M. A., Gale, M. M., Woodford, M. H. & Casperd, N. M. (1970). Comparative studies on the fatty acid composition of wild and domestic meats. *Int. J. Biochem.*, **1,** 295–305.

Crawford, M. A., Casperd, N. M. & Sinclair, A. J. (1976). The long-chain metabolites of linoleic and linolenic acids in liver and brain in herbivores and carnivores. *Comp. Biochem. Physiol.*, **54B,** 395–401.

Crawford, M. A., Hassam, A. G. & Rivers, J. P. W. (1978). Essential fatty acid requirements in infancy. *Am. J. Clin. Nutr.*, **31,** 2181–5.

Croset, M., Sala, A., Folco, G. & Lagarde, M. (1988). Inhibition by lipoxygenase products of TXA_2-like responses of platelets and vascular smooth muscle: 14-hydroxy from 22:6n-3 is more potent than 12-HETE. *Biochem. Pharmacol.*, **37,** 1275–80.

Doll, R. & Peto, R. (1981). The causes of cancer: quantitative estimates of avoidable risks of cancer in the United States today. *J. Nat. Cancer Inst.*, **66,** 1191–1308.

Dunstan, G. A., Sinclair, A. J., O'Dea, K. & Naughton, J. M. (1988). The lipid content and fatty acid composition of various marine species from southern Australian coastal waters. *Comp. Biochem. Physiol.*, **91B,** 165–9.

Dyerberg, J. & Bang, H. O. (1979). Haemostatic function and platelet polyunsaturated fatty acids in Eskimos. *Lancet,* **2,** 433–5.

Dyerberg, J., Bang, H. O. & Hjorne, N. (1975). Fatty acid composition of the plasma lipids in Greenland Eskimos. *Am. J. Clin. Nutr.*, **28,** 958–66.

Dyerberg, J., Bang, H. O., Shefferson, E., Moncada, S. & Vane, J. R. (1979). Eicosapentaenoic acid and prevention of thrombosis and atherosclerosis. *Lancet,* **2,** 117–9.

Eaton, S. B. & Konner, M. (1985). Paleolithic nutrition. A consideration of its nature and current implications. *New Engl. J. Med.*, **312,** 283–9.

Edington, J., Geekie, M., Carter, R., Benfield, L., Fisher, K., Ball, M. & Mann, J. (1987). Effect of dietary cholesterol on plasma cholesterol concentration in subjects following a reduced fat, high fibre diet. *Brit. Med. J.*, **294,** 333–5.

Emken, E. A., Rohwedder, W. K., Adlof, R. O., Rakoff, H. & Gulley, R. M. (1987). Metabolism in humans of *cis*-12, *trans*-15-octadecadienoic acid relative to palmitic, stearic, oleic and linoleic acid. *Lipids,* **22,** 495–504.

English, R. (1987). *Towards better nutrition for Australians.* Report of the Nutrition Taskforce of the Better Health Commission, Commonwealth Department of Health, Australian Government Publishing Service, Canberra, pp. 51–9.

Evans, L. T. (1975). Crops and world food supply, crop evolution and the origin of crop physiology. In *Crop Physiology: Some Case Histories;* Ed. L.T. Evans. Cambridge University Press, pp. 1–22.

Faber, M., Benade, A. J. S. & van Eck, M. (1986). Dietary intake, anthropometric measurements and blood lipid values in weight training athletes (body builders). *Int. J. Sports Med.*, **7,** 342–6.

Fiennes, R. N. T.-W., Sinclair, A. J. & Crawford, M. A. (1973). Essential fatty acid studies in primates. Linolenic acid requirements of capuchins. *J. Med. Primatol.*, **2,** 155–69.

Fischer, S. & Weber, P. C. (1984). Prostaglandin I_3 is formed *in vivo* in man after dietary eicosapentaenoic acid. *Nature,* **307,** 165–8.

Fischer, S., Weber, P. C. & Dyerberg, J. (1986). The prostacyclin/thromboxane balance is favourably shifted in Greenland Eskimos. *Prostaglandins,* **32,** 235–41.

Fischer, S., Vischer, A., Preac-Mursic, V. & Weber, P. C. (1987). Dietary docosahexaenoic acid is retroconverted in man to eicosapentaenoic acid, which can be quickly transformed to prostaglandin I_3. *Prostaglandins,* **34,** 367–75.

Fisher, E. A., Blum, C. B., Zannis, V. I. & Breslow, J. L. (1983). Independent effects of dietary saturated fat and cholesterol on plasma lipids, lipoproteins, and apolipoprotein E. *J. Lipid Res.,* **24,** 1039–48.

Fogerty, A. C., Evans, A. J., Ford, G. L. & Kennett, B. H. (1986). Distribution of omega-6 and omega-3 fatty acids in lipid classes in Australian fish. *Nutr. Rep. Int.*, **33,** 777–86.

Gibson, R. A. (1988). The effect of diets containing fish and fish oils on disease risk factors in humans. *Aust. NZ J. Med.*, **18,** 713–22.

Glatthaar, C., Welborn, T. A., Stenhouse, N. S. & Garcia-Webb, P. (1985). Diabetes and improved glucose tolerance: a prevalence estimate based on the Busselton 1981 Survey. *Med. J. Aust.,* **143,** 436–40.

Gortner, W. A. (1975). Nutrition in the United States, 1900–1974. *Cancer Res.*, **35,** 3246–53.

Greenberg, R. A. (1984). Meat—the last roundup. In *Agriculture and Human Nutrition,* ed. H. Glasson & R. Wise. NSW Department of Agriculture, Sydney, pp. 44–54.

Grundy, S. M. (1986). Comparison of monounsaturated fatty acids and carbohydrates for lowering cholesterol. *New Engl. J. Med.*, **314,** 745–8.

Grundy, S. M. (1987). Monounsaturated fatty acids, plasma cholesterol, and coronary heart disease. *Am. J. Clin. Nutr.*, **45,** 1168–75.

Hansen, A. E., Wiese, H. F., Boelsche, A. R., Haggard, M. E., Adam, D. J. D. & Davis, H. (1963). Role of linoleic acid in infant nutrition. Clinical and chemical study of 428 infants fed on milk mixtures varying in kind and amount of fat. *Paediatrics*, **31** (Supplement 1, part 2), 171–92.

Hansen, H. A. (1986). The essential nature of linoleic acid in mammals. *TIBS*, **11**, 263–5.

Harris, W. (1989). Fish oils and plasma lipid and lipoprotein metabolism in humans: a critical review. *J. Lipid Res.*, **30**, 785–807.

Hegsted, D. M., McGandy, R. B., Myers, M.L. & Stare. F. J. (1965). Quantitative effects of dietary fat on serum cholesterol in man. *Am. J. Clin. Nutr.*, **17**, 281–95.

Hepburn, F. N., Exler, J. & Weihrauch, J. L. (1986). Provisional tables on the content of omega-3 fatty acids and other fat components of selected foods. *J. Am. Dietet. Assoc.*, **86**, 788–93.

Higgs, G. A. (1985). The effects of dietary intake of essential fatty acids on prostaglandin and leukotriene synthesis. *Proc. Nutr. Soc.*, **44**, 181–7.

Holman, R. T. (1968). Essential fatty acid deficiency. *Prog. Chem. Fats Other Lipids*, **9**, 279–348.

Holman, R. T. (1970). Biological activities of and requirements for polyunsaturated fatty acids. *Prog. Chem. Fats Other Lipids*, **9**, 607–82.

Holman, R. T., Caster, W. O. & Weise, H. (1964). The essential fatty acid requirements of infants and the assessment of their dietary intake of linoleate by serum fatty acid analysis. *Am. J. Clin. Nutr.*, **14**, 70–5.

Holman, R. T., Johnson, S. B. & Hatch, T. F. (1982). A case of human linolenic acid deficiency involving neurological abnormalities. *Am. J. Clin. Nutr.*, **35**, 617–23.

Horrobin, D. F. (1988). Prostaglandin E_1: physiological significance and clinical use. *Wiener Klinische Wochen.*, **100**, 471–7.

Horrobin, D. F. & Huang, Y.-S. (1987). The role of linoleic acid and its metabolites in the lowering of plasma cholesterol and the prevention of cardiovascular disease. *Int. J. Cardiol.*, **17**, 241–55.

Katan, M. B. & Beynen, A. C. (1987). Characteristics of human hypo- and hyperresponses to dietary cholesterol. *Am. J. Epidem.*, **125**, 387–99.

Keys, A., Menotti, A., Karvonen, M. J., Aravanis, C., Blackburn, H., Buzina, R., Djordjevic, B. S., Dontas, A. S., Fidanza, F., Keys, M. H., Kromhout, D., Nedeljkovic, S., Punsar, S., Seccareccia, F. & Toshima, H. (1986). The diet and 15-year death rate in the Seven Countries Study. *Am. J. Epidem.*, **124**, 903–15.

Kinsella, J. E. (1987). Effects of polyunsaturated fatty acids on factors related to cardiovascular disease. *Am. J. Cardiol.*, **60**, 236–326.

Kulmacz, R. J., Sivarajan, M. & Lands, W. E. M. (1986). Measurement of the incorporation of orally administered arachidonic acid into tissue lipids. *Lipids*, **21**, 21–5.

Lasserre, M., Mendy, M., Spielman, D. & Jacotot, B. (1985). Effects of different dietary intakes of essential fatty acids and on C20:3n-6 and C20:4n-6 serum levels in human adults. *Lipids*, **20**, 227–33.

Ledger, H. P. (1968). Body composition as a basis for a comparative study of some East African mammals. *Symp. Zool. Soc. Lond.*, **21**, 289–310.

Lee, T. H., Hoover, R. L., Williams, J. D., Sperling, R. I., Ravalese, J., Spur, B. W., Robinson, D. W., Corey, E. J., Lewis, R. A. & Austin, K. F. (1985). Effect of dietary enrichment with eicosapentaenoic and docosahexaenoic acids on in vitro neutrophil and monocyte leukotriene generation and neutrophil function. *New Engl. J. Med.*, **312,** 1217–24.

Lokesh, B. R., German, B. & Kinsella, J. E. (1988). Differential effects of docosahexaenoic acid and eicosapentaenoic acid on suppression of lipoxygenase pathway in peritoneal macrophages. *Biochem. Biophys. Acta,* **958,** 99–107.

Martinez, M. & Ballabriga, A. (1987). Effects of parenteral nutrition with high doses of linoleate on the developing human liver and brain. *Lipids,* **22,** 133–8.

Mattson, F. H. & Grundy, S. M. (1985). Comparison of effects of dietary saturated, monounsaturated, and polyunsaturated fatty acids on plasma lipids and lipoproteins in man. *J. Lipid Res.*, **26,** 194–202.

Mead, J. F. (1968). The metabolism of the polyunsaturated fatty acids. *Prog. Chem. Fats Other Lipids,* **9,** 161–92.

Meehan, B. (1977). Hunters by the seashore. *J. Hum. Evolut.*, **6,** 363–70.

Mehta, J., Lopez, L. M. & Wargovich, T. (1987). Eicosapentaenoic acid: its relevance in atherosclerosis and coronary artery disease. *Am. J. Cardiol.,* **59,** 155–9.

Morita, I., Saito, Y., Change, W. C. & Murota, S. (1983). Effects of purified eicosapentaenoic acid on arachidonic acid metabolism in cultured aortic smooth muscle cells, vessel walls and platelets. *Lipids,* **18,** 42–9.

Naughton, J. M., O'Dea, K. & Sinclair, A. J. (1986). Animal foods in traditional Aboriginal diets: polyunsaturated and low in fat. *Lipids,* **21,** 684–90.

Needleman, P., Raz, A., Minkes, M. S., Ferrendelli, J. A. & Sprecher, H. (1979). Triene prostaglandins: prostacyclin and thromboxane biosynthesis and unique biological properties. *Proc. Natl. Acad. Sci.,* **79,** 944–8.

Nestel, P. J., Connor, W. E., Reardon, M. F., Connor, S., Wong, S. & Boston, R. (1984). Suppression by diets rich in fish oil of very low density lipoprotein production in man. *J. Clin. Invest.,* **74,** 82–9.

Neuringer, M. & Connor, W. E. (1986). n-3 fatty acids in the brain and retina: evidence for their essentiality. *Nutr. Rev.*, **44,** 285–94.

Neuringer, M., Connor, W. E., Anderson, G. J. & Reisbick, S. (1989). Are omega-3 fatty acids necessary for normal brain and retinal development. *Proc. Nutr. Soc. Aust.*, **14,** 1–12.

Nichaman, M. Z. & Hamon, P. (1987). Low-fat, high carbohydrate diets and plasma cholesterol. *Am. J. Clin. Nutr.,* **45,** 1155–60.

O'Dea, K. (1984). Marked improvement in carbohydrate and lipid metabolism in diabetic Australian Aborigines after temporary reversion to traditional lifestyle. *Diabetes,* **33,** 596–603.

O'Dea, K. & Sinclair, A. J. (1982). Increased proportion of arachidonic acid in plasma lipids after two weeks on a diet of tropical seafood. *Am. J. Clin. Nutr.,* **36,** 868–72.

O'Dea, K. & Sinclair, A. J. (1985). The effects of low fat diets rich in arachidonic acid on the composition of plasma fatty acids and bleeding time in Australian Aborigines. *J. Nutr. Sci. Vitaminol.,* **31,** 441–53.

O'Dea, K. & Spargo, R. M. (1982). Metabolic adaption to a low carbohydrate–high protein ('traditional') diet in Australian Aborigines. *Diabetol.*, **23**, 494–8.

O'Dea, K., Spargo, R. M. & Akerman, K. (1980). The effect of transition from traditional to urban lifestyle on the insulin secretory response in Australian Aborigines. *Diabetes Care*, **3**, 31–7.

O'Dea, K., Spargo, R. M. & Nestel, P. J. (1982). Impact of westernization on carbohydrate and lipid metabolism in Australian Aborigines. *Diabetol.*, **22**, 148–53.

O'Dea, K., Steel, M., Naughton, J., Sinclair, A. J., Hopkins, G., Angus, J., He, G.-W., Niall, M. & Martin, T. J. (1988*a*). Butter-enriched diets reduce arterial prostacyclin production in rats. *Lipids*, **23**, 234–41.

O'Dea, K., White, N. G. & Sinclair, A. J. (1988*b*). An investigation of nutrition related risk factors in an isolated Aboriginal community in northern Australia: advantages of a traditionally oriented lifestyle. *Med. J. Aust.*, **148**, 177–80.

O'Dea, K., Traianedes, K., Chisholm, K., Leyden, H. & Sinclair, A. J. (1990). Cholesterol-lowering effect of a low fat diet enriched with lean beef is reversed by the addition of beef fat. *Amer. J. Clin. Nutr.* (In Press.)

Pietinen, P. & Huttunen, J. K. (1987). Dietary fat and blood pressure—a review. *Europ. Heart J.*, **8**, Supplement B, 9–17.

Pilbeam, D. (1984). The descent of hominoids and hominids. *Sci. Am.*, **250**, 84–96.

Posatti, L. P., Kinsella, J. E. & Watt, B. K. (1975). Comprehensive evaluation of fatty acids in food 1. Dairy products. *J. Am. Dietet. Assoc.*, **66**, 482–8.

Ramesha, C. S., Gronke, R. S., Sivarajan, M. & Lands, W.E.M. (1985). Metabolic products of arachidonic acid in rats. *Prostaglandins*, **29**, 991–1008.

Renaud, S. (1985). Dietary fatty acids and platelet function. *Proc. Nutr. Soc. Aust.*, **10**, 1–13.

Rivett, D. E., Tucker, D. J. & Jones, G. P. (1983). The chemical composition of seeds from some Australian plants. *Aust. J. Agric. Res.*, **34**, 427–32.

Rizek, R. L., Welsh, S. O., Marston, R. M. & Jackson, E. M. (1983). Levels and sources of fat in the U.S. food supply and in diets of individuals. In *Dietary Fats and Health*, eds E.G. Perkins & W.J. Visek. American Oil Chemist's Society, Champaign, IL, pp. 13–43.

Salem, N., Kim, H.-Y. & Yergey, J. A. (1986). Docosahexaenoic acid: membrane function and metabolism. In *Health Effects of Polyunsaturated Fatty Acids in Seafoods*, eds A.P. Simopoulos, R.R. Kifer, & R.E. Martin. Academic Press, pp. 263–317.

Samuelsson, B., Dahlen, S. E., Lindgren, J. A., Rouzer, C. A. & Serhan, C. N. (1987). Leukotrienes and lipoxins: structures, biosynthesis and biological effects. *Science*, **237**, 1171–6.

Schonfeld, G., Patsch, W., Rudel, L. L., Nelson, C., Epstein, M. & Olsen, R. L. (1982). Effects of dietary cholesterol and fatty acids on plasma lipoproteins. *J. Clin. Invest.*, **69**, 1072–80.

Senti, F. R. (1985). Health aspects of dietary *trans* fatty acids. *Fed. Am. Soc. Exp. Biol.*, Bethesda, MD, pp. 27–34.

Simopoulos, A. P. (1989). Summary of NATO Advanced Workshop on dietary omega-3 and omega-6 fatty acids: biological effects and essentiality. *J. Nutr.*, **119,** 521–8.

Sinclair, A. J. & Crawford, M. A. (1975). The incorporation of radioactive polyunsaturated fatty acids into the liver and brain of the developing rat. *Lipids*, **10,** 175–84.

Sinclair, A. J. & O'Dea, K. (1987*a*). The lipid levels and fatty acid compositions of the lean portions of Australian beef and lamb. *Food Technol. Aust.*, **39,** 228–31.

Sinclair, A. J. & O'Dea, K. (1987*b*). The lipid levels and fatty acid compositions of the lean portions of pork, chicken and rabbit meats. *Food Technol. Aust.*, **39,** 232–3, 240.

Sinclair, A. J., Slattery, W. J. & O'Dea, K. (1982). The analysis of polyunsaturated fatty acids in meat by capillary gas–liquid chromatography. *J. Sci. Food Agric.*, **33,** 771–6.

Sinclair, A. J., O'Dea, K. & Naughton, J. M. (1983). Elevated levels of arachidonic acid in fish from northern Australian coastal waters. *Lipids,* **18,** 877–81.

Sinclair, A. J., O'Dea, K., Dunstan, G., Ireland, P. D. & Niall, M. (1987). Effects on plasma lipids and fatty acid composition of very low fat diets enriched with fish or kangaroo meat. *Lipids,* **22,** 523–9.

Snow, B. J., Brand, J. C. & Nathan, G. (1987). The glycemic index of traditional Pima Indian staples: a population at high risk from diabetes. *Proc. Nutr. Soc. Aust.*, **12,** 99.

Sommerfeld, M. (1983). *Trans* unsaturated fatty acids in natural products and processed foods. *Prog. Lipid Res.*, **22,** 221–33.

Spady, D. K. & Dietschy, J. (1982). Dietary saturated triacylglycerols suppress hepatic low density lipoprotein receptor activity in the hamster. *Proc. Natl. Acad. Sci.*, **82,** 4526–30.

Sprecher, H. (1977). Biosynthesis of polyunsaturated fatty acids and its regulation. In *Polyunsaturated Fatty Acids,* eds W.-H. Kunau & R.T. Holman. American Oil Chemists Society, Champaign, IL, pp. 1–18.

Stamler, J., Wentworth, D. & Neaton, J. D. (1986). Is the relationship between serum cholesterol and risk of premature death from coronary heart disease continuous or graded? *J. Am. Med. Assoc.*, **256,** 2823–8.

Steel, M. S., Naughton, J. M., Hopkins, G. W., O'Dea, K. & Sinclair, A. J. (1987). Effects of different types of dietary fat on haemostatic function and plasma and aortic fatty acid composition in rats. *Proc. Nutr. Soc. Aust.*, **12,** 108.

Thorburn, A. W., Brand, J. C. & Truswell, A. S. (1987*a*). Slowly digested and absorbed carbohydrate in traditional bushfoods: a protective factor against diabetes? *Am. J. Clin. Nutr.*, **45,** 98–106.

Thorburn, A. W., Brand, J. C., O'Dea, K., Spargo, R. M. & Truswell, A. S. (1987*b*). Plasma glucose and insulin responses to starchy foods in Australian Aborigines: a population now at high risk for diabetes. *Am. J. Clin. Nutr.*, **46,** 282–5.

Tinoco, J. (1982). Dietary requirements and functions of α-linolenic acid in animals. *Prog. Lipid Res.*, **21,** 1–46.

Traianedes, K., Chisholm, K. C., O'Dea, K. & Sinclair, A. J. (1987). The cholesterol-lowering effect of low fat diets rich in lean beef is reversed by adding beef fat to the diet. *Proc. Nutr. Soc. Aust.*, **12,** 107.

Vergroessen, A. J. (1977). Physiological effects of dietary linoleic acid. *Nutr. Rev.*, **35,** 1–5.

von Schacky, C. & Weber, P. C. (1985). Metabolism and effects on platelet function of the purified eicosapentaenoic and docosahexaenoic acids in humans. *J. Clin. Invest.*, **76,** 2446–50.

Whyte, M., Nestel, P. & MacGregor, A. (1977). Cholesterol metabolism in Papua New Guineans. *Europ. J. Clin. Invest.*, **7,** 53–60.

Willis, A. L. (1981). Nutritional and pharmacological factors in eicosanoid biology. *Nutr. Rev.*, **39,** 289–301.

Wise, P. H., Edwards, F. M., Thomas, D. W., Elliott, R. B., Hatcher, L. & Craig, R. (1976). Diabetes and associated variables in the South Australian Aboriginal. *Aust. NZ J. Med.*, **6,** 191–6.

Woodward, D. R., Macphail, M. & King, H. (1987). A history of the diet and health of Tasmanian Aborigines. In *Epidemiology in Tasmania,* Ed. H. King. Brolga Press, Canberra, pp. 233–58.

Yamanaka, W. K., Clemans, G. W. & Hutchinson, M. L. (1981). Essential fatty acid deficiency in humans. *Prog. Lipid Res.*, **19,** 187–215.

Chapter 2

Dietary Fat and Human Health

G. ROSE
London School of Hygiene and Tropical Medicine, University of London, UK

BACKGROUND

Since the days of Hippocrates it has been recognised that environment and life-style affect the incidence and distribution of common diseases; and a principal route by which they do this is through the food we eat.

Throughout most of his history, man's chief dietary problem has been undernutrition, particularly of energy (calories); and in much of the world this is still true. To be poor meant to be thin, and this was reflected in the view that the greatest dietary privilegc was 'to eat off the fat of the land'. That situation has now changed, the problems of undernutrition having been largely replaced in developed countries by the problems of overnutrition and nutrient imbalance, so that overweight is now commoner than underweight. This is due partly to improved food supply and purchasing power, but also to a large and continuing decline in energy requirements, as advances in transport and mechanisation increasingly remove the need for physical activity.

The energy density of fat (calories per gram) is more than double that of carbohydrate or protein, and this differential is increased by the lower amounts of water and fibre in fatty foods. This means that the energy yield of the modern high-fat diet is high in relation to its filling power, which tends to promote overweight; thus people who change to a low-fat diet tend to lose weight spontaneously. In this way a high fat intake is one of a number of factors contributing to obesity, with all its attendant health problems (principally musculoskeletal disorders, diabetes and cardiovascular disease). This is a non-specific effect of an augmented energy intake in the face of declining needs.

In addition to its non-specific effect on energy balance, a high intake of fat also has potent effects on metabolism. These are specific, and they depend not only on the total fat intake but also on its particular fatty acid composition. There is evidence that these disturbances of bodily fat metabolism are the necessary precursor of the epidemic of coronary heart disease which in recent decades has so dominated mortality throughout the industrialised countries of the world. This evidence and its implications will be reviewed in some detail.

Finally some correlations have been found between fat consumption and the incidence of cancers of the breast and colon. Although far less complete at present than the evidence linking fats with heart disease, this might turn out to be important and it too will be reviewed, though more briefly.

THE CHANGING WESTERN DIET

The history of nutrition is a dangerous discipline, especially if it involves international comparisons. Standardised national food surveys are of comparatively recent origin; they are conducted differently in different countries (and often not at all); and from time to time within one country their methods are changed. Moreover they do not measure wastage within the household, and this may vary over place and time. For example, in a recent survey most young people said that they cut the visible fat off meat, but we do not know if this is a new phenomenon. Even greater uncertainties apply to the differences based on national estimates of food entering the supply chain.

It is widely believed—though with little quantitative evidence—that the typical working-class diet of the 19th and early 20th centuries was high in complex carbohydrates and fibre and relatively low in fat. Because most animals were fed low energy diets (mainly grass), meat would have been leaner and its fat would have been less saturated. One supposes therefore that those who were not rich generally had a low total fat intake, with a high P:S ratio.

In the UK the government's National Food Survey has been maintained with much the same techniques since 1952 (Derry & Buss, 1984). At that time the esitmated mean intake of fat was 94 g/person/day; by

1970 this had risen to around 120 g, when a progressive and continuing decline set in, the 1986 figure being 98 g.

These figures for total fat intake are an oversimplification, in the first instance because they conceal major changes in the nature of the fats eaten. Table 1 shows for USA, Australia and UK an irregular but on the whole downwards trend in the proportion of fat derived from animals. US Department of Agriculture figures for food *supply* suggest a continuing rise in fat supply over the period 1909–81, largely reflecting a growth in mono- and polyunsaturates (Epstein, 1983). On the other hand US Nationwide Food *Consumption* Surveys report a 17% decline in total fat intake over the period 1965 to 1977. The common ground between the two kinds of data is that the intake of polyunsaturates has risen considerably in the last 20 years.

Table 1
Animal fats as a percentage of total fat intake (after Dwyer & Hetzel, 1980)

Year	*USA*	*Australia*	*UK*
1935	65	82	76
1950	65	79	60
1958	61	81	66
1962	68	84	66
1966	66	76	65
1970	62	75	65
1974	57	73	64

Table 2 gives more detail for recent UK experience, showing (for meat fat) some shift away from 'beef, lamb and pork', accompanied by much larger declines for milk and butter. As a result the total intake of saturated fatty acids fell by about 10% in five years. At the same time however total energy intake also fell (related to declining physical activity), and as a result the fat and saturated fat intakes as a proportion of total energy remained almost the same. It is supposed that it is these proportionate values rather than the absolute intakes which best indicate the health effects, although there is little direct evidence on this point.

The fatty acid composition of the diet can be assessed much more accurately by analysis of fat biopsy specimens, which can be taken easily and

Table 2
Average weekly intakes of selected fat-containing foods in Britain (UK National Food Surveys: Ministry of Agriculture, Fisheries and Food)

	1980	*1984*	*1986*
Beef, lamb, pork (g)	475	365	375
Poultry (g)	189	205	207
Meat products (g)	443	442	447
Milk: whole (litre)	2·36	2·05	1·73
reduced fat (litre)	—	0·19	0·40
Cheese (g)	110	109	118
Butter (g)	115	81	64
Margarine (g)	108	116	116
Energy (Kcal)	2230	2060	2070
Fat (g) (% energy)	106(42·6)	97(42·3)	98(42·6)
Satd fatty acids (g) (% energy)	47(18·8)	42(18·3)	41(17·7)

almost painlessly with a small needle from the abdominal wall. Table 3 gives the results of a recent international comparison. In Finland and Scotland (with very high mortality from heart disease) the P:S ratio is about half that in Italy (where the heart disease mortality is much lower). The distribution of individual fatty acids reflects the Italians' higher consumption of oils and lower consumption of meat.

The pattern of fat intake is capable of rapid change in response to public interest and health education. In a World Health Organization trial men were advised to eat less fat and to substitute some animal fats by

Table 3
Fatty acid composition (% and standard error) of adipose tissue from men in Finland, Scotland and Italy (Riemersma *et al.*, 1986)

	Finland	*Scotland*	*Italy*
All saturates	38·5(0·5)	36·3(0·6)	25·9(0·4)
All polyunsaturates	8·7(0·3)	10·2(0·3)	15·0(0·5)
P:S	0·23(0·01)	0·30(0·01)	0·58(0·02)
Stearic	7·0(0·5)	6·4(0·2)	3·5(0·1)
Oleic	45·0(0·3)	46·4(0·4)	54·3(0·7)
Linoleic	7·4(0·3)	8·8(0·3)	13·4(0·5)

polyunsaturated margarines and oils (Kornitzer & Rose, 1985). After four years of intervention in the Italian section of the trial, the P : S ratio of body fats had risen by 31%. Surprisingly, linoleate content had remained constant but arachidonate had increased by 42%, illustrating that the biological effects of dietary advice may be unpredictable.

CORONARY HEART DISEASE

Nature of the Problem

The problem is not new: it has been identified in the mummified body of a Chinese princess who died suddenly 4000 years ago! What is new is its mass occurrence. In Western countries it accounts for about one quarter of all deaths; many more suffer from symptoms or non-fatal attacks; and few of us survive to old age with healthy arteries. Measuring the risk factors can identify different levels of risk; but even among the group with the lowest estimated risk, the single commonest cause of death is coronary heart disease. It is a mass problem, and we must therefore look for mass causes, and for means to control them.

The incidence varies widely, even among industrialised countries: in Japan the rate is one-sixth of that in America or Britain, and it is not rising; but when Japanese migrate to America, their children acquire American coronary rates, implying that environmental rather than genetic factors must underlie the incidence differences. This view is supported by the speed and variety of changes within individual countries. In Eastern European countries, coronary mortality is rising at more than 5% a year, whereas in the USA and Australia it has fallen by 40% in 20 years (the equivalent of an average gain of two years in life expectancy).

Dietary Fat, Blood Cholesterol and Cardiovascular Disease

The evidence linking fats, cholesterol and heart disease goes back to three observations made about 80 years ago, soon after it had become possible to measure the blood cholesterol concentration. Clinicians then realised that patients (such as diabetics) with raised blood cholesterol were subject to premature atherosclerosis. In the laboratory Anitschkov found that by feeding an omnivorous diet to rabbits he could produce cholesterol deposits in artery walls (Anitschkov, 1933). And de Langen noted that blood cholesterol levels were lower among Javanese eating traditionally than in those who adopted the fattier Western diet (Keys, 1967). Such obvservations laid the basis for an argument which in essence is now firmly substantiated, although still in process of continuing elaboration and development:

1. Dietary fat regulates blood lipid levels.
2. Blood lipid levels (especially cholesterol) determine the risk of atherosclerotic diseases.
3. Control of blood lipids reduces the risk of heart attack.

The evidence has continued to come (as it began) from clinical, laboratory and epidemiological sources.

Blood Cholesterol Regulation

For many years it was thought that it was the total fat intake which determined the cholesterol level, until Groen *et al.* (1952) showed the varying effects of different fatty acids. Controlled experiments in metabolic wards led to the 'Keys' formula' for predicting serum cholesterol changes following a change in dietary fats and cholesterol (Grande *et al.*, 1972):

$$\Delta Chol = 0{\cdot}039\Delta Z + 0{\cdot}031(2\Delta S - \Delta P)$$

where: $\Delta Chol$ = Serum cholesterol change (mmol/litre)
ΔZ = $\sqrt{\text{Change in cholesterol intake (mmol/day)}}$
ΔS = Change in percent energy from C_{12-16} saturated fatty acids
ΔP = Change in percent energy from polyunsaturated fatty acids

From this forumla it can be seen that the cholesterol level is raised by shorter chain fatty acids, though not by stearic or longer chain acids; and it is lowered by polyunsaturated fatty acids, but with only half the potency of the cholesterol-raising effects of saturates. The effects of particular fatty acids are still being debated (Reiser *et al.*, 1985). Dietary cholesterol is thought to make some contribution to the blood cholesterol level; but it is not large, and it probably depends on the particular dietary context. Most of the body's cholesterol is synthesised in the liver, and any increase in dietary cholesterol is largely compensated by diminished biosynthesis. The regulatory effect of dietary fats on the blood cholesterol level is achieved by altered rates both of biosynthesis and of biliary excretion of cholesterol degradation products.

The regulation of the blood cholesterol level is evidently an active homeostatic mechanism designed to produce an appropriate internal balance between cholesterol and fatty acids. Cell membrane liquidity, which is crucial to transport across cell membranes and to cell division, is found to depend on the balance between cholesterol concentration and the P:S ratio of fatty acids in the phospholipids of the cell wall. The P:S ratio passively reflects the pattern of dietary fat, and so it seems that the maintenance of the proper P:S–cholesterol balance has to depend on cholesterol variations. The system evolved to cope with much higher P:S ratios and lower saturated fat intakes than are prevalent nowadays. The high blood cholesterol levels of Western man can be seen as an active adaptation to this unusual diet. So far as cellular biochemistry is concerned, the adaptation seems successful: but it is achieved at the price of blood cholesterol levels which arteries cannot tolerate.

At a population level it is found that the wide differences in national diets closely match their mean serum cholesterol levels: Keys (1970) reported a correlation coefficient of 0·9 between mean serum cholesterol and mean percentage of energy from saturated fats, the slope of the regression being similar to that seen in dietary experiments on individuals. In cross-sectional studies in individuals it has proved difficult to identify any relation between diet and serum cholesterol, partly because of variablity in both dietary and lipid measurements, but chiefly because there are large genetically-determined differences in the set-points around which individuals regulate their cholesterol levels.

Most published work in this area is based on measurements of the total serum cholesterol. The critical component in causing arterial damage is carried on a low density lipoprotein (LDL–cholesterol), which accounts for about 85% of the total cholesterol. The fraction carried on high density lipoprotein (HDL–cholesterol) has a quite different significance, being a measure of the rate of clearance of cholesterol from arteries and other tissues. Fat intake is not an important determinant of the HDL fraction.

Blood Cholesterol and Heart Disease

The incidence of coronary heart disease (CHD) in populations correlates closely with both the mean serum cholesterol level and the mean proportion of energy derived from saturated fats. Keys (1970) found a correlation coefficient in each case of 0·8. In every population where CHD is common, the mean serum cholesterol level always exceeds 5·17 mmol/litre (200 mg/dl) with a correspondingly high saturated fat intake. Above that level, incidence varies widely, due to the influence of many other factors, both known and unknown. Below that level—as in Japan—smoking, hypertension and obesity may be common, but not coronary heart disease.

A high national intake of saturated fat, mirrored by high serum cholesterol levels, stands in a unique relation to coronary heart disease: it is the one essential factor for mass disease. Atherosclerosis (involving cholesterol deposition in artery walls) is the primary underlying pathology, and atherosclerosis does not develop—in animals or man—unless there is an excess of circulating cholesterol; but if that excess is present then other factors, not by themselves sufficient, become major aggravating causes.

The effect is illustrated in Fig. 1 (Shaper, 1980). This emphasises the role of the P:S ratio, whereas earlier studies emphasised the intake of saturated fat. In the past these two have been so highly correlated that their effects can hardly be separated: it is only in recent years that some populations have increased their dietary P:S ratio whilst still maintaining a very high total fat intake.

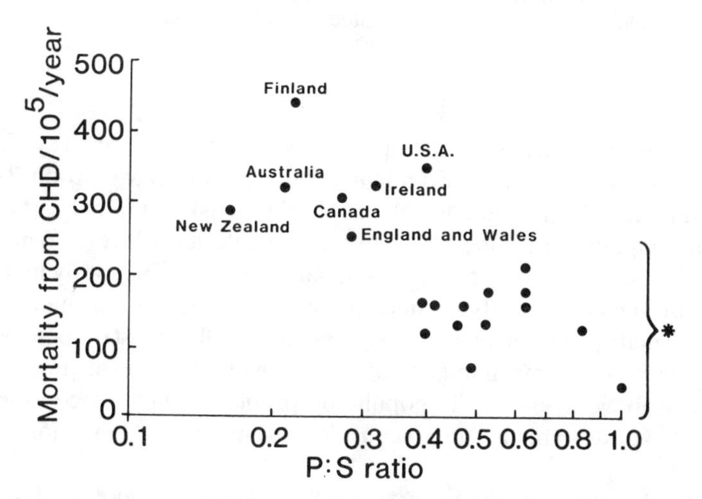

Fig. 1. National dietary P:S ratio and mortality from coronary heart disease in men aged 45–54 in 20 countries. (*Austria, Belgium, Denmark, France, W. Germany, Japan, Italy, Israel, Netherlands, Norway, Sweden, Switzerland, Venezuela.)

Within populations, wherever this critical mean cholesterol level is exceeded there is a strong and consistent relation between individual cholesterol levels (reflecting genes as much as diet) and the risk of heart attack. This is shown in Fig. 2 (Rose & Shipley, 1986). The rise in risk is curvilinear, but the important point is that there is no evident threshold: within Western populations, virtually everyone's cholesterol level seems to be too high. The whole population distribution has been shifted up into the risky range.

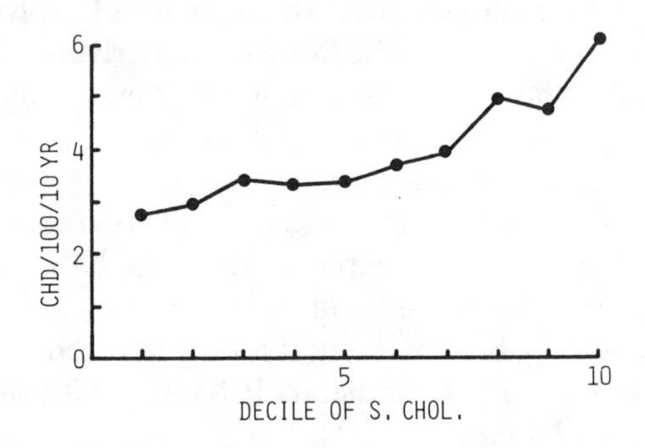

Fig. 2. Ten-year mortality from coronary heart disease according to decile of serum cholesterol in the Whitehall Study of 17718 male civil servants aged 40–64.

This observation has major implications for preventive strategy. Figure 3 is based on data from the Framingham Study (Kannel & Gordon, 1970), and it relates the individual cholesterol-related risk of a fatal heart attack to the population frequency of different cholesterol levels. When the level is above 8 mmol/litre the individual's risk is conspicuously high; but because such individuals are uncommon, they generate less than 10% of the population burden of cholesterol-related deaths. Most of these deaths arise from the large numbers of people with cholesterol levels at or slightly above average: the population problem is not the conspicuous minority of deviants but the inconspicuous majority, all of whom are at risk.

This means that the main problem for prevention is not the high-risk minority, with exceptional eating habits or unhealthy family background, but 'Mr and Mrs Average': the main underlying cause of the coronary epidemic is the average national diet.

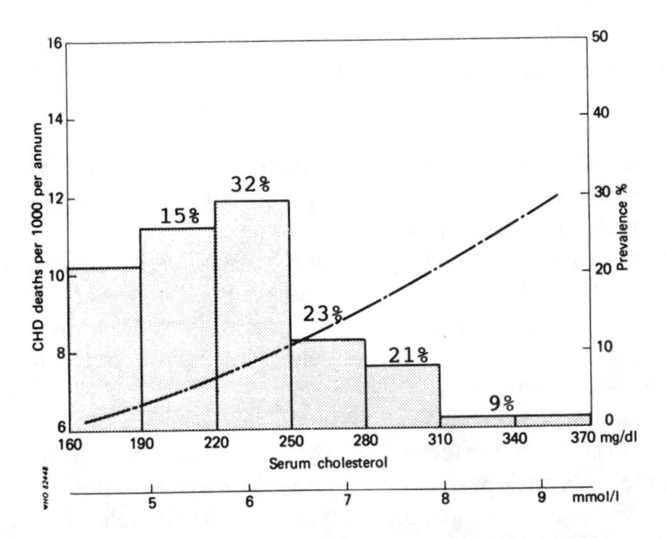

Fig. 3. Prevalence distribution (histogram) of serum cholesterol concentrations related to coronary heart disease mortality (interrupted line). The number above each column is the percentage of cholesterol-attributable deaths (men, aged 55–64).

Other Mechanisms

For many years aetiological thinking on heart disease was dominated by the saturated fat/cholesterol/atherosclerosis theory. This theory is now firmly established, but there are other ways also in which dietary fat may influence heart disease.

An acute heart attack is usually initiated by a blood clot forming on top of long-standing atherosclerosis. It has emerged that the plasma concentrations of certain clotting factors (Factor 7 and fibrinogen) are important risk factors (Meade *et al.*, 1986). The concentration of Factor 7 responds rapidly to changes in dietary fat—apparently an effect of total fat intake rather than of specific fatty acids, probably mediated by the neutral fat transport system (plasma chylomicrons and triglycerides). It implies that a high total fat intake could be risky, regardless of its nature.

Another active line of enquiry concerns a specific role for certain polyunsaturated fatty acids. The stickiness of vascular endothelium and platelets, which initiates clot formation, is controlled by prostaglandin synthesis, for which n-3 fatty acids form a substrate. A high intake of, for example, eicosapentaenoic acid (abundant in fish oils) inhibits the natural clotting tendency; but there is at present no direct evidence on how this affects coronary thrombosis. Trials are in progress. Indirect evidence from a Dutch study (Kromhout *et al.*, 1985) shows that coronary mortality is inversely related to fish consumption, but it did not seem that the protection was particularly related to fatty fish.

A more striking association has emerged between coronary mortality and a low intake of linoleic acid, as measured by fat biopsies. In certain populations this could explain some puzzling variations in mortality (Riemersma *et al.*, 1986), as well as helping to explain the occurrence of heart attacks in individuals.

These differences in linoleic acid intake are so clear as to suggest that polyunsaturates may protect against coronary disease more directly than by simple saturate-sparing or cholesterol-lowering effects; but the mechanism of such a protection, and in particular its relation to thrombosis, remains to be worked out.

Finally, the possibility must be mentioned that unnatural fatty acids produced by hydrogenation may have adverse biological effects. The use in margarine manufacture of cheap marine or vegetable oils can yield considerable quantities, for example, of *trans* isomers or of lightly-unsaturated longer chain fatty acids which are rare in traditional fat sources, the effects of which are unpredictable. Thomas (1975) reported correlations between consumption of hydrogenated fats and coronary mortality rates.

Evidence from Prevention

More than 20 years ago a National Diet–Heart Study was set up in the USA to assess the feasibility of a randomised controlled trial of the effects of a change in dietary fat on the incidence of heart attacks. It was concluded that such a trial was feasible, but that it would need 47 000 high cholesterol participants; no one was prepared to fund such a costly study, and so the definitive dietary trial will never be done.

A large number of smaller controlled trials have since been completed, with generally favourable results; but taken individually only one (the Oslo Heart Study: Hjerrmann *et al.*, 1981) achieved statistical significance. Recently Peto and colleagues (pers. comm.) have undertaken a regression analysis of all the published randomised trials of cholesterol-lowering interventions. Overall there is a very highly significant relationship between fall in serum cholesterol and reduction in heart attacks: a

10% fall in cholesterol corresponds to a 24% reduction in heart attacks. The relationship is similar in both dietary and drug trials, and the gradient corresponds to that reported in observational population studies.

Earlier in the chapter reference was made to the large changes in coronary mortality occurring in various countries, and to the weakness of historical nutrition data which might help to explain them (Epstein, 1983; Marmot, 1985). In the USA and Australia—the two countries with the longest-running and largest declines—this decline has involved all ages, men and women, and (in the USA) blacks and whites to a broadly similar extent. The environmental change which best parallels such a widespread decline seems to be the rise in dietary P:S ratio (Dwyer & Hetzel, 1980). A decline in saturated fat intake is less clear, although in the USA, Finland and Norway the mean serum cholesterol level in the population has fallen (by 10% or more in Finland and Norway). Conversely the alarming increase in coronary rates in eastern Europe seems to have accompanied a large *fall* in the dietary P:S ratio (Cooper, 1981).

FATS AND CANCER

Whereas the evidence linking fats with heart disease is of many different kinds, the link with cancers rests mainly on cross-population correlations with total fat intake (Armstrong & Doll, 1975; Liu *et al.*, 1979; Joossens & Geboers, 1985). Some of these correlations are however rather striking, particularly for cancers of the breast and colon (Figs 4(a) and 4(b), from Knox, 1977). There is also a correlation in Japan between a rising fat intake and a rising incidence of breast cancer.

If these correlations reflected a causal relation then they would be extremely important, since they relate to some of the commonest cancers; but at present the evidence is incomplete.

MEAT FAT

In all developed countries meat is a major source both of total and saturated fat intake. It is however much harder to quantify than the amounts derived from dairy products (the other main source of saturated fats). The composition of dairy fat is more or less constant under current farming conditions; and essentially all that is produced enters the market and is ultimately consumed. For meat fats, however, there is a wide but poorly

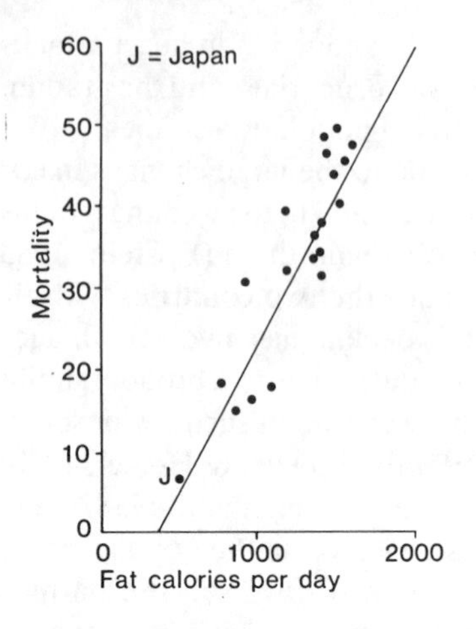

Fig. 4(a). Fat intake and death from breast cancer (aged 55–64) in different countries.

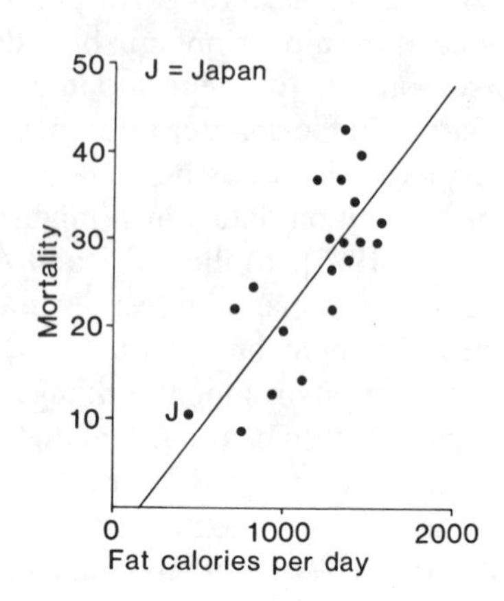

Fig. 4(b). Fat intake and death from cancer of the colon (aged 55–64) in various countries.

documented variation both in the product as marketed and in the proportion that subsequently gets eaten.

The total fat content of meat as produced depends on feeding and marketing practices; a variable and uncertain amount is then trimmed by butchers and disposed of by various routes; and finally there are substantial losses in cooking and in trimming of visible fat on the plate. The chemical composition of meat fat also varies, but this is more readily assessed. There are predictable differences in composition between different meat sources. To some extent carcass fat passively reflects the animal's feed: for example pigs reared on low-energy diets have less saturated fat and those fed high levels of oils have a fatty acid profile resembling that of the feed. This leads to important compositional differences between meat in developing and developed countries, but the variations within developed countries are currently not very important.

The data in Table 4 come from the most recent edition of the UK *Mc Cance and Widdowson* food tables (Paul & Southgate, 1978) and are (mostly) averages, based on laboratory analyses of a substantial number of supposedly representative samples. In the last decade however there have been large declines in average fat content of meat, particularly affecting pork and bacon. Trimming of visible fat, both in the kitchen and

Table 4
Fat content of various meats and meat products (Paul & Southgate, 1978)

Item	*g fat/100 g*
Bacon rashers, fried	22–45
Bacon rashers, grilled	19–36
Beef, sirloin, roast	21
Beef, rump steak, grilled	27
Chicken, roast	4–14
Lamb, leg, roast	26
Pork, chops, grilled	24
Pork, chops, grilled, lean only	11
Pork, leg, roast	20
Pork, leg, roast, lean only	7
Pork, sausages, grilled/fried	24
Veal, fillet, roast	11
Canned corned beef	12
Canned ham and pork, chopped	24
Pork pie	27
Beefburger, fried	17
Salami	45

on the table, has also increased, so for all these reasons the current contribution of meat to total fat intake is less than it appears, though still substantial.

Table 5 describes the typical fatty acid composition of different meats. Lamb is the most saturated, chicken much less so. The effect on serum cholesterol is thought to depend largely on the C_{14} and C_{16} saturated acids, stearic acid ($C_{18:0}$) having little effect (although still contributing to

Table 5
Fatty acid composition (%) of meat fats (dressed carcass, raw) (after Paul *et al.*, 1980)

	Saturated (excluding stearic)	*Monounsaturated*	*Polyunsaturated*
Beef	45(32)	51	4
Chicken	36(28)	48	16
Lamb	53(32)	42	5
Pork	43(29)	49	8

other biological effects). When stearic acid is excluded, the variation in content of the remaining saturates becomes unimportant.

IMPLICATIONS FOR NATIONAL FOOD POLICY

After many years of controversy there has now emerged a widespread agreement that excess dietary fat is a major cause of the current epidemic of coronary heart disease. In most developed countries the problem is not of a misguided minority of overeaters but of a faulty national diet, characterised by excess calories, excess saturated fat, and too low a P:S ratio.

In the World Health Organization Expert Committee report 1982 we concluded that:

> the balance of evidence indicates sufficient assurance of safety and a sufficient probability of major benefits to warrant action at the population level. . . . A population average value (of serum cholesterol) of under 5·17 mmol/litre (200 mg/dl) is likely to be associated with no more than a moderate frequency of coronary heart disease.

Such a population level of cholesterol is likely to be achieved when the national diet contains 10% or less of its energy as saturated fatty acids, and this figure was therefore adopted as a generally desirable goal. Given the likely composition of typical fat sources, this implied a goal of about 30% of total energy as fat; but this figure could be higher in countries where olive oil, for example, is popular.

The general principles of these WHO recommendations have been widely accepted. Obviously any one country may choose to work to a more realistic goal, achievable in a small number of years, rather than the 'ideal' WHO goals. In the UK, for example, it has been accepted that the national diet should aim to contain no more than 15% of energy as saturated fat and 35% as total fat (Committee on Medical Aspects of Food Policy, 1984), implying decreases of 25% and 17% respectively. (High risk individuals, or course, need more stringent guidelines.) These recommendations have been widely endorsed by professional bodies and by the food industry.

Official advice is clear on the need to reduce the intake of saturated fat, and it is most unlikely that any new evidence could upset this recommendation (unless it were to make it more specific in relation to particular fatty acids). The emphasis hitherto has been on the reduction in saturates, and a fall in total fat intake and some substitution of polyunsaturates have been seen simply as adjuvants to this end. Recent evidence, however (reviewed earlier in this chapter) is increasingly supporting both of these changes as perhaps desirable in themselves. Further evidence could well lead to advice on benefits from particular polyunsaturated sources.

A substantial fall in total fat intake is not popular, and at present the achievement of a reduction in saturates requires a considerable substitution of polyunsaturates. The emerging result in a number of countries is a high total fat, high-polyunsaturates diet. Countries that have moved in this direction are enjoying a decline in coronary mortality; but it is a diet of which we have little previous experience, and it will be important to monitor its health effects. So far, all seems well.

The dietary guidelines proposed on health grounds could be achieved in various ways. The current sources of saturated fat in the British diet are roughly dairy 40%, meat 30% and others 30%. To achieve a 25% reduction in the total is likely to need some contribution from each of these sources; but the balance is negotiable.

From the point of view of food producers the big issue is whether this reduction in saturated fat consumption will come from a reduced volume of dairy and meat production or from a changed constitution of the marketed products. The desired reduction is in fact achievable within the present production volumes, if there were major qualitative changes (leaner meat, and more low-fat dairy products)—*provided the excess animal fats are taken out of the human food cycle*.

The way that health advice and public choice will move in the coming years will depend substantially on the extent to which the food producers and distributors facilitate and encourage the reduction in saturated and total fat by qualitative changes.

REFERENCES

Anitschkov, N. N. (1933). Experimental arteriosclerosis in animals. In *Arteriosclerosis*, ed. E. V. Cowdry. Macmillan, New York, p. 271.

Armstrong, B. & Doll, R. (1975). Environmental factors and cancer incidence and mortality in different countries, with special reference to dietary practices. *Int. J. Cancer,* **15,** 617–31.

Committee on Medical Aspects of Food Policy (1984). *Diet and Cardiovascular Disease*, Report of the Panel on Diet in Relation to Cardiovascular Disease. HMSO, London.

Cooper, R. (1981). Rising death rates in the Soviet Union—the impact of coronary heart disease. *New Engl. J. Med.,* **304,** 1259–65.

Derry, B. J. & Buss, D. H. (1984). The British National Food Survey as a major epidemiological resource. *Br. Med. J.*, **288,** 765–7.

Dwyer, T. & Hetzel, B. S. (1980). A comparison of trends of coronary heart disease mortality in Australia, USA and England and Wales with reference to three major risk factors—hypertension, cigarette smoking and diet. *Int. J. Epidemiol.*, **9,** 65–71.

Epstein, F. H. (1983). Coronary heart disease—geographical differences and time trends. In *Atherosclerosis VI,* ed. F. G. Schettler, A. M. Gotto, G. Middelhoff *et al.* Springer Verlag, Berlin.

Grande, F., Anderson, J. T. & Keys, A. (1972). Diets of different fatty acid composition producing identical cholesterol levels in man. *Am. J. Clin. Nutr.*, **25,** 53–60.

Groen, J., Tjiong, B. K., Kamminga, C. E. & Willebrands, A. F. (1952). The influence of nutrition, individuality and some other factors, including various forms of stress, on the serum cholesterol; an experiment of nine months duration in 60 normal human volunteers. *Voeding,* **13,** 556–87.

Hjermann, I., Velve Byre, K., Holme, I. & Leren, P. (1981). Effects of diet and smoking intervention on the incidence of coronary heart disease. *Lancet,* **ii,** 1303–10.

Joosssens, J. V. & Geboers, J. (1985). *Diet, Cancer and Other Diseases, in Diet and Human Carcinogenesis.* Proceedings of the 3rd annual symposium of the European Organisation for Cooperation in Cancer Prevention Studies, Aarhus, Denmark, June 19–21, 1985. Excerpta Medica, Amsterdam.

Kannel, W. B. & Gordon, (T. Eds) (1970). Section 26. Some characteristics related to the incidence of cardiovascular disease and death: Framingham Study, 16-year follow-up. US Government Printing Office, Washington, DC.

Keys, A. (1967). Blood lipids in man—a brief review. *J. Am. Diet. Assoc.*, **51,** 508–16.

Keys, A. (Ed.) (1970). Coronary heart disease in seven countries. *Circulation,* **41,** Suppl. 1.

Knox, E. G. (1977). Foods and diseases. *Br. J. Prev. Soc. Med.*, **31,** 71–80.

Kornitzer, M. & Rose, G. (1985). WHO European collaborative trial of multifactorial prevention of coronary heart disease. *Preventive Medicine,* **14,** 272–8.

Kromhout, D., Bosschieter, E. B. & Coulander, C. de L. (1985). The inverse relation between fish consumption and 20-year mortality from coronary heart disease. *New Engl. J. Med.*, **312,** 1205–9.

Liu, K., Stamler, J., Moss, M., Garside, D., Persky, V., Soltero, I. (1979). Dietary cholesterol, fat, and fibre, and colon-cancer mortality. *Lancet,* **ii,** 782–5.

Marmot, M. (1985). Interpretation of trends in coronary heart disease mortality. *Acta Med. Scand.*, Suppl. 701, 58–65.

Meade, T. W., Stirling, Y., Thompson, S. G. *et al.* (1986). An international and interregional comparison of haemostatic variables in the study of ischaemic heart disease. *Int. J. Epidemiology,* **15,** 331–6.

Ministry of Agriculture, Fisheries and Food. (Annual). Household food consumption and expenditure. HMSO, London.

Paul, A. A. & Southgate, D. A. T. (1978). *McCance and Widdowson's The Composition of Foods,* 4th edn. HMSO, London.

Paul, A. A., Southgate, D. A. T. & Russell, J. (1980). *First Supplement to McCance and Widdowson's The Composition of Foods. Amino Acid Composition, Fatty Acid Composition*. HMSO, London.

Reiser, R., Probstfield, J. L., Silvers, A. *et al.* (1985). Plasma lipids and lipoprotein response of humans to beef fat, coconut oil and safflower oil. *Am. J. Clin. Nutr.*, **42,** 190–7.

Riemersma, R. A., Wood, D. A., Butler, S. *et al.* (1986). Linoleic acid content in adipose tissue and coronary heart disease. *Br. Med. J.*, **292,** 1423–7.

Rose, G. & Shipley, M. (1986). Plasma cholesterol concentration and death from coronary heart disease: 10 year results of the Whitehall study. *Br. Med. J.*, **293,** 306–7.

Shaper, A. G. (1980). Dietary prevention at home and abroad. *Postgrad. Med. J.*, **56,** 593–6.

Thomas, L. H. (1975). Mortality from arteriosclerotic disease and consumption of hydrogenated oils and fats. *Br. J. Prev. Soc. Med.*, **29,** 82–90.

World Health Organization (1982). Prevention of coronary heart disease. Report of a WHO Expert Committee, Technical Report Series 678. WHO, Geneva.

Chapter 3

Consumer Attitudes to Fat in Meat

JUDITH WOODWARD & VERNER WHEELOCK
Food Policy Research Unit, University of Bradford, UK.

INTRODUCTION

The shaping of an individual's attitudes and beliefs towards food is a complex and fluid process. It is influenced by many different factors including cultural, social and economic dimensions; the knowledge about, and availability of certain foods; organoleptic characteristics and perceptions of quality. Consumer attitudes towards fat in meat are no less complex and need to be examined within this broad context.

In recent years a major development in the food chain in the UK and other Western countries has been the expansion of supermarket retailing. This has provided consumers with a very wide range of foods from which to choose so that it is now very easy to alter the pattern of food purchasing. Hence, markets have become increasingly volatile. Furthermore, the retailers and manufacturers are constantly monitoring changes in public attitudes and perceptions towards food. With their huge buying power they can ensure that their suppliers develop the products which match the demands of the consumer.

As a consequence, consumer interest in food has been stimulated, and is reflected by the growing demand for high quality food and for product information. This phenomenon is illustrated by the current developments in healthy eating as the public become aware of the results of recent scientific evidence on the links between diet and health. These trends have been reinforced in the UK by the publicity given to several reports, including that prepared by the Committee on Medical Aspects of Food Policy (COMA, 1984) on *'Diet and Cardiovascular Disease'*, which was quickly accepted by the British Government. This report made recommendations on how the nation's diet should be changed with particular emphasis on the need to reduce fat intake.

This chapter looks at the various issues affecting consumer choice in relation to fat in meat and the implications for the rest of the food system. Most of the material considered has been drawn from data on British consumers. However, there are similarities in the experience of other Western industrialised countries and so reference is also made to information from New Zealand, the United States, West Germany, Belgium, and Finland.

Agriculture

Since World War II, the main objective of agricultural policy in the United Kingdom and in the European Community (EC) has been to increase production of virtually all commodities in order to gain self-sufficiency (Fallows & Wheelock, 1983). As a consequence of this policy, output has expanded dramatically in the UK (Fig. 1) and in the EC (Table

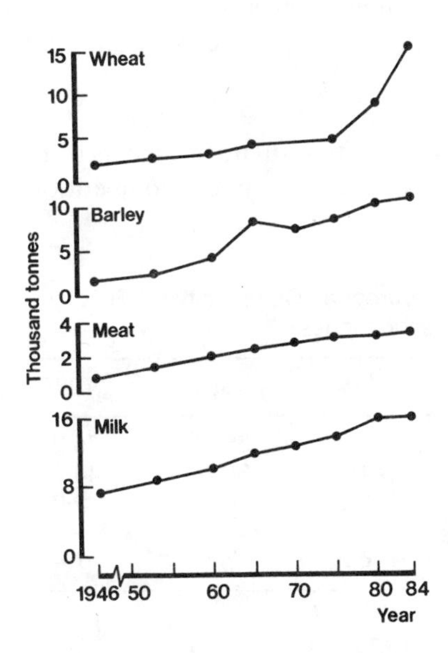

Fig. 1. Changes in the output of selected commodities in the UK, 1946–84. (Source: MAFF, 1967–1985).

1), where self-sufficiency for certain commodities was as follows:

Sugar	(1985/86)	124%
Wine	(1985/86)	108%
Cereals	(1985/86)	112%
Butter	(1984)	129%
Total meat	(1986)	102%
Pork	(1986)	107%
Beef	(1986)	103%
Sheep and goats	(1986)	84%
Poultry	(1986)	98%

Source: Eurostat (1988)

The implication is that there are now substantial surpluses for several commodities. However, production has continued to increase because the Common Agricultural Policy (CAP) effectively guarantees a market and price for the producers' output, irrespective of total demand. Approximately 70% of EC production, including beef and veal, pigmeat, most dairy products, and certain fish products is subject to internal price support as well as external protection. The cost to the community of such a scheme has risen steadily in recent years although objections to it are based not only on cost but on other grounds, such as (Howarth, 1985):

- It is inequitable between small and large producers.
- There is wide variation in support between producers of different commodities.
- It is ineffective.

The effect of such support for agriculture is to insulate producers from the market place. As support and protection declines, competition

Table 1
Production of selected commodities in the European Community (EUR 10) in recent years (thousand tonnes)

	1981	*1982*	*1983*	*1984*	*1985*
Common wheat	50·2	49·7	55·7	55·2	70·2
Barley	41·4	39·4	41·4	36·1	44·2
Butter	2·0	2·1	2·3	2·1	2·0
Skimmed milk	45·9	48·9	53·1	49·0	—
Beef/veal	7·0	6·7	6·9	7·5	7·4

European Commission (1987).

increases. Hence, farmers are having to attend more closely to consumer requirements and adjust their production accordingly. Some of the changes in consumption will not be important as far as agriculture is concerned because they are transitory or can be accommodated during the processing stage (for example, the removal of food additives). However, there will be others which are likely to have a significant impact on the amount and/or composition of agricultural commodities required.

In these circumstances, it is becoming imperative that the amount and composition of agricultural produce should match the demands of the consumer and the market place. Furthermore, because the market is changing and agriculture cannot respond quickly, it is becoming necessary to anticipate how consumer attitudes are likely to develop in future.

Retailers

At the retailer and consumer end of the UK food system the major development has been the growth of the supermarkets.

Today, the top five multiples have expanded to dominate the retail scene; Sainsbury, Tesco, Dee, Argyll, and Asda account for 52% of grocery sales, whilst the 10 biggest multiple retailers account for 70% of grocery sales (Institute of Grocery Distribution, 1987). This development has significantly influenced, and has been influenced by, changes in food processing and manufacturing. The major retailers have grown by offering convenience and variety together with value for money. There is intense competition between retailers and every effort is made to tune in to the demands of the consumer. This applies especially to fresh food including meat, as well as private-label products. When supermarket selling began, low price was regarded as the main selling attribute. In recent years, however, quality has become a major factor in the retailers' decision to stock, and in the consumers' decision to purchase.

The buying power of supermarket retailers enables them to exert considerable influence over their suppliers—ordering to detailed specification. The nature of the specifications is largely dependent upon the retailer's view of future customer demand. It follows, therefore, that there is considerable interest in obtaining information on how consumer attitudes to foods are changing. Major market surveys are regularly conducted to determine the impact of a variety of factors on consumer food choice.

Socioeconomic Factors Affecting Consumer Choice

During the past 30 years in the UK, there has been a significant increase in the proportion of married women in paid employment, from approximately 20% in the 1950s to approximately 70% in the 1980s. Family structure has also changed, with a decline in family size. Between 1971 and 1982 the average number of persons per household fell from 2·91 to 2·63.

Changes in lifestyle have also occurred; there is increased interest in leisure activities; people are less prepared to spend long hours producing meals and are more adventurous in their choice of foods. Furthermore, there has been an increase in affluence for the majority of the population, which enables consumers to take factors other than price into consideration when purchasing food with the result that there has been a growth in demand for value added goods such as convenience meals and speciality foods.

Surveys conducted by 'Harris International Marketing' indicate that price is no longer the dominant factor given by consumers as the reason for shopping at a particular site. In 1980, it was found that 55% of those interviewed identified 'price' as the most important reason for choosing a store compared with 30% who identified 'convenience' (i.e. accessibility). By 1985, 'convenience' (59%) had become the main reason and 'price' was only cited by 35% of the respondents (Anon. 1985).

However, it would be wrong to assume that this picture is universal, since the growth of single parent families (an increase from 8% in 1971/3 to 12% in 1981/3) and unemployment (from 0·6 million in 1974 to 3 million in 1984) has made many households significantly poorer during the last decade. For such households, price remains the dominant influence on food choice. A survey into the food circumstances, attitudes and consumption of people on low incomes revealed that 'good value for money' was the dominant factor affecting food choice for 75·8% of respondents, whilst nutrition and health considerations were important for 53·6% of respondents. Only 6·4% of those interviewed mentioned a food being 'different and unusual' as an important factor influencing food choice (Lang *et al.*, 1984).

Since attitudes to food are changing it follows that the traditional conservative attitudes to food choice are being eroded: consumers have become much more adventurous. This is largely due to the impact of the media, particularly television; the influx of ethnic foods into the UK; and the ability of the food manufacturer and retailer to respond rapidly and encourage new tastes and trends in consumer food choice.

The increasing availability of a wide range of innovative, exotic, or ethnic products acts to put further pressure on the more traditional relatively plain foods, which may have characteristics no longer perceived as desirable nor, in some cases, acceptable.

An example of this change in attitude towards a product is provided by the traditional British sausage. Sausage consumption began to decline in 1984 after three years of static sales. In 1985 consumption fell by 2%. This decline has been attributed to the increase in health awareness amongst consumers and the move towards a lower dietary intake of fat (Food Research Association, 1986). Unfavourable media publicity in 1985 had a particularly damaging effect on consumption, leading to a short term slump in sales of up to 15%.

Manufacturers responded to the diet and health concerns by producing a variety of 'low fat' and 'additive free' sausages. In 1984, Bowyers and Walls introduced low fat sausages, followed in subsequent years by other brands. Today, low fat sausages have stabilised at between 5% and 10% of total market consumption. In addition, the fat content of the 'ordinary' sausage produced by the major manufacturers has been reduced. The demand for new interesting products has resulted in the provision of an extended range of 'flavours' (Anon., 1987).

Demand for Quality

Changes in taste and an increased interest in food and food issues have led to the perception of quality gaining in importance as a factor affecting food choice. Perception of quality is largely subjective, being influenced by:

- — organoleptic characteristics such as taste and texture
- — visual characteristics
- — novelty value
- — image
- — peer group approval

Quality itself can be defined as 'A measure of the desirable characteristics of a product'. Some aspects of quality prevalent in the UK at present are:

- healthiness
- non-processed
- additive free
- naturalness
- freshness

Certain characteristics of foods which are deemed as positive quality attributes will be transitory in their appeal. Others, however, remain as a desirable attribute for long periods of time, resulting in a change in the nature of products offered to the consumer. Such changes in attitude towards quality have affected the whole food industry including the meat industry.

MEAT CONSUMPTION

Over the last 10–20 years a number of interesting features can be detected with regard to meat consumption in the EC (Frank, 1987) (Fig. 2a–d). In Italy, Greece, Spain and Portugal, consumption of meat is rising steeply. In Belgium, Ireland, and the Netherlands (Fig. 2b) it appears to have reached a plateau. However, since the late 1970s, the consumption of meat has fallen in France, West Germany (Fig. 2c), Denmark (Fig. 2d) and the United Kingdom (Fig. 2a).

In general there has been a growth in demand for pigmeat and for poultry, whilst consumption of beef and sheep meat has declined. Relative price is believed to be the major factor responsible for this change, although other influences, for example nutritional value, also play a part.

In the UK, meat consumption has slowly declined since the mid 1960s. Between the periods 1965–69 and 1980–84, consumption fell by approximately 4·7 kg per head per year (Frank, 1987). However, such a general decline in total meat consumption conceals considerable variation between the meat species.

Beef

Shortly after the end of rationing in 1954, consumption of beef in the UK increased, but began to decline in the late 1960s. In 1975 beef accounted for 54% of household consumption of carcass meat. However, by 1980 this had dropped to 49% (Ministry of Agriculture, Fisheries and Food,

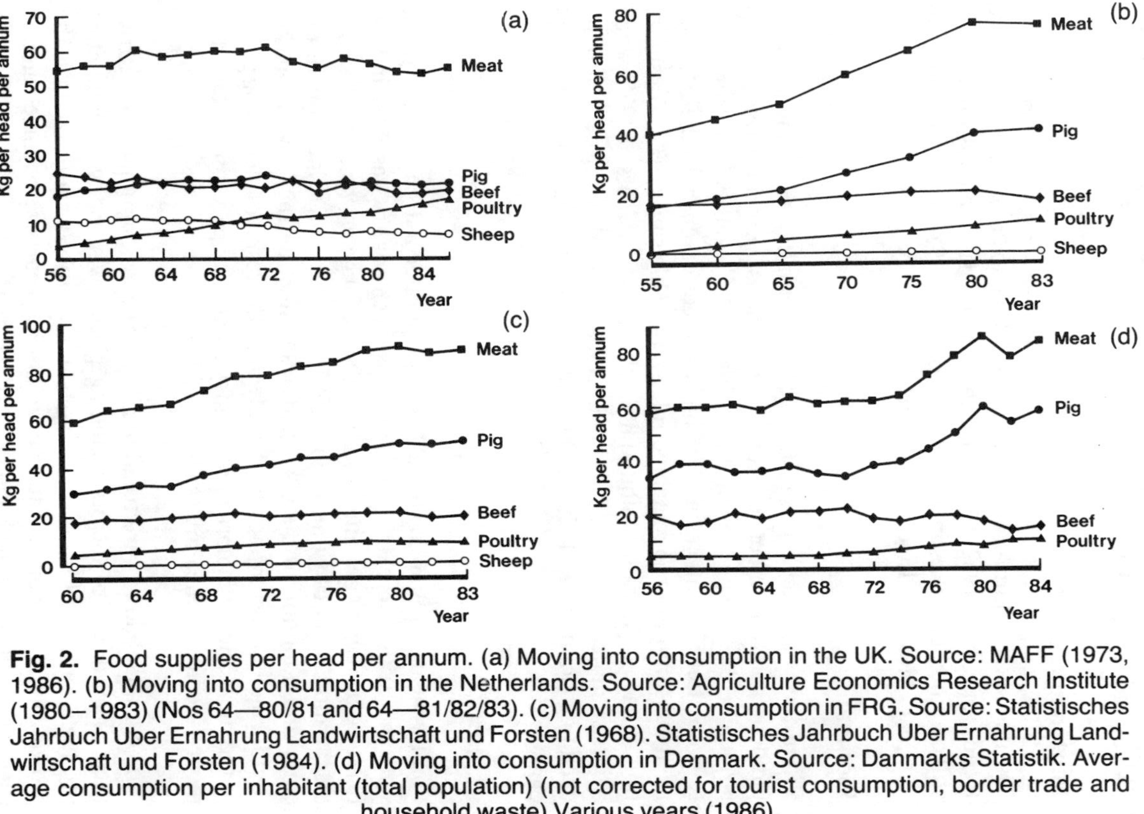

Fig. 2. Food supplies per head per annum. (a) Moving into consumption in the UK. Source: MAFF (1973, 1986). (b) Moving into consumption in the Netherlands. Source: Agriculture Economics Research Institute (1980–1983) (Nos 64—80/81 and 64—81/82/83). (c) Moving into consumption in FRG. Source: Statistisches Jahrbuch Uber Ernahrung Landwirtschaft und Forsten (1968). Statistisches Jahrbuch Uber Ernahrung Landwirtschaft und Forsten (1984). (d) Moving into consumption in Denmark. Source: Danmarks Statistik. Average consumption per inhabitant (total population) (not corrected for tourist consumption, border trade and household waste) Various years (1986).

1982). Consumption of beef appears to have increased slightly since 1983/4 and currently consumption is approximately 19·1 kg per head per year (Meat and Livestock Commission, 1987).

Mutton and Lamb

The demand for mutton and lamb in the UK has experienced a downward trend from about 1962. Since 1977 the decline has been less rapid but is still continuing. The estimated consumption for 1986 was 6·4 kg per head per year (Meat and Livestock Commission, 1987).

Pigmeat

When the war ended, production of pork, poultry meat and eggs expanded rapidly to meet the demand for animal protein. Consumption gradually rose to a plateau in 1965. Between 1976 and 1979 consumption revived after a decline caused by higher prices on entry to the EC. Pigmeat consumption has remained relatively stable since 1979. Consumption of bacon and ham exhibited little change in the years to 1972, when a persistent decline set in. This fall in demand correlates closely with the declining popularity of the traditional British breakfast of bacon and eggs.

Poultry

Poultry meat consumption in the UK has exhibited a strong upward trend since the 1950s. More than seven times as much poultry was eaten in 1986 (17·1 kg *per capita* per annum) as in 1954 (2·3 kg *per capita* per annum) (Meat and Livestock Commission, 1987). Even the price increases at EC entry and adoption of the CAP had little effect on the upward market trend. Higher consumption has largely been the result of increasing prices of red meat and falling prices of poultry meat. The perception of poultry meat as relatively 'healthy' is also important.

The gradual downward trend in UK meat consumption can be attributed to a variety of factors including price, availability of competing products, perceived value for money and quality. A major factor is the growing awareness that meat is not an essential part of a balanced diet. A survey carried out by the Food Policy Research Unit at Bradford University into consumer attitudes to diet and health in 1986 revealed that 63·2% of the respondents felt that meat was not necessary for a meal to be well balanced, and 81·3% of the respondents agreed that meals without meat could be just as substantial as meals including meat (Wright, 1987).

A second important factor contributing to a decline in meat consumption is the steady growth in the number of people who eat meat rarely or not at all. In 1984 this group accounted for approximately 4·0% of the population, whereas in 1987 it had risen to about 6·0% (3·5 million people) (Gallup, 1987). The importance of this change is that it has encouraged manufacturers to develop non-meat meals. These appeal to the non-meat eater, as well as being attractive to people who eat meat, but also enjoy a non-meat meal. This increased availability of products as protein alternatives acts to increase the competition for meat.

The decline in meat consumption, or the shift from one meat species to another is also attributable to perceptions of quality. One major factor in the perception of quality for meat is the lean : fat ratio. Attitudes towards fat in meat depend on a number of issues such as general organoleptic characteristics, perceived wastefulness, and health.

There has been a general shift away from fat consumption on taste/waste grounds for many years. However, in recent years the relationship with health has become more important and is now regarded as the main issue for many people.

DIET AND HEALTH

The existence of a causal relationship between food consumption patterns and health has been known for many years. Periodically, medical advisors in the UK have been commissioned to review the state of knowledge and to provide recommendations to government. Historically, such reviews have been produced by the medical profession for the medical profession and/or government—and as such have attracted little public attention. This is no longer the case in the UK, as nutritional issues move rapidly from the academic into the public arena.

Since the mid 1970s, nutrition and other aspects of food and health have attracted considerable media attention. Inevitably, mass media coverage of such issues can only present a snapshot of the deliberations rather than the considered evaluation of all the available material on the subject in question. In addition, consumers are constantly subjected to the marketing output of the various food manufacturers and retailers, much of which incorporates or alludes to the role of diet in the achievement of good health. One consequence of this shift in emphasis to the public discussion of the issues, is undoubtedly a heightened public awareness of the relationship between diet and health.

Medical Reports

The 1980s saw the publication in the UK of two particularly influential reports relating to the links between diet and disease. The first of these was a report to the National Advisory Committee on Nutrition Education (NACNE, 1983).

The NACNE Report

The NACNE commissioned an *ad hoc* group to review the existing reports relating to diet and health issues. As part of its remit the group produced a comprehensive series of guidelines on the dietary changes proposed. These guidelines were to be used as a basis for programmes of nutrition education. The resultant report was published as a discussion document by the Health Education Council (HEC) in September 1983.

The NACNE working group proposed that the average UK diet should be altered along the following lines:

- Reduce total fat consumption, especially saturated fat.
- Increase consumption of polyunsaturated fatty acids.
- Reduce sugar consumption.
- Increase fibre consumption.
- Reduce salt consumption.
- Moderate alcohol consumption

Although the NACNE Report was the result of an initiative to improve nutrition education and essentially not concerned with official policy on food and agriculture, the actual publication of the report was the centre of controversy and publicity. The results of this publicity helped to push the diet–health issue into the limelight and to make it a topical area of discussion for people outside the medical profession.

The proposals contained in the report were unprecedented in the UK because they actually quantified the nutritionally desirable amounts of dietary components such as total fat, for the British population. Previous quantified reports were limited to single medical topics or were concerned with recommended intakes of energy or particular nutrients (Fallows & Gosden, 1985). Official publications such as *Eating for Health* were not fully quantified (Department of Health and Social Security, 1978).

Shortly after the NACNE Report was published, the official DHSS report on *Diet and Cardiovascular Disease* was published (COMA, 1984).

COMA Report

The recommendations contained in the COMA report were very similar to those made by the NACNE working group. Furthermore the report was quickly accepted by the Government. Announcing the report in the House of Commons, the Secretary of State for Social Services, Norman Fowler, said (Hansard, 1984):

> I welcome this report. Heart attacks kill more men under 65 than any other disease—30 000 each year in England and Wales alone. This is an important report based on a very careful study of all the available evidence on the relationship between diet and health.

The 1984 COMA report did not restrict itself to diet, but also made recommendations on how the changes in food consumption could be facilitated. These included:

> Those responsible for health education should inform the general public of the recommendations and how to implement them. In particular, advice should be given on how to construct diets and regulate physical activity in order to minimise the risk of cardiovascular disease and avoid obesity. (Para. 2.3.1)

To producers, manufacturers and distributors of food and drink, and caterers:

> The percentage by weight of fat and of saturated fat, polyunsaturated and *trans* fatty acids in butter, margarine, cooking fats and edible oils should be printed on the container or wrapping in which they are sold. Consideration should be given to providing in addition (i.e. not in place of) uniform and more simple labelling codes to enable the general public to distinguish easily between fats and oils with low or high fat contents of saturated fatty acids. (para. 2.4.1.)

The report also recommended that detailed information on fat composition be provided wherever practical, for all foods with a

> fat content of more than 10% by weight, or which are major contributors to fat intake

This information should be included on the label of prepacked food wherever possible. Those which are not prepacked should have the information displayed prominently at retail outlets. Similar information should be provided by caterers (Para. 2.4.2).

The Government was also recommended to take appropriate action to improve the quality of nutrition education for the general population as well as in schools (Para. 2.6.1)
In addition:

> Consultation should take place between the relevant government departments and producers, manufacturers and distributors of food and drink and caterers which will lead to legislation and to Codes of Practice to improve public knowledge of the composition of foods . . . and lead to the provision of alternative preparations of some foods with lower content of saturated and of *trans* fatty acids and/or common salt. (Paras 2 .4.2, 2.4.3, 2.4.5.).

For years there had been arguments in scientific circles, in government and in industry about the relationship between diet and health and in particular about the role of dietary fat in the development of heart disease. Although it had been officially recognised by many expert committees that it would be advantageous to make changes in diet as part of a programme to reduce the incidence of the degenerative diseases, the UK had been slow to respond. The importance of the 1984 COMA report and its acceptance by the Government was in effect a watershed. It meant that there was a significant weakening in the position of those who opposed the introduction of dietary recommendations. Much more importantly it acted as a signal to those who were sitting on the sidelines. As a result some of the more progressive elements immediately moved into action. For example, within weeks of the release of the COMA report, Tesco, a major multiple food retailer, had taken the decision to implement a Healthy Eating Campaign. By January 1985, the Campaign was launched with the appearance of booklets on healthy eating in Tesco stores.

It would be a mistake to assume that the NACNE and COMA reports triggered the interest in healthy eating in the UK. Data from the National Food Surveys show that demand for certain foods, which may be considered as indicators for healthy eating, such as polyunsaturated margarine, low fat spreads, and wholemeal breads, although relatively small, had actually been growing for years. The significance of the impact of the COMA and NACNE reports is that they highlighted healthy eating as an important issue in contemporary society. The concept was rapidly moved from the fringes into the mainstream of food marketing.

Consumer Attitudes to Diet and Health

Consumer reaction to, and attitudes about such issues as the diet and health debate are constantly being assessed by social surveys. Such surveys are commissioned by commercial interests and independent research groups. The findings of such research, together with statistics on food consumption trends, have revealed a growing interest in 'healthy eating' and perhaps more importantly a growth in the number of consumers putting such knowledge into practice by making specific changes in their dietary habits.

Since this research is repeated on a regular basis it is becoming clear that the trend towards a healthier diet is now well established and is being facilitated by the increased availability of a choice of 'healthy' products. It can no longer be dismissed as a passing fad.

A survey into the consumer climate for red meat, in the United States, revealed that over 80% of respondents (n = 1211) agreed that 'these days, nutrition is as much a concern as price in buying food'. However, 45% agreed that they were more concerned with taste than nutrition when buying food. Of respondents in 1985, 68% agreed strongly that it was important to limit the fat in one's diet—even if they were not concerned about weight control, and 26% of respondents were considering or had already cut down on the amount of meat they were eating, for health reasons (Yankelovich *et al.*, 1985).

Group discussions held with consumers in May and September 1986 revealed the level of interest in diet and health topics in England (Wright & Slattery, 1986). Participants in the discussions spoke about a range of health problems, including heart disease, diabetes, obesity, arthritis, cancer, dental problems and allergies especially in children. Concern about heart disease had led to some avoidance of foods considered to be high in fat and cholesterol. Participants also mentioned that they were replacing red meat with poultry and fish, and changing from butter to polyunsaturated margarines and low fat spreads. White meat or fish were each widely viewed as preferable to red meat as part of a healthy diet.

Participants had also noticed the increasing variety of cuts of meat being made available, and that there was a move towards less fatty meat being on offer. Here is a typical comment (Wright & Slattery, 1986):

> You can get cuts of meat now that you wouldn't have heard of 20 years ago, some of these don't have much fat on, and make you think they are better for you.

A postal survey undertaken by the Food Policy Research Unit into consumer attitudes to diet and health, elicited further responses on the subject (Wright, 1987). Out of a sample of 576 respondents, 82% felt that they had started eating some foods because they considered them to be healthy. Whilst virtually all the respondents (98%) felt that 'people were a lot more conscious of health these days'. Another important aspect revealed in the survey was the changing attitude towards the importance of red meat in the diet. As many as 36% of respondents agreed with the attitudinal statement that a healthy diet does not contain red meat.

Though the trend towards healthy eating in New Zealand is less advanced than that in the UK (Hamilton, personal communication), meat has gained a negative image there. Red meat does not conform to the current demand for what has become known as 'light', 'white' food. Examples of such foods are chicken, fish and white wine. In addition, red meat products are losing popularity because of the view that they contain a high saturated fat content—which is considered to be unhealthy. In a survey of the attitudes of meat purchasers (consumers) toward meat and meat products, approximately 39% of those interviewed indicated they had a low consumption of meat (28% of the respondents only bought red meat once every 2–4 weeks). This group regarded chicken and cheese as having variety, interest, health and value for money (Food and Industry, 1987).

Nutritionists in West Germany, in common with their counterparts in other Western industrialised countries, are actively encouraging a trend towards a healthy diet. This includes a reduction in fat intake (Ernahrung Heute, 1984). Social research has revealed similar attitudes towards meat as those held by consumers in other countries. Desirable characteristics are that meat should be healthy, low fat, tender and juicy, have a good flavour and must have a price appropriate to the product (Rewerts & Folkers, 1985).

In 1986, a survey was conducted into the attitudes and beliefs of the Finnish population towards healthy eating, and in particular towards meat and meat products, and milk and milk products (Poytakivii, 1986). Of those interviewed (n = 498), 31%, 19% and 45% ate beef, pork and sausage respectively, several times a week. When asked which foods they felt should be dropped in order to create a healthy diet, 63% said that red meat should be maintained in the diet and 20% said that consumption should be increased. However, 70% said that sausage consumption should be reduced. By comparison, 77% and 43% of the respondents said that fish and chicken consumption respectively should be increased.

In the Finnish survey, both pork and sausage were perceived as unhealthy and containing a lot of fat. Beef was perceived as a 'light food' together with fish, milk and milk products. Beef was also seen as a healthy food. In contrast sausage was considered by those interviewed to be unhealthy (87%), a 'heavy food' (82%), having a high fat content (81%), containing poor ingredients (60%) and harmful substances and additives (63%). However 45% of those interviewed ate sausage several times a week. It can be speculated that other factors of importance to consumers were price, ease of preparation and 'organoleptic qualities'. Such results highlight the necessity of weighing the different factors when attempting to understand consumer perception and actions.

Consumer awareness of the links between diet and health has manifested itself as a determinant of food purchase in many different product areas. Concern however, usually relates to the presence or absence in a product of a particular food component such as fat or salt, because these components are often the focus of both media coverage and new product development.

We can look at one of these components—dietary fat—from the perspective of the recommendations in the NACNE and COMA reports, as well as from that of the consumer.

FAT IN THE DIET

The Health Dimension

Dietary fat has been associated with several conditions of ill health. The links are in most cases indirect, and not necessarily causal. The major issues relate to cardiovascular disease (CVD) (DHSS, 1974; COMA, 1984); obesity (Royal College of Physicians, 1983) and cancer (Committee on Diet, Nutrition and Cancer, 1982). High fat diets have also been associated with the development of diabetes and gallstones.

Many reports have been published which attempt to review progress in research on the relationship between nutrition and cardiovascular disease (Truswell, 1983). These include the reports of NACNE and COMA as well as the more recent one by the British Medical Association (1986). Invariably they recommend dietary changes which it is hoped will reduce the incidence of CVD amongst the population. Much of the dietary advice concentrates on fats and recommends lower intakes of both total fat and saturated fatty acids than is typical for countries such as the UK.

The COMA panel (1984) recommended that fats in total should supply only 31–35% energy in the diet. This represents a 17–25% reduction compared with the National Food Survey value of 42% of energy from fats, at that time.

COMA went on to recommend that if the current balance between saturated and polyunsaturated fatty acids is maintained (P:S ratio 0·23) then the advised reduction in total fats should be 25%. If a proportion of the saturated fatty acids is replaced by polyunsaturated fatty acids (P:S ratio 0·45) then a smaller reduction in total fats is deemed to be acceptable.

The World Health Organisation expert group (World Health Organisation, 1982) advised a reduction in fat intake to just 30% of energy of which no more than one-third should be supplied from saturated fatty acids.

Fat is the dietary component which contains the greatest quantity of energy per unit weight:

Fat	37 kJ/g (9 kcal/g)
Carbohydrate	16 kJ/g (3·75 kcal/g)
Protein	17 kJ/g (4 kcal/g)

This means that a reduction in fat consumption has a proportionately greater effect on total energy intake than does the removal of the same quantity of protein or carbohydrate. The Royal College of Physicians in their report on obesity (Royal College of Physicians, 1983) recommended that the availability of foods with a lower fat content should be increased and their use encouraged in weight control programmes.

Consumer Awareness of and Attitudes to Fat in the Diet

Publicity regarding the diet, health and fat issues is being given in a number of ways: via media coverage, by health education programmes and by the food industry supplying information on the diet and health subject as part of their marketing strategy. In addition, the increase in availability of innovative products marketed as 'healthy' has compounded public awareness of such issues.

The intensity of promotions extolling the nutritional benefits of food components such as polyunsaturated fat has both stimulated public awareness and reinforced consumer attitudes. In a pilot survey conducted in 1985, people were asked to name those foods which they considered to be main sources of fat in the diet (Fallows & Gosden, 1985, Woodward,

1985). Reponses to this question revealed a prevalence towards interpreting 'main sources' as foods 'with a high fat content'. For example cream which has a high fat content was frequently cited even though the contribution to the fat in the diet is negligible for most consumers. This exposes a simplification of the health messages into 'good' and 'bad' foods.

In all, 34 different foods were identified as main sources of fat. Those mentioned most frequently are given in Table 2, which shows that meat was considered to be a main source of fat by 42% of respondents.

Table 2
Consumers' view of the main sources of fat in the diet

Food	*Total responses (n = 132)*	*Percentage Total respondents*
Butter	78	59
Cheese	73	55
Meat	56	42
Milk	45	34
Margarine	24	18
Bacon	21	16

Woodward (1985).

Table 3
Consumer attitudes to dietary fat

Attitudinal statement	*Percentage respondents agreeing (n = 576)*	
	1985	*1986*
Saturated fat is better for you than polyunsaturated fat	12	8
Animal fat is saturated fat	—	58
I don't think fat is any good for you at all	54	42
Even if fat were no good for you I would still eat it	59	49
All animal fats are bad for you	—	40

Wright (1987).

In a postal survey (Wright, 1987) conducted in 1986 (Table 3) with 576 respondents, only 8% agreed that saturated fat is better for you than polyunsaturated fat; 58% considered that animal fat is saturated fat and 40% believed that all animal fats are bad for you. Nevertheless 49% of the respondents said that even if fat was no good for them, they would still eat it. This represented a reduction on the 59% of respondents who agreed with that statement the previous year using exactly the same sample. However, in 1986, 42% agreed with the statement 'I don't think fat is any good for you at all' whereas in 1985, 54% of respondents had taken this view, representing a moderation of the extreme standpoint on 'dietary fat'.

Consumer Attitudes towards Fat in Meat

Consumer awareness of and attitudes towards dietary fat in general have necessarily shaped attitudes towards the fat content of meat and meat products. Consumers are known to dislike the fat in meat for three main reasons: taste, wastefulness and health.

Research undertaken by the Food Policy Research Unit at the end of 1986 and early 1987 ($n = 584$) revealed that consumers are reducing their meat consumption for reasons of health and price (Woodward, 1987). Of those who ate meat 33% stated that they were eating meat less frequently than they used to: a similar value (35%) was found for meat products.

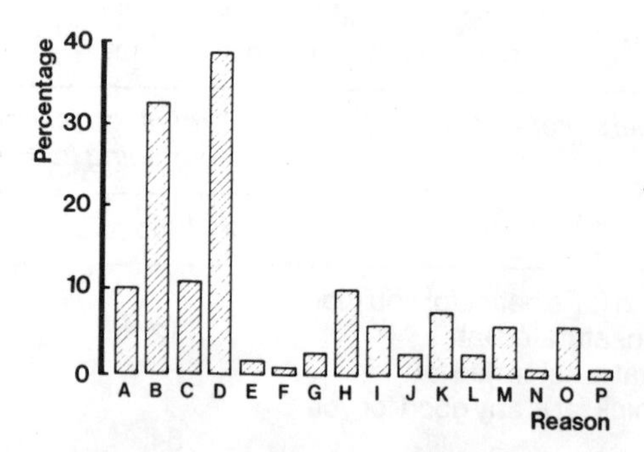

Fig. 3. Reasons given for eating less meat (spontaneous) ($n = 121$). A, Faster food; B, health; C, lifestyle change; D, cost (high); E, additives; F, ingredients; G, not enough fat; H, preparation unpleasant; I, calories high; J, medical advice; K, alternatives; L, too much fat; M, smaller appetite; N, adverse publicity; O, taste; P, animal welfare. (Source: Woodward, 1987.)

Figure 3 shows that the main reasons given for this change in meat consumption were cost (38%) and health (32%).

When the respondents were presented with eight preselected reasons for eating less meat, general health was chosen by 51% of respondents. This was followed by cost (45%) as being an important factor in causing them to reduce their meat consumption (Fig. 4).

Figure 5 shows that 24% of respondents (n = 167) stated that they were eating fewer meat products for reasons of health; 14% because of cost; 11% for taste-related reasons and 7% because of meat product contents.

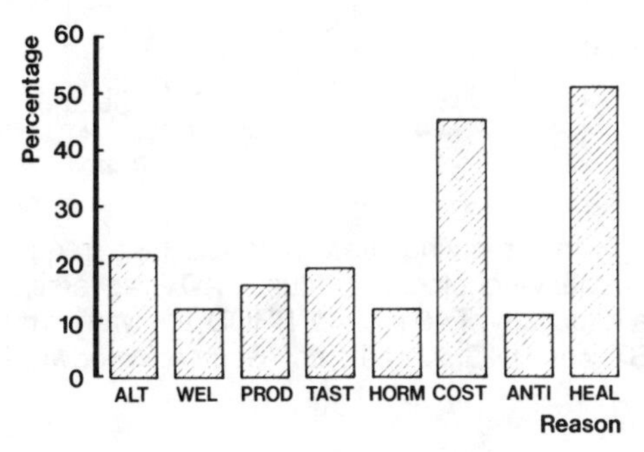

Fig. 4. Reasons given for eating meat less frequently (prompted) (n = 167). ALT, More alternatives; WEL, animal welfare; PROD, production methods; TAST, taste; HORM, growth hormones; COST, cost of meat; ANTI; antibiotics; HEAL, general health. (Source: Woodward, 1987.)

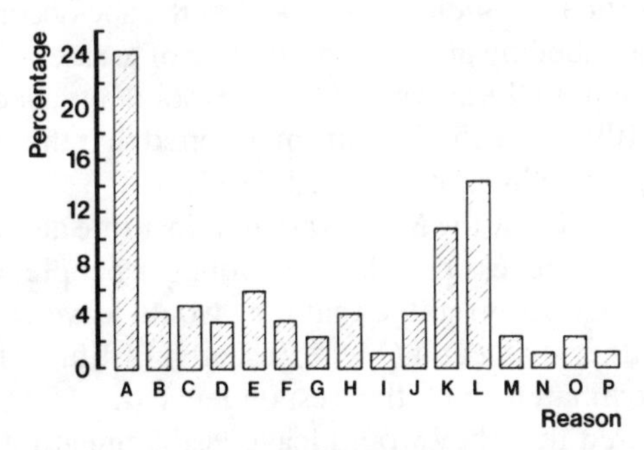

Fig. 5. Reasons given for eating meat products less frequently (spontaneous) (n = 167). A, Health; B, additives; C, fat included; D, smaller appetite; E, contents; F, more alternatives; G, calorific value; H, adverse publicity; I, prefer non-meat meals; J, lifestyle; K, taste; L, cost; M, animal welfare; N, convenience; O, prefer fresh meat; P, moral. (Source: Woodward, 1987.)

When the respondents were presented with nine preselected reasons for eating fewer meat products, 'too fatty' was chosen by 50% of the respondents, 49% stated 'health' and 27% cited reasons of 'cost' (Fig. 6).

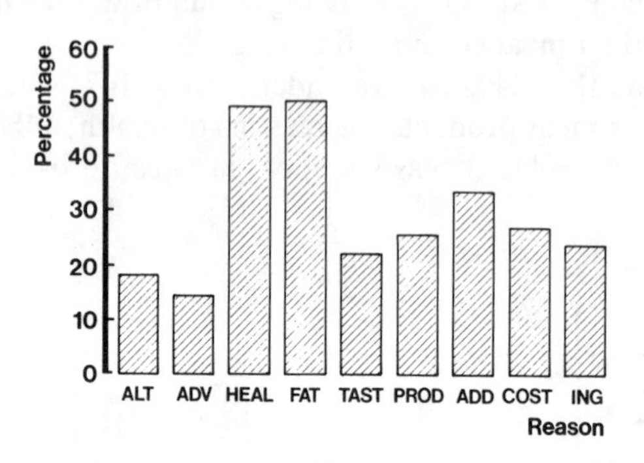

Fig. 6. Reasons given for eating meat products less frequently (prompted) $n = 167$). ALT, Alternative choice of products; ADV, adverse publicity; HEAL, general health; FAT, too fatty; TAST, taste; PROD, the way it's made; ADD, additives; COST, cost; ING, ingredients. (Source: Woodward, 1987.)

When asked what they would like to see labelled on a meat product, 11% ($n = 514$) stated that they would like to see fat percentages included (Fig. 7). Fat percentages were mentioned most frequently after additives (27%) and ingredients (16%) across the socioeconomic groups from C1 to E. Fat labelling also appeared to be of greatest importance to the 45–54 age group. In the survey, 18% of respondents aged 45–54; 13% aged 25–34 and 10% aged 25–44 years mentioned that they would like to see fat percentage labelling (Woodward, 1987).

Research undertaken by the Meat Research Institute also revealed that meat fat appeared to be less popular with younger people. Over 60% of the under sixteen age group stated that they would leave visible meat fat on their plates rather than eat it (Meat Research Institute, 1975).

Though approximately half the respondents ($n = 500$ households) in this survey stated that they would leave visible meat fat on the plate rather than eat it, reactions varied towards the fat of different meat species. Forty-six percent of respondents said that they would leave beef fat on the plate; 44% would reject pork fat and 52% would leave visible fat from lamb on their plates rather than eat it.

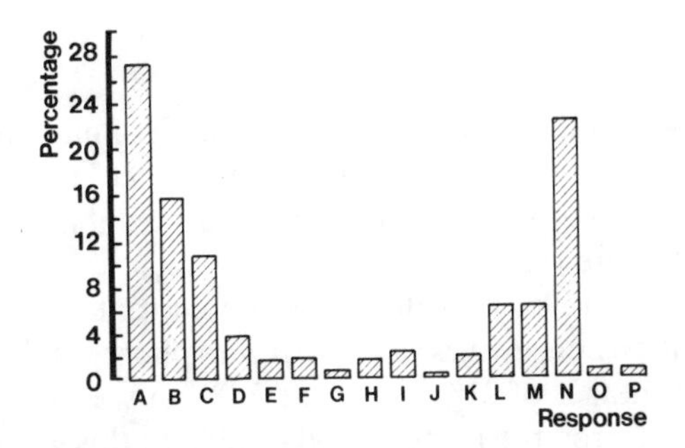

Fig. 7. Items which should be included on meat product labels (spontaneous) $n = 514$). A, Additives; B, ingredients; C, fat percentage; D, sell by date; E, calorific value; F, type of meat; G, cereal; H, hormones; I, water; J, percentage lean; K, nutritional information; L, meat percentage; M, salt; N, fresh; O, other. (Source: Woodward, 1987.)

These different perceptions of the meat species have been noted in several countries. A survey of 100 consumers in Edinburgh (Shearer *et al.*,1986) showed that 69% felt that beef was leaner than lamb, and pork and beef were both considered to be fattier than chicken by virtually all those interviewed.

In New Zealand, a recently conducted survey revealed that 43% of all adults questioned over the past two years had reduced their fat intake by trimming visible fat off meat, and using less fat in cooking (Hamilton, personal communication).

In Belgium, consumers surveyed generally felt that some meats were healthier than others (4·4 on a scale of 1 (strongly disagree) to 5 (strongly agree)). There was also the view that pork and mutton were particularly fattening (Vandercammen & Viaene, 1988).

In a West German survey into attitudes towards pork, 14% of the respondents ($n = 285$) said that they did not eat pork for the following main reasons: pork was not considered to be 'good for you' (36%); they were generally eating little or no meat (29%); they wanted better quality than that which pork provided (9%); and pork was considered to be too fatty/had too many calories (6%).

It therefore appears that different attitudes towards fat in meat are not only dependent upon such factors as socioeconomic group, age, and geographical location, but also on consumer perceptions of the different meat species.

Beef

Beef is the most popular (by choice) of the four common meat species in the UK, and beef fat is preferred whether hot or cold to lamb fat. However, as with the other meat species, those cuts exhibiting a high ratio of visible lean to fat are the most popular.

Brayshaw *et al.*, (1967)—well before the current debate on diet and health—presented housewives (n = 976) with photographs of steaks with visible fat percentages of 21%; 26% and 34%. It was found that 36% of respondents preferred the leanest steak—13% of this group expressed a preference for an even higher lean : fat ratio. Only 15% of respondents preferred the steak exhibiting 34% visible fat. Six percent of respondents stated that they would prefer the steak to have no visible fat at all.

Almost a decade later, consumer research (Baron & Carpenter, 1976) revealed that consumers consistently rejected beef with a high fat content, and that the ratio of lean to fat in beef was the most important factor influencing consumer purchases.

In 1982, visitors attending the Royal Smithfield Show were asked to indicate their preference from four samples of raw wing rib of beef which differed principally in fatness (Dransfield, 1983). The study was designed to repeat a ballot undertaken at the show in 1955 (Pomeroy, 1956). Thus it was hoped that any changes in preference might be identified. The four ribs of beef had visible fat percentages of 48%; 40%; 37%; and 30% with lean : fat ratios of 1·05; 1·36; 1·52 and 2·22.

In 1982 the leanest joint was most popular (60% respondents) (n = 1880) with the fattest joint the least popular (4% respondents). By contrast, in 1955, the middle range joints had been the most popular (see Table 4).

Table 4.
Preference for beef joints of different fatness (n = 1880)

Average fatness of joint (%)	Percentage responses	
	1955	*1982*
30	12	60
37	17	12
40	62	23
48	6	4

Dransfield (1983).

Table 5
Percentage of consumers who ranked either of the two fattier joints as their first choice

A By age

Age (years)	*Percentage respondents*	
	1955	*1982*
Under 30	66	26
30–39	61	28
40–49	70	25
50 and over	75	33

B By region

Region	*Percentage respondents*	
	1955	*1982*
East	74	25
North East	77	31
East Midlands	71	29
West	70	27
London	67	32
Home Counties	67	29
South	67	30
West Midlands	67	29
Scotland	61	17
Wales	59	23
North West	55	35
Overseas	—	17
Unspecified	65	31

C By occupation

Occupation	*Percentage respondents*	
	1955	*1982*
Farming	75	27
Butcher	61	49
Others	64	25
Unspecified	65	26

Dransfield (1983)

A shift in preference towards lean meat across the complete age range was revealed in the study (Table 5(A)), with an alteration in preference being revealed by those participants who would have been in the under 30 age group had they taken part in 1955. These participants, now in the 50 or over age group, appear to prefer leaner meat now, despite the fact that a greater percentage of them prefer the fattier joints, compared with those in the younger age groups.

A shift in preference away from the fattier joints was also revealed across the various regions of the UK. Just over a third (35%) or less of participants in each of the regions preferred the fattier joints. Less than a quarter of participants from Scotland, Wales, East England, and overseas chose one of the two fattier joints in preference to the leaner ones (Table 5(B)).

It is interesting to note that butchers taking part in the exercise in 1982 preferred fattier beef than those participants engaged in farming or other occupations (Table 5(C)). Butchers had changed their views less than other groups during the 27-year period.

Fat levels in meat have traditionally been higher in the USA than in Europe because of differences in feeding systems and meat industry attitudes. However recent national surveys have shown that US consumers dislike high levels of fat and would buy more beef if it was leaner. As a result, beef cuts are now trimmed of fat to a much greater extent than previously before sale to consumers. (Savell *et al.*, 1988).

Lamb

Mutton and lamb are experiencing the most consistent decline in popularity amongst UK consumers of all the meat species. This is again due to several factors including price, seasonality and wastefulness (a high bone and fat content) (Baron *et al.*, 1973). Lamb fat is also the least popular of the red meat fats due to its 'greasy' nature and is equally unpopular whether hot or cold. However, interestingly, pork is often cited by consumers as being fattier than lamb. Research undertaken by the Meat and Livestock Commission (Kempster, 1979) showed that considerably more fat is produced yearly in lamb carcasses than is required by UK consumers (approximately 30 000 tonnes). A major problem with lamb carcasses is that much of the fat is hidden and so cannot be trimmed off during conventional butchery. Overfat lamb joints are therefore more likely to be presented to consumers.

Pigmeat

Pork is becoming increasingly popular with consumers, though bacon consumption is in decline. The perceived higher fat content of bacon is a negative factor (Woodward, 1985) whereas pork is considered by some to be a 'white meat' with relatively high levels of unsaturated fatty acids and a low fat content. As with lamb and beef, consumers demand pork with a high lean : fat ratio (Baron & Carpenter, 1976).

The pig industry in the UK has been most in tune with customer requirements for lean meat, price incentives reflecting this having encouraged a steady reduction in the fat percentage of the average pig carcass during the past thirteen years (Meat and Livestock Commission, 1985*a*).

REGULATIONS AND LEGISLATION CONCERNING FAT IN MEAT

As can be seen from the above, there is a clear preference being expressed by a significant proporiton of consumers in several countries for products which are low in fat. The 'healthy eating' messages promoted largely via the media have acted to compound and accelerate an existing dislike for fat, by adding the diet and health dimension.

Consumers are becoming increasingly interested in issues such as health, and in issues relating to the quality of foods. As a part of this, there is the desire for more information regarding the composition of foods which would enable them to make informed choices.

The COMA report, as part of its recommendations concerned with the reduction of fat in the UK diet, proposed that a system of fat labelling be introduced on foods. This was based on the premise that consumers should be given the means with which to effect a change in diet if they desired to do so.

As a result of the recommendations made by the COMA panel regarding the provision of information relating to fat, and the composition of foods (Paras 2.4.1, 2.4.2, 2.6.2 COMA, 1984) MAFF held preliminary discussions with representatives of the food industry. The results of these discussions were circulated in a paper entitled *Labelling of Fat and Other Nutrients*. The paper indicated that in response to industry proposals, the UK government would be willing to consider the introduction of a more comprehensive system of nutrition labelling than just fat labelling.

Nutrition Labelling and Fat Labelling

The Government proposed that there should be:

(a) The introduction of legislation for mandatory fat labelling.
(b) Guidelines, either voluntary or mandatory, for the standardisation of the approach to nutrition labelling.

It should be noted that by this time some food manufacturers were introducing a form of nutrition labelling on products marketed as 'healthy' partially to promote the 'positive' attributes of the product.

In addition to such consultations with the industry, MAFF in collaboration with the Consumers Association (CA) and the National Consumers Council (NCC) commissioned a consumer survey, the results of which were published in June 1985 (CA, MAFF, and NCC, 1985).

Just prior to that, in March 1985, MAFF set out government proposals for the implementation of the recommendations outlined in the COMA report—particularly those regarding fat labelling (Hansard, 1985).

The main proposal was for a statutory requirement for foods (whether prepacked or loose) making a significant contribution to the fat intake in the diet to be labelled with total fat and saturated fatty acid (SFA) content. Yellow fats, edible oils and fats would include any *trans* fatty acids in the declaration of SFA content.

MAFF also released proposed voluntary guidelines for a comprehensive system of nutrition labelling. These guidelines were prepared taking into account the earlier work of the Food Standards Committee (FSC) (MAFF, 1985) and the Food Standards Programme of the Codex Alimentarius Commission. It was proposed that there be no limit on the amount of information given on the label. However, such information would have to conform to a prescribed format.

The final guidelines were published in July 1987 (MAFF, 1987) and consist of a 'staged' approach developed from earlier drafts and the Codex Alimentarius Commission work. The stages are as follows:

- A basic nutritional profile (energy, protein, carbohydrate, fat) as agreed internationally within Codex.
- A progressively more detailed breakdown of the main nutrients may be given to show first saturates plus *trans* fatty acids, then sugars, with sodium and fibre in addition.

A compulsory system was not possible within the framework of EC law—the introduction of nutrition labelling unilaterally by one Member State could be construed as a non-tariff barrier to trade.

In February 1986, draft proposals for *Fat Content of Food (Labelling) Regulations* (MAFF, 1986) were circulated for comment. Essentially these proposed that all foods containing more than 0·5% fat (with a small number of exemptions) should be labelled or marked with a declaration of the actual fat content and the SFA content (expressed as 'saturates') given in grams per 100 grams (100 millilitres) of the food product.

As the proposals made were for mandatory action for labelling of fat content, the development of the regulations necessarily follows the processes common to all secondary legislation (Fallows, 1986). Discussions on mandatory fat labelling are continuing with the EC Commission and with other Member States. At present the EC is also drawing up proposals with respect to nutrition labelling (Freckleton, 1987).

The Meat Product Regulations

A second set of regulations directly affecting the meat industry, which take into account recommendations to reduce dietary fat, are the meat product regulations. In 1975 MAFF began the legislative review procedure for meat products by calling for evidence from any interested parties. The brief given to the FSC was to review the whole range of meat products and to recommend a system of control appropriate to the changing market environment.

The existing system of control for meat products based upon product-specific compositional standards had been recognised as becoming outdated for a variety of reasons including complexity of the regulations, the perceived restriction on product innovation and changing market requirements (Woodward, 1986).

In the light of the submissions received, the FSC considered that the existing regulations on the main meat products were generally satisfactory from the point of view of the consumer, and the reputable manufacturer. However, various areas of concern were highlighted including:

- Rigidity of regulations restricting product innovation.
- Problems with interpretation of certain sections of the regulations for enforcement authorities.
- Concern that the proportion of fat permitted in the meat content with existing regulations was too high.

In 1980 the findings and recommendations of the FSC were published in the *Report on Meat Products* (MAFF, 1980).

The general philosophy of the FSC and MAFF was to move away from detailed compositional standards and towards a system which would combine a system of compositional standards for well established meat products (e.g. sausages) with a system of informative labelling.

Previous meat product regulations had used minimum meat content as the basis of control by compositional standards. This was continued to a certain degree with the adoption of a declaration of minimum meat content as the basis of informative labelling. The final regulations which were introduced in 1984 are seen to represent this philosophy in the following labelling requirements which apply specifically to meat products:

(a) Products which are ready for delivery to the consumer or to a catering establishment in any quantity must carry a declaration indicative of one or more of the following:
    1. Minimum meat content.
    2. Minimum corned meat content.
    3. Maximum added water content.
(b) Methods of declaration will differ according to the type for product, but will normally be on product labels in or in immediate proximity to the list of ingredients or the name of the product.
(c) A notice in immediate proximity to the product is required for products not prepacked or prepacked for direct sale (Meat and Livestock Commission, 1985*b*).

As will be seen, the original intention to label according to 'lean meat content' was later modified so that 'meat content' was finally chosen as the basis for declaration (the definition of meat being '. . . the flesh and the fat, skin, rind, gristle and sinew in amounts naturally associated with the flesh used') (Meat Products and Spreadable Fish Products Regulations, 1984).

Meat Content Declaration

The 1980 *Report on Meat Products* took account of the 1974 DHSS recommendations that total fat in the diet, especially saturated fat, should be reduced (DHSS, 1974). The same report stated that the existing regulations allowed the inclusion of too much fat in the meat content of meat products. Hence the view was taken that it would be desirable that any control on the meat content of meat products should be framed to limit the amounts of fat and connective tissue which could count as meat whilst not placing any restrictions on the total amount of meat that could be incorporated in the product. One way of doing this is to prescribe minimum standards in terms of lean meat. When the FSC put forward the recommendation for lean meat declaration they also stated that:

> as consumers begin to take notice of, and understand such declarations we consider that they will increasingly exercise a choice and influence the composition of products in relation to their conception of what is desirable and the price they are prepared to pay

Consumer groups and enforcement authorities were in favour of declaration by lean meat content, stating 'lean meat' would be a concept more easily understood by those consumers who were concerned with quality and product comparison, though it was argued that the definition of lean meat would have to be tightened up considerably.

However, there was strong opposition from the food industry to this proposal. They argued that the declaration of 'lean meat content' presented analytical difficulties. This was accepted and the final regulations contained the provision that 'percentage total meat content' should be declared rather than 'percentage lean meat'.

In practice, the problems of analysis perceived by the food industry were overcome by manufacturers producing products in direct response to a market demand for low fat foods. These are often labelled with 'lean meat' and 'fat' content. Two examples of the ways in which this was achieved are given below:

Product A
Ingredients: Lean Pork (min. 40%), Water, Rusk, Pork Fat (max. 15%), Salt, Milk Protein, Polyphosphate, Preservative E223, Flavour Enhancer 621, Spices, Antioxidant E301, Sodium Citrate, Colour 128, herbs. Minimum 52% Pork.

Product B
Ingredients: Pork (min. 50%, min. lean 38%, max. fat 12%), Water, Rusk, Starch, Salt. Ingredients less than 1%: Spices, Herbs, Dextrose, Polyphosphates, Antioxidants E304, E307, Preservative E223, Colour 128.

Thus, although the meat products regulations have not met the original aim of reducing the fat content of meat products through control over 'lean meat', voluntary labelling may achieve the same end in the market place. A system of self regulation of product quality, to be imposed by the meat products industry, is also under development.

DISCUSSION

The enormous growth in EEC agricultural output since World War II combined with the recent accumulation of surpluses has had a profound impact on the food system and on the associated political and economic climate. Worries about food shortages have been largely replaced by concerns over the costs of producing, storing and disposing of surpluses. Inevitably state financial support for agriculture will be reduced: as this occurs, farmers will be exposed more and more to market forces.

At the same time consumers are becoming much more critical and demanding in their attitudes towards food and markets are changing rapidly. This creates real problems for primary producers because it can take years to adjust farming systems while consumers can change their purchasing patterns overnight.

If producers are to cope effectively, it is essential for them to become aware of the factors which influence food consumers, and how they are changing, in order to anticipate how demands for different foods are likely to evolve in the future. Understanding how consumer concerns about health can affect the choice of food is an extremely important part of the process of awareness.

Within scientific circles, debates about the relationship between diet and the degenerative diseases have raged since the 1960s (Truswell, 1983). More recently however, there has been a growing belief that diet is an important contributory factor in the development of many diseases including coronary heart disease, obesity, diabetes, gallstones and certain cancers. Hence, people are being advised that certain alterations to the regular pattern of consumption would help to avoid or delay the onset of these diseases. The COMA report spelled out some of the steps that could be taken to facilitate and encourage such change in the general population, which can be summarised as follows.

- Agriculture — The encouragement of the production of leaner animals by removing subsidy payments for animals graded as too fat.
- Education — The introduction of health promotion campaigns in hospitals and the community, and education in schools.
- Labelling — The provision of better information about the composition of food products through fat content labelling.

- Manufacturers — The encouragement of 'innovative' 'healthy' products which are low in fat, high in fibre, low in sugar and low in salt.

In the years since the COMA report was published there have been developments in all of the above areas along the lines recommended, and there is no doubt that these are having an impact on food purchasing patterns.

Studies on attitudes to meat show that there is a substantial body of consumers who do not like fat and that the preference for lean meat and lean meat products is growing. When these changes in attitude were first detected, taste was the dominant consideration. More recently this has been augmented by health concerns.

Because the contemporary consumer has such a wide range of choice, unsatisfactory quality of meat or meat products will lead to lower sales. It is certainly recognised by retailers that consumers are demanding low fat fresh meat and meat products. The provision of information on the meat product label facilitates the choice of low fat products.

Surplus fat can be trimmed before sale but this involves extra labour costs and in any case fat is extremely expensive to produce. There can be little doubt that market forces will filter back to producers and economics will favour the production of leaner meat animals.

Trends towards leaner animals have already begun with pigs and a similar pattern is emerging with cattle and sheep. Any steps which will facilitate this change, whether by legislation or price changes which stimulate modifications in production or processing should therefore be encouraged.

REFERENCES

Agricultural Economics Research Institute (LEI). *Periodical Reports*, no. 64–80/81 and 64-81/82/83, Ls Den Haag, Netherlands.

Anon. (1985). The major changes in shopping habits; prices give way to shoppers convenience. *Grocer*, 30 November, 50.

Anon. (1987). Health and diet take a hold in a tale of two meaty products. *Meat Trades Journal*, 23 July.

Baron, P. J. & Carpenter, E. M. (1976). *A Review of Consumer Attitudes and Requirements for Meat*, Report no. 23. Department of Agricultural Marketing, University of Newcastle upon Tyne.

Baron, P. J., Cowie, W. J. G., Hughes, D. R. & Lesser, D. (1973). *Housewives Attitudes to Meat*, Report no. 16. Department of Agricultural Marketing, University of Newcastle upon Tyne.

Brayshaw, G. H., Carpenter, E. M. & Perkins, R. J. (1967). *Consumer Preferences for Beef Steaks,* Report no. 2. Department of Agricultural Marketing, University of Newcastle upon Tyne.
British Medical Association (1986). *Diet, Nutrition and Health.* British Medical Association, London.
Committee on Diet, Nutrition and Cancer (1982) *Diet, Nutrition and Cancer.* Assembly of Life Sciences, National Research Council, National Academy Press, Washington DC.
Committee on the Medical Aspects of Food Policy (1984). *Diet and Cardiovascular Disease,* Department of Health and Social Security. HMSO, London.
Consumers' Association, Ministry of Agriculture, Fisheries and Food, National Consumer Council (1985). *Consumer Attitudes to, and Understanding of Nutrition Labelling* (Two Volumes). HMSO, London.
Danmarks Statistik (1986). *Average Consumption per Inhabitant (Total Population) (Not Corrected for Tourist Consumption, Border Trade, and Household Waste), various years.* Denmark.
Department of Health and Social Security (1974). *Diet and Coronary Heart Disease,* Report on Health and Social Subjects No. 7. HMSO, London.
Department of Health and Social Security (1978). *Prevention and Health: Eating for Health.* HMSO, London.
Dransfield, E. (1983). *Consumer Preference for Beef—Smithfield Show Results.* Institute of Meat Bulletin, No. 120, 4–5.
Ernahrung Heute (1984). Aus Dem Ernahrung Bericht 1984 Der Deutschen Gelleschaft Fur Ernahrung, Frankfurt.
European Commission, *Agricultural Prices Proposals, 1987/1988,* Green Europe Newsflash 39, Brussels.
Eurostat (1988) *Agriculture-Statistical Yearbook.* Office des publications des Communautes europeenes, Luxembourg.
Fallows, S. J. (1986). *Deciding Our Food Laws* Food Policy Research Unit, University of Bradford. Horton Publishing, Bradford.
Fallows, S. J. & Gosden, H. (1985). *Does the Consumer Really Care?* Food Policy Research Unit, University of Bradford. Horton Publishing, Bradford.
Fallows, S. J. & Wheelock, J. V. (1983). A policy for food self sufficiency in the UK. *Agricultural Administration,* **13,** 165–77.
Food and Industry (1987). *National Food and Nutrition Policy Needed in New Zealand,* Vol. III, No. 5.
Frank, J. (1987). *European Food Consumption and Expenditure Patterns, UK and International Food Consumption Patterns, Volume 4,* Food Policy Research Unit, University of Bradford. Horton Publishing, Bradford.
Freckleton, A. (1987). *Nutrition Labelling: The Role of Government, Food Industry, Consumer Advocates and Consumers in the Development of Nutrition Labelling.* PhD thesis, University of Bradford.
Gallup, (1987). Survey conducted for the Realeat Company.
Hansard (1984). Written answer, sixth series, volume 63, volumn 665.
Hansard (1985). Written answer, sixth series, Volume 75, columns 79-80.
Howarth, R. W. (1985) *Farming for Farmers?* Institute of Economic Affairs. Hobart Paperback 20, London.

Institute of Grocery Distribution (1987). *Food Retailing*. Institute of Grocery Distribution, Watford.

Jefcoate, H. M. (1986). *Meat and Meat Products in the UK*. Food Market Updates No. 17. Leatherhead Food Research Association, Leatherhead.

Kempster, A. J. (1979). Variation in the carcase characteristics of commercial British sheep with particular reference to over fatness. *Meat Science* **3,** 199–208.

Lang, T., Andrews, H., Bedale, C. & Hannon, E. (1984) *Jam Tomorrow? A report of the first findings of a pilot study on the food circumstances, attitudes and consumption of 1100 people on low incomes in the North of England.* Food Policy Unit, Manchester Polytechnic.

Meat and Livestock Commission (Economics and Livestock Marketing Services) (1985*a*). *Pig Year Book*. Meat and Livestock Commission, Bletchley

Meat and Livestock Commission (1985*b*). *A guide to the meat products and spreadable Fish Products Regulations, 1984*. Meat and Livestock Commission, Bletchley.

Meat Industry Research Institute of New Zealand (1985). *Nutritive Analysis of Lamb–The Mirinz Study*. Mirinz, Hamilton, New Zealand.

Meat Industry Research Institute of New Zealand (1986). *Nutritive Analysis of Beef—The Mirinz Study*. Mirinz, Hamilton, New Zealand.

Meat Products and Spreadable Fish Products Regulations (1984). Statutory Instrument, 1984/1566. HMSO, London.

Meat Research Institute (1975) *Consumer Reaction to Visible Fat on the Plate.* Biennial Report, Meat Research Institute, Bristol.

Ministry of Agriculture, Fisheries and Food. *Annual Reviews of Agriculture (Formerly Annual Review and Determination of Guarantees)* for years 1967 to 1985. HMSO, London.

Ministry of Agriculture, Fisheries and Food (1973). Estimates of food supplies moving into cosumption in the UK. *Trade and Industry*, 30 August, 459-66.

Ministry of Agriculture, Fisheries and Food (1977). Food consumption in 1976 recovers from low levels of 1975. *Trade and Industry*. 18 November.

Ministry of Agriculture, Fisheries and Food (1980, 1986). *Estimates of Food Supplies Moving into Consumption in the UK,* Food Facts no. 9 (1980), Food Facts no. 12 (1986).

Ministry of Agriculture, Fisheries and Food, Food Standards Committee (1980). *Report on Meat Products.* HMSO, London.

Ministry of Agriculture, Fisheries and Food (1985). Press Release, 13 March. MAFF, London.

Ministry of Agriculture, Fisheries and Food (1986). *Proposed Fat Content of Food (Labelling) Regulations,* MAFF communication, 13 February. MAFF, London.

Ministry of Agriculture, Fisheries and Food (1987). *Guidelines on Nutritional Labelling*. MAFF, London.

Ministry of Agriculture, Fisheries and Food (Annual). *Household Food Consumption and Expenditure: Annual Report of the National Food Survey Committee.* HMSO, London.

National Advisory Committee on Nutrition Education (1983). *A Discussion Paper on Proposals for Nutritional Guidelines for Health Education in Britain.* Health Education Council, London.

Office of Fair Trading (1985). *Competition and Retailing.* HMSO, London.

Otto, E. (1986). *Zum Fettgehalt Bei Rind—Und Schweinfleisch,* Ernahrungsforschung 31 (1986) l, 24-25, DDR.

Pomeroy, R. W. (1956). *Consumer Preferences for Beef.* Institute of Meat Bulletin, No. 14, 3–9.

Poytakivii, R. (1986). Mielipiteet Uhasta Ja Makkarasta, Smoven Gallup Oy, Finnish Meat Information Centre, Helsinki.

Rewerts, I. & Folkers, D. (1985). Die Qualitat Von Schweinfleisch Aus Der Sicht Der Verbraucher, Aid-Verbraucherdienst, 30 (1985) Heft 4, West Germany.

Royal College of Physicians (1983). *Obesity*, reprinted from the *Journal of the Royal College of Physicians,* **17,** 1.

Savell, J. W., Cross, H. R., Hale, D. S. & Beasley, L. (1988). *National Beef Market Basket Survey.* Meat Research Brief. Texas A&M University, College Station, Texas.

Shearer, J., Burgess, G & English, P. R., (1986). A study of consumer attitudes to fat in meat. *Animal Production,* **42,** 458.

Statistisches Jahrbuch Uber Ernahrung Landwirtschaft und Forsten. (1968), Verlag Paul Parey, Hamburg, West Germany.

Statistisches Jahrbuch Uber Ernahrung Landwirtschaft und Forsten. (1984), Landwirtschaftsverlag GmbH, Munster-Hiltrop, West Germany.

Truswell, A. S., (1983). The development of dietary guidelines. *Food Technology in Australia,* **35,** 498–502.

Vandercammen, G. & Viaene, J. (1988). Consumer Attitudes Towards Red Meat and Poultry, Consumer Attitudes and Requirements for Food, Report 2. EEC Contract V1/4741/84-NL (Prec/764), Department of Agricultural Marketing, State University of Ghent.

Woodward, J. (1985). Consumer Attitudes to Meat and Meat Products, Unpublished Report. Food Policy Research Unit, University of Bradford.

Woodward, J. (1986). *Regulating the Meat Product Industry*, Food Policy Research Unit, University of Bradford. Horton Publishing, Bradford.

Woodward, J. (1988). *Consumer Attitudes to Meat and Meat Products with regards to Composition and Labelling.* Food Policy Research Unit, University of Bradford, Horton Publishing, Bradford.

World Health Organisation Expert Committee. Prevention of Coronary Heart Disease, Technical Report Series no. 678, 1982. World Health Organisation, Geneva

Wright, G. (1987). *Milk and the Consumer*, Food Policy Research Unit, University of Bradford. Horton Publishing, Bradford.

Wright, G. & Slattery, J. (1986). *Talking about Healthy Eating,* Food Policy Research Unit, University of Bradford.

Yankelovich, Skelly & White (1985). *The Consumer Climate for Red Meat, an Executive Summary.* Prepared for the American Meat Institute, the National Live Stock and Meat Board, New York.

Chapter 4

Metabolic and Endocrine Control of Adipose Tissue Accretion*

H. J. MERSMANN‡
US Department of Agriculture
Roman L. Hruska, US Meat Animal Research Center, ARS, Clay Center, Nebraska, USA

INTRODUCTION

Until recently, producers of cattle, pigs and sheep as sources of palatable protein for human consumption were not concerned with the extent of fat accretion concomitant with the accretion of muscle mass. Reasons for a low interest in reducing fat deposition included: the assumption that intramuscular fat (marbling fat) in bovine muscle is important for palatability; the relatively low cost of feeds in the USA, Canada and some countries of the European community; and little concern about the effects of fat on human health. Most of these factors have been radically changed in the last decade so that today there is a major impetus to produce a lean muscle product for human consumption.

The belief that intramuscular fat improves palatability (mainly juiciness and tenderness) applies particularly to cattle and is most strongly held in the USA. The US attitude to marbling fat has developed for a number of

*This chapter is dedicated to the persons who taught me experimental biology; C. A. Privitera, D. R. Strength, H. L. Segal and H. C. Stanton.

‡Present address: USDA/ARS Children's Nutrition Research Center, Baylor College of Medicine, Houston, Texas 77030, USA.

reasons, for example, comparisons of diverse breeds such as Angus and Brahman, at different stages of chronological or physiological development, have often led to the conclusion that marbling fat improves tenderness. The US penchant for dry cookery (usually grilling) to extremes of doneness (Savell *et al.*, 1987) is also a major factor in the belief that intramuscular fat improves tenderness because the fat acts as a buffer to maintain some juiciness when large portions of water are lost from the meat during cooking. Nevertheless, it is difficult to objectively demonstrate a strong association of intramuscular fat with palatability in beef (Campion & Crouse, 1975; Crouse *et al.*, 1978).

The high cost of feed for finishing animals has become a major factor in agriculturally affluent countries like the USA and Canada only in recent times. Even though feed costs fluctuate dramatically, they now represent the major costs of animal production so that the portion used to produce excess fat is expensive to the producer and the consumer. Pigs are sold on a grading system that rewards lean animal production in many European countries and Canada but there has been no widespread acceptance of this practice in the USA (Anon., 1982). A decade ago, cattle in the USA were sold on a grading system that rewarded carcass fatness. The cattle grading system has since been changed several times to reflect increasing consumer interest in leaner meat.

Finally, consumer preference for lean meat, primarily because of perceptions of the harmful effects of excessive animal fat to human health, is probably the most important factor encouraging change toward the production of lean animals. At the same time consumers are realizing that purchase of retail meat cuts with considerable subcutaneous or intermuscular fat is economically undesirable because the fat is discarded rather than eaten. Consequently, the retailer or slaughter plant operator must absorb the cost of physically removing excess fat from the retail product or reward the producer for production of lean animals. In any case, the present and future goal of the animal industry will be the efficient production of a palatable lean muscle product for human consumption.

This chapter will address the biology of the characteristic cell of adipose tissue, the adipocyte, as it pertains to the accretion of fat in growing cattle, pigs and sheep. The approach will be to present the anabolic and catabolic pathways associated with fat accretion in this cell coupled with endocrine, genetic, nutritional and pharmacological control of those pathways. Finally, some opportunities to modify animal growth through manipulation of the adipose tissue cell will be discussed.

METABOLIC PATHWAYS

Comprehensive reviews document the literature on anabolic and catabolic pathways in adipose tissue in general (Vernon & Clegg, 1985) and specifically in adipose tissue from cattle and sheep (Vernon, 1981) and pigs (Mersmann, 1986*a*). In addition, a book on adipose tissue biology containing chapters on cell proliferation, adipocyte metabolism and adipocyte culture systems contains information on the adipose tissues of cattle, pigs, and sheep (Hausman & Martin, 1987). Consequently, only overall metabolic patterns will be discussed with emphasis on literature following these reviews.

Anabolism

The aspect of adipocyte anabolism most studied is *de novo* synthesis of long chain fatty acids. This process provides a mechanism for storing excess energy from dietary carbohydrate as fat but also is a mechanism to provide nonessential fatty acids for phospholipid synthesis when dietary fat intake is low. In nonruminant species, glucose or another dietary hexose sugar is generally the major carbon precursor for this synthesis. Although other metabolites may serve as carbon donors for long chain fatty acid biosynthesis in porcine adipose tissue systems *in vitro*, the function of such donors has seldom been established *in vivo* (Mersmann, 1986*a*). Glucose is metabolized (Fig. 1) via the glycolytic pathway to pyruvate which enters the mitochondrion, is decarboxylated to a two carbon moiety,

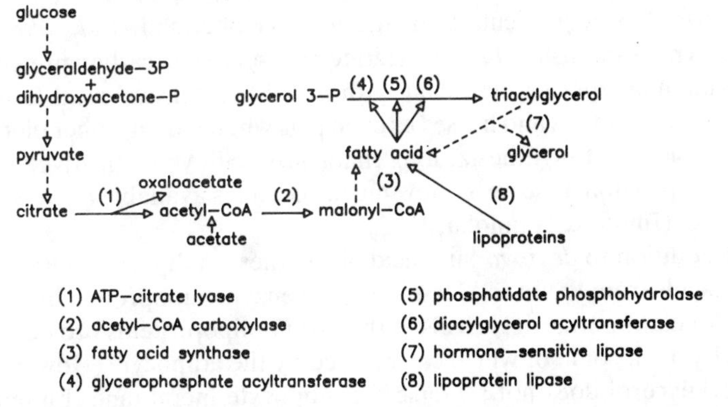

Fig. 1. Anabolic and catabolic pathways of adipose tissue lipid metabolism.

acetyl-CoA, which is condensed with oxaloacetate (four carbons) to form citrate. Citrate can traverse the mitochondrial membrane and is cleaved in the cytosol by ATP-citrate lyase (i.e. citrate cleavage enzyme) to form cytosolic acetyl-CoA and oxaloacetate. The cytosolic acetyl-CoA is carboxylated to form malonyl-CoA by acetyl-CoA carboxylase, the enzyme that in many cases may provide a regulatory step to control fatty acid synthesis. Malonyl-CoA is the carbon donor which provides two carbon units for sequential polymerization by fatty acid synthase to form long chain fatty acids. The primary product of fatty acid synthase is usually the saturated 16 carbon fatty acid, palmitate. Reducing equivalents needed for synthesis of fatty acids are potentially provided by four dehydrogenase enzymes, glucose-6P and 6P-gluconate dehydrogenases associated with the pentose phosphate pathway of glucose metabolism, malic enzyme (NADP-malate dehydrogenase) and cytosolic NADP-linked isocitrate dehydrogenase. Much of the hydrogen, estimated to be 60–90% in porcine adipose tissue (Rosebrough & Steele, 1986), 50–100% in ovine adipose tissue (Yang *et al.*, 1982*a*; Smith & Prior, 1986) and 30–100% in bovine adipose tissue (Smith, 1983; Smith & Crouse, 1984; Smith & Prior, 1984) is supplied by the pentose phosphate pathway dehydrogenases whereas the contribution of malic enzyme, at least in some circumstances, may be questioned (Conover, 1987).

In ruminant species, glucose is conserved and the major carbon donor appears to be acetate (Hanson & Ballard, 1967; Ingle *et al.*, 1972; Yang & Baldwin, 1973*a*; Bauman, 1976; Yang *et al.*, 1982*b*; Smith, 1984; Vernon *et al.*, 1985*a*). Lactate could be an important carbon donor under some circumstances (Prior & Jacobson, 1979; Prior *et al.*, 1981; Whitehurst *et al.*, 1981; Yang *et al.*, 1982*a*; Smith & Crouse, 1984; Smith & Prior, 1986), particularly in the fetus (Robertson *et al.*, 1981). Although not established *in vivo*, acetate may serve as a carbon precursor in nonruminant species, particularly those with extensive cecal fermentation such as rabbits and horses or even pigs when fed high fiber diets. Pigs can use acetate to synthesize long chain fatty acids very effectively *in vitro* but the question arises as to how much acetate is available *in vivo* for this purpose (Imoto & Namioka, 1983).

In addition to *de novo* fatty acid biosynthesis, adipocytes may acquire fatty acids from the blood plasma via the enzyme lipoprotein lipase (Fig. 1). Circulating triacylglycerols in the form of lipoproteins provide a pool of fatty acid, some of which can be used by the adipocyte. However, the triacylglycerol does not traverse the adipocyte membrane. Lipoprotein lipase is produced by adipocytes and migrates to the surface of the capillary endothelial cell where fatty acids are hydrolyzed from triacylglycerol molecules (Nilsson-Ehle *et al.*, 1976). The long chain free fatty acid can then traverse membranes and thus becomes available to the adipocyte. In addition, plasma free fatty acids bound to albumin are available to the adipocyte. Thc lipoprotein lipase mechanism is important in providing not only dietary fatty acids, the source of essential fatty acids as well as one source of nonessential fatty acids, but also in redistributing lipoprotein fatty acids.

The role of lipoprotein lipase in accretion of fat in cattle, pigs and sheep has not been studied extensively (Kris-Etherton & Etherton, 1982). Enzyme activity fluctuates in ovine (Haugebak *et al.*, 1974*a*) and porcine (Enser, 1973; Steffen *et al.*, 1978) adipose tissue in accord with at least some of the periods of fat deposition. The lack of emphasis on this enzyme and source of fatty acids may be partially justified because these three species usually are fed diets low in fat so that in these circumstances the *de novo* synthesis pathway is probably more important. However, the role of lipoprotein lipase will become more significant as new dietary sources are used for pigs and ruminal bypass is utilized to deliver more long chain fatty acids to ruminant species. The enzyme is one of the earliest enhanced in the development of adipocytes as demonstrated in rats (Hietanen & Greenwood, 1977) and in fetal pigs (Hausman, 1982; McNamara & Martin, 1982). Because of its role in early development of adipocytes and probably later in their differentiation and growth, more emphasis should be placed on control of fat accretion by this enzyme in the meat-producing species. There is very little information regarding lipoprotein lipase activity in adipose tissue from cattle, pigs, or sheep manipulated endocrinologically, nutritionally or pharmacologically.

The major depot lipid in the adipocyte is triacylglycerol. The major pathway for synthesis of this molecule involves the esterification of long chain fatty acids to the three carbon precursor, glycerol 3-P (Fig. 1). Two fatty acids are initially esterified by one or two enzymes (glycerol phosphate acyltransferase) to form phosphatidic acid which is dephosphorylated by phosphatidate phosphohydrolase to yield diacylglycerol that is finally esterified with a third fatty acid by diacylglycerol acyltransferase to yield triacylglycerol. This pathway has not been studied extensively in adipose tissue of cattle, pigs or sheep. The overall esterification process has been examined by using radiolabelled long chain fatty acids as precursors in adipose tissue from pigs (Etherton & Allen, 1980). Much less is known regarding the individual steps of the pathway although some rudimentary knowledge has emerged, especially regarding porcine adipose tissue (Mersmann, 1986*a*). Although triacylglycerol biosynthesis is stimulated by insulin and inhibited by adrenergic hormones in rat adipose tissue *in vitro* (Saggerson, 1985), insulin stimulation remains to be demonstrated in porcine adipose tissue and adrenergic hormone inhibition only occurs under specific incubation conditions (Rule *et al.*, 1987).

Catabolism

The degradation of triacylglycerol or lipolysis in adipose tissue is initiated by hormone-sensitive lipase and when completed results in the production of three fatty acids and glycerol (Fig. 1). The fatty acids can enter the adipocyte fatty acid pool and can be reesterified or be transported to the exterior, i.e. plasma *in vivo* or in the incubation medium *in vitro* (Fain, 1980). Lipolysis is controlled by a cascade type regulatory system beginning with activation of a membrane-bound hormone-receptor (Fig. 2). The hormone-receptor complex, through interaction with a GTP-binding protein (G_s) activates adenylate cyclase to produce intracellular cAMP that in turn binds to the regulatory subunit of protein kinase to free, i.e. activate, the catalytic subunit that phosphorylates hormone-sensitive lipase (Fig. 2). Phosphorylated or activated hormone-sensitive lipase begins the cleavage of fatty acids from triacylglycerol. The regulatory cascade can be negatively controlled by receptors that interact with an inhibitory GTP-binding protein, G_i that inhibits adenylate cyclase and,

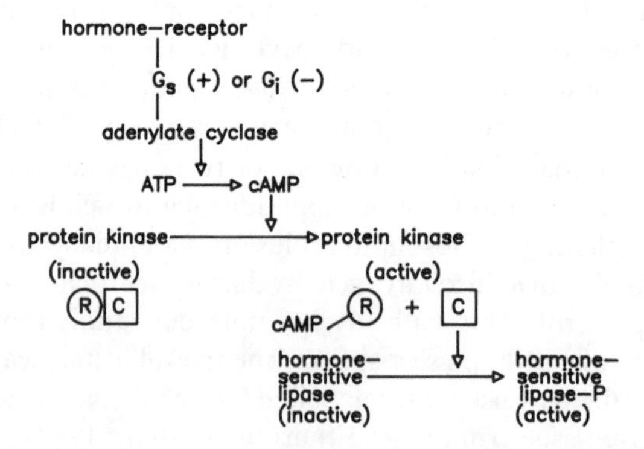

Fig. 2. Activation cascade for adipose tissue lipolysis.

consequently, cAMP accumulation. Another negative control may be exerted by stimulation of the phosphodiesterase enzyme that cleaves cAMP to lower the concentration of this intracellular mediator. Many hormones and other biologically active materials stimulate or inhibit adipose tissue lipolysis (Table 1). Rat adipose tissue is controlled *in vitro* by essentially all of these agents whereas other species tend to have a more restricted spectrum of potential control mechanisms.

Table 1
Potential effectors for lipolysis

Positive	*Negative*
β-Adrenergic agonists	β-Adrenergic antagonists
α-Adrenergic antagonists	α_2-Adrenergic agonists
Glucagon	Insulin
Somatotropin	Prostaglandin E_2
Adrenocorticotropin	Adenosine
Thyrotropin	Phosphodiesterase stimulators
Phosphodiesterase inhibitors	Nicotine (*in vitro*)
Nicotine (*in vivo*)	

A modest amount of information is available concerning lipolysis in porcine adipose tissue including preliminary studies on some of the cascade enzymes and tissue cAMP concentrations (Mersmann, 1986*a*; Hu *et al.*, 1987). However, most studies only measure rates of lipolysis *in vitro* in adipose tissue obtained from animals manipulated endocrinologically, genetically, nutritionally, or pharmacologically. The investigations in ruminant adipose tissue also center on attempts to assess the lipolytic status of adipose tissue in animals by measuring metabolic activity *in vitro* (Sidhu *et al.*, 1973; Yang & Baldwin, 1973*b*; Metz *et al.*, 1974; Pothoven *et al.*, 1975; Etherton *et al.*, 1977; Pike & Roberts, 1980; Jones & Marchello, 1983; Prior *et al.*, 1983; Smith *et al.*, 1983, 1984; DiMarco *et al.*, 1986). The implication of many of these studies is that the lipolytic process, assessed as rates *in vitro*, does not fluctuate significantly in many situations using experimental animals with diverse adipose tissue accretion. This lack of association of lipolytic rates *in vitro* with fat accretion rates *in vivo* is observed in studies with cattle, pigs and sheep.

There is some evidence for control of fat deposition by lipolytic activity from studies *in vivo*. Lean Pietrain pigs mobilize more fat than fatter

Large White pigs during norepinephrine infusion, and insulin infusion is less antilipolytic in Pietrain than in Large White pigs (Wood *et al.*, 1977). Plasma free fatty acid concentrations are lower in two different obese strains of pigs compared to lean pigs during fasting (Bakke, 1975; Mersmann & MacNeil, 1985). If plasma concentration during fasting reflects the lipolytic capacity *in vivo*, then decreased lipolysis may contribute to the obesities.

Metabolic Rates *In Vitro*

Whole organ metabolism studies via perfusion techniques or arterial–venous concentration differences coupled with blood flow measurements are difficult to do in cattle, pigs and sheep because adipose tissue is distributed diffusely in the animal body with no single artery and vein entering and leaving a tissue depot. The tailhead fat depot in fat-tailed sheep is an exception (Khachadurian *et al.*, 1966). Metabolism of tissues or individual cell types is studied *in vitro* because of the problems associated with isolation of adipose tissue for metabolic assessment *in vivo*. There are a number of difficulties associated with measurement of metabolism *in vitro* that should be recognized.

Firstly, the tissue preparation used *in vitro* may influence the results. Tissue slices or pieces represent the closest model to the intact tissue *in vivo* but have several negative attributes. There are other cell types present , e.g. connective tissue cells, blood vessels, reticulocytes and nerves as well as extracellular collagen that might interfere with observations attributed to adipocytes; there is no circulatory perfusion to distribute substrates, oxygen, metabolites and CO_2; some cells are cut during the slicing or other preparative procedures.

These disadvantages are eliminated with isolated adipocytes because a single cell type is studied and each cell is bathed totally in the medium, increasing the surface area for exchange with the medium. However, the negative aspects of isolated adipocytes are also considerable e.g.: cell breakage during preparation, especially in the large cells found in finishing cattle, pigs, and sheep; and proteolytic damage to cell surface receptors and structural membrane proteins caused by the enzymes used to isolate adipocytes from the tissue.

Secondly, the incubation conditions used for measurement *in vitro* can markedly modulate the qualitative patterns and especially the quantitative rates observed as indicated in the following:

(a) A choice must be made about the salt composition of the medium, the buffer composition and the medium pH. Usually this is done with little empirical evidence to support the choice and there is much variation between laboratories.

(b) A choice has also to be made concerning the substrate(s) to be used for carbon flux studies and although it is recognized that the use of ^3H_2O to measure fatty acid biosynthesis is not confounded by intracellular pool size as for the various carbon precursors (Jungas, 1968), water as a substrate does not indicate the flux of individual carbon precursors when several are present simultaneously, as expected *in vivo*.

(c) Regardless of the choice of radiolabelled substrate, a concentration of carbon precursor must be chosen; the choice is between an optimal concentration, i.e. one yielding rates approaching apparent maximal velocity, or some presumed physiological concentration. Unfortunately, because intracellular metabolite concentration is not always known and subcellular compartmentalization essentially unkown, plasma concentrations are assumed to represent relevant intracellular concentrations. The choice then is to use optimal concentrations that yield velocities expressing capacity of the tissue or enzyme, or to attempt to approximate physiological concentrations that yield highly variable rates because these concentrations are usually in a range where velocity changes rapidly with minor fluctuations in substrate concentration.

(d) Bovine serum albumin is a component of the incubation medium for isolated adipocytes but also, at times, for tissue slices *in vitro*. It has been demonstrated (Walton & Etherton, 1986) and confirmed (Mersmann & Hu, 1987) that the commercial source and even the batch of albumin used may profoundly change the metabolic rates and even the qualitative nature of the response of adipose tissue. For example, insulin only stimulates *de novo* fatty acid synthesis *in vitro* with specific albumin preparations.

(e) With broken cell preparations, the dilution of enzymes and other intracellular substituents as well as disruption of spatial integrity that occurs upon tissue homogenization, may yield artifactual results. An example of these effects is the synthesis of triacylglycerol by homogenates of porcine adipose tissue wherein the proportion of total lipid synthesis, represented by triacylglycerol synthesis, varies widely with the homogenate concentration in the assay *in vitro* (Rule *et al.*, 1988*a*,*b*).

(f) Extrapolation of methodology from another tissue or another species without attempts to ascertain relevant conditions for the enzyme or metabolic pathway being studied can lead to erroneous results. For example, conditions for assay of ovine and caprine adipose tissue acetyl-CoA carboxylase are different from those of rat adipose tissue (Vernon & Taylor, 1986). Conditions of assay must be tailored to the species and tissue being investigated (e.g. Mersmann *et al.*, 1973; Steffen *et al.*, 1979; Mersmann & Hu, 1987) to produce relevant results.

Thirdly, the method used to measure the end product of adipose tissue metabolic flux may bias the results. Recently, in adipose tissue lipolysis studies *in vitro*, the intracellular free fatty acid pool has been examined in addition to the free fatty acids in the medium (McNamara *et al.*, 1985; McNamara & Hillers, 1986). Because the tissue pool is sizeable, reinterpretation of some lipolytic results may be necessary as more information accumulates about this pool. The procedure chosen for extraction of the product and for separation of precursor from product may greatly affect the observed process by selective extraction of individual lipid classes or by loss of specific lipids. For example, with long chain fatty acids as substrates for glycerolipid biosynthesis, the usual washing procedures remove much of the phospholipid fraction, potentially one of the major products (Rule *et al.*, 1988*b*).

Finally, in ruminant and porcine adipose tissue *in vitro*, measurements of catabolic rates exceed those of anabolic rates, i.e. fatty acid biosynthesis (Table 2). Thus these rates do not represent the metabolic status of the tissues *in vivo* because in growing animals rapidly depositing fat, anabolism must exceed catabolism. The reason for excessive catabolism *in vitro* is unknown, although poor oxygenation of cells may be the cause; one attempt to include a fluorocarbon oxygen carrier in the incubation medium did not result in net anabolic rates (Mersmann & Hu, 1987). Measurement of triacylglycerol synthesis rates *in vitro* yields large anabolic activity but when extrapolated to the whole animal, excessive estimates of fat deposition are obtained (Mersmann, 1986*b*).

In summary, metabolic rates *in vitro* must be interpreted cautiously; the extent to which they represent the status of adipose tissue *in vivo* is unknown. An example of the disparity of apparent rates *in vitro* and *in vivo* is the measurement of changes in plasma free fatty acid concentration *in vivo* and the rates of lipolysis *in vitro* in adipose tissue taken by biopsy from the same animals; the rates *in vitro* are poor indicators of the plasma

Table 2
Adipose tissue metabolic rates *in vitro*[a]

Species[b]	*Anabolism*		*Catabolism*	
	Fatty acid synthesis	*Triacylglycerol synthesis*	*Basal lipolysis*	*Stimulated lipolysis*
Pig	10	500	100	300
Cattle and sheep	16	1200	100	300

[a]These data are amalgamated from several research reports and present relative rates of fatty acid synthesis, incorporation into triacylglycerol or release during lipolysis *in vitro*. Calculations were based on a 16 carbon fatty acid. The basal lipolytic rate was arbitrarily chosen to represent 100%.
[b]Several references were used for pig adipose tissue values (Etherton & Allen, 1980; McNamara *et al.*, 1985; Mersmann, 1986*b*) and for the combined cattle and sheep data (Ingle *et al.*, 1972; Pothoven *et al.*, 1975; Etherton *et al.*, 1977; Jaster & Wegner, 1981; Vernon, 1981, 1982; Jones & Marchello, 1983; Prior *et al.*, 1983; Smith *et al.*, 1984; DiMarco *et al.*, 1986; Etherton & Evock, 1986; McNamara & Hillers, 1986; Smith & Prior, 1986).

changes observed in these acute experiments (Mersmann, 1986*b*). Measurement of metabolic or enzymatic rates *in vitro* should not be abandoned because most of what we currently know about adipose tissue metabolism and most of what we learn in the near future will come from in-vitro studies. Measurement of metabolism *in vitro* probably is valid much of the time to provide qualitative information about a tissue, e.g. substrate selectivity or hormone effects. Although measurement of adipose tissue metabolism *in vitro* will continue to be valuable it must be recognized that at times, the results, especially rate measurements, will be misleading regarding in-vivo function.

ADIPOSE TISSUE DEVELOPMENT

Because the mass of adipose tissue in an animal at any particular time is dependent on the number and the size of adipocytes, the accretion of fat may be regulated by control of hyperplasia or of hypertrophy. The topics of adipocyte differentiation and growth have been reviewed (Hausman *et al.*, 1980; Leat & Cox, 1980; Hausman, 1985; Vernon, 1986). Adipocyte hyperplasia has been studied much less than the hypertrophic process in large animals for several reasons. Firstly, much of cell proliferation in this tissue occurs in the fetus precluding many approaches to the problem. Secondly, large animal destruction is costly and most adipose tissue depots are not as readily accessible or defined for total dissection as in some laboratory species. Thirdly, the cost of radiolabelled precursors to study proliferation becomes prohibitive except for the most preliminary studies in the youngest animals (Kirtland & Gurr, 1980; Hausman & Kauffman, 1986*a*). Fourthly, no marker for the adipocyte is defined in any species at this time, i.e. no specific protein for which an antibody can be generated or any other characteristic to define developing adipocytes has been identified (Hausman *et al.*, 1980). Consequently, there is no way to determine the precursor cell and so early developing adipocytes can only be defined as cells that have multiple lipid droplets. This is not an absolute characteristic because many cell types accumulate some lipid (Mersmann *et al.*, 1975*a*); however, when a cell is observed with five or more small lipid droplets in an anatomical position that will produce adipose tissue, a reasonable assumption is that the cell is an early multilocular adipocyte (Hausman & Richardson, 1982). There is little problem identifying more mature multilocular adipocytes when further accumulation of lipid to form many droplets occurs and these coalesce to produce very large lipid accumulations. The characteristic unilocular adipocyte with its essentially round shape, peripheral nucleus and cytoplasmic space, and large central major lipid droplet is unmistakable.

Because of the problems of estimating hyperplasia, we know little of its role in adipose tissue accretion in cattle, pigs or sheep. There have been estimates of adipocyte number in growing animals but these are universally flawed because they utilize determination of cell number at one defined depot site with extrapolation to the entire animal using the dissectible fat mass (e.g. Anderson & Kauffman, 1973; Etherton *et al.*, 1982). Accretion of fat in these animals, especially in regard to the contiguous subcutaneous and intermuscular fat depots, is extremely complex so that it is not possible to sample one or even a few sites and then extrapolate to the whole animal (Gurr *et al.*, 1977; Mersmann & MacNeil, 1986). In conclusion, we do not know the role of hyperplasia in the accretion of fat in cattle, pigs or sheep.

Growth and further differentiation of cells that can be identified as adipocytes have been examined in the red meat-producing species with greatest emphasis on the pig (Hermans, 1973; Mersmann *et al.*, 1975*a*; Vodovar *et al.*, 1977; Hausman, 1985; Hausman & Kauffman, 1986*b*). In the porcine fetus, multilocular adipocytes are observable first at about 50–65% of the gestational period; they continue to increase in number and accumulate more lipid. At least in the pig, only multilocular cells are present before birth. Very shortly after birth there is rapid accumulation of adipose tissue stores in this species with formation of unilocular cells. Individual cells continue to fill with triacylglycerol so that the size of unilocular cells increases to $> 150\ \mu m$ (Anderson & Kauffman, 1973; Allen, 1976; Wood *et al.*, 1978). There appears to be a limit to cell size (cells up to 250 μm have been observed) which is species dependent and after an unknown fraction of the cells reach or approach the limit, more cells are added to continually increase the adipose tissue depots (Allen, 1976). Because the precursor cell is not identifiable, it is not clear whether continued accretion of fat involves recruitment and filling of adipocytes that have already differentiated or increased hyperplasia to produce more differentiated cells. In rats, after the postpartum hyperplastic activity, there is a drastic reduction in adipocyte cell division (Greenwood & Hirsch, 1974). However, hyperplasia is observed when older rats experience very high dietary energy intakes (Bertrand *et al.*, 1984; Miller *et al.*, 1984). It is not known whether this type of observation can be extrapolated to any of the red-meat producing species because there appear to be major species variations in the differential roles of hyperplasia and hypertrophy for fat accumulation in several laboratory species (DiGirolamo & Mendlinger, 1971).

Energy intake is a major factor in the control of fat deposition so that increased energy intake above requirements for maintenance and in growing animals for muscle, bone and internal organ growth results in adipocyte hypertrophy. Hypertrophy has been documented during growth of cattle (Robelin, 1981; Cianzio *et al.*, 1985), pigs (see Mersmann, 1986*a*, for review) and sheep (Haugebak *et al.*, 1974*b*; Hood & Thornton, 1979). There are certainly important genetic effects on fat deposition as dramatically seen in genetically obese rodents (York, 1985) as well as pigs (Hetzer & Harvey, 1967; Standal *et al.*, 1973). Genetic effects causing diverse patterns of fat deposition during growth are also observed in different breeds within a species; an extreme example of an obese breed is the feral Ossabaw pig (Martin *et al.*, 1972). The literature on obese pig types has recently been reviewed (Mersmann, 1990). Because a large portion of apparent divergence in body composition results from comparing breeds with different maturation rates and sometimes different stature (Hood & Allen, 1977), valid compositional comparisons can only be made on animals that have achieved a constant percentage of compositional maturity. Dismissing the pragmatic implications for meat production, it is not biologically valid to compare body composition in an Angus and Simmental steer or in a Pietrain and Landrace barrow at the same body weights.

There is major positive and negative regulation of hypertrophy and probably hyperplasia of adipose tissue by a variety of endocrine and pharmacological substances. As subsequently discussed, these effects are observable also in fetuses in which pituitary and central nervous system control of adipose tissue differentiation are evident (Hausman *et al.*, 1987; Ramsay *et al.*, 1987*a,b*). Whole body fat accretion can be modulated by such mechanisms so that manipulation of animals raised for meat production is possible to yield a more acceptable and efficiently produced product.

NUTRITIONAL CONTROL

As already indicated, adipose tissue deposition is dependent on energy intake. In many husbandry situations with animals raised to produce muscle mass as a commercial endpoint, overfeeding increases the accumulation of an unwanted product, fat (Mersmann, 1987*a*). The practice of feeding *ad libitum* arose because of the availability of cheap, high-quality feedstuffs and because labour for feeding is expensive. Alternative feeding schemes arose in locales where feedstuffs were less available. Regardless, feeding at less than *ad libitum* levels (limit-feeding) yields efficient production of lean muscle mass (Lister, 1976; Gurr *et al.*, 1977; Henry, 1977; Prior *et al.*, 1977; Byers, 1982). The amount of fat is reduced by limit-feeding in the intramuscular depot of cattle in which large amounts of fat accumulate (Smith *et al.*, 1977) and in the subcutaneous, perirenal and intermuscular depots of cattle (Smith *et al.*, 1977) and pigs (Henry, 1985). Limit-feeding with animals grouped in pens is not perfected. A pen of animals fed a fixed amount per day will be composed of an aggressive group that has *ad libitum* intakes, a group that has the desired intake reduction and a group that has impaired growth because intake is reduced beyond the desired level. Animals can be fed individually when grouped in a common pen by utilization of electronically controlled feeders such that each animal will receive only its own ration for the day. The cost effectiveness of such systems remains to be determined. Limit-feeding is applicable to pigs and to cattle and sheep when placed on high-energy diets but is usually of little interest when ruminant species are fed on pasture.

Ultimately, feed intake is controlled by the central nervous system so that knowledge of satiety signals to the brain may be useful in controlling feed intake. It is not clear what the peripheral satiety signals are but many, such as glucose, fat and endocrine substances, e.g. cholecystokinin have been proposed. More clear are the ways in which the central nervous system centers for satiety or hunger are manipulated (Houpt *et al.*, 1979; Baile *et al.*, 1983; Della-Fera & Baile, 1984; Houpt, 1984). Thus, it is known that injection of a number of substances directly into the brain (usually into the third ventricle), will increase or decrease feed intake. There has been limited success with administration of various materials peripherally because of the problem of peripheral isolation from the brain by the blood–brain barrier. Anorectic agents could be used to control the *ad libitum* intake of animals to achieve the desired level of limit-feeding.

During animal growth, there is a major shift in the partitioning of dietary component intake, including energy, between muscle and adipose tissue (Byers, 1982; Berg & Walters, 1983; Etherton & Walton, 1986). Early in development the muscle mass is accreted to a much larger extent than the adipose tissue mass. As development proceeds, adipose tissue accretion accelerates to eventually become a major sink for energetic intake. An understanding of the trigger(s) for this shift in dietary component and especially energy partitioning would allow control of excess fat deposition as the animal matures. This approach, although complex on a theoretical basis because of the intricacies of endocrine and metabolic factors controlling the change in pattern of dietary component distribution, may be useful on a pragmatic basis by manipulation of a single hormonal material as evidenced by sex steroid, somatotropin or adrenergic agonist-induced changes in body composition.

ENDOCRINE CONTROL

There is considerable evidence that many endocrine substances influence fat deposition (Weekes, 1983; Muir, 1985; Brockman & Laarveld, 1986; Mersmann, 1987*a*; Tucker & Merkel, 1987). Some of the evidence is from extreme experimental situations such as extirpation of the endocrine gland followed by observations of changes in body composition; sometimes this is coupled with replacement therapy of the endocrine substance. The regulation of postnatal adipose tissue growth is almost certainly not the result of the presence or absence of an endocrine substance but of the interaction of varied concentrations of hormone at the tissue level coupled with changes in the receptivity of the target tissue to the endocrine substance. Furthermore, several endocrine substances are present during postnatal growth so that complex endocrine regulation occurs involving interactions between several hormones. Much of the evidence accrued regarding endocrine influence on adipose tissue is from studies of tissue preparations or cells *in vitro*. These approaches allow isolation of a single endocrine substance with a single tissue or cell type so that interpretation is clear. However, the effects *in vitro* must be verified *in vivo* to be certain they reflect physiological observations and not pharmacological phenomena or artifacts imposed by changes in tissue responsiveness when removed from the animal. Hormonal regulation of adipose tissue metabolism has been reviewed recently (Saggerson, 1985). There are a number of reviews regarding endocrine manipulation of animal growth and all have some information about control of fat deposition (Lu & Rendel, 1976; Lister, 1980; Trenkle, 1981; Meissonnier, 1983; Etherton & Kensinger, 1984; Roche & O'Callaghan, 1984; Schanbacher, 1984; Buttery, 1985; Brockman & Laarveld, 1986; Roche & Quirke, 1986; Mersmann, 1987*a*).

Sex Steroids

The most obvious endocrine influence on adiposity is the sex of the animal (Seideman *et al.*, 1982). In mammals, the intact male is more muscular and less fat than the intact female whereas the castrated male is more fat than the male and in some species, the female. Administration of exogenous estrogens to castrated male cattle and sheep produces improved gain and feed efficiency coupled with leaner carcasses (Preston, 1975; Lu & Rendel, 1976; Galbraith & Topps, 1981; Meissonnier, 1983; Schanbacher, 1984; Roche & Quirke, 1986). The mechanism of this estrogenic effect is not known but may result from elevated plasma growth hormone concentrations (Trenkle, 1976). For unknown reasons, exogenous estrogenic compounds do not influence body composition in pigs (De Wilde & Lauwers, 1984; Roche & Quirke, 1986). There are estrogen receptors in rat adipose tissues (Gray *et al.*, 1981) but these receptors have yet to be demonstrated in adipose tissue from meat-producing species. Acute exposure of bovine or ovine (Etherton *et al.*, 1977) or of porcine (Mersmann, 1984*a*) adipose tissues to estrogens *in vitro* does not alter lipolytic or lipogenic rates. However, manipulation of the sex hormone status of living animals does influence adipose tissue metabolism in pigs (Mersmann, 1984*a*) and cattle (Prior *et al.*, 1983). Although not studied in depth, at least some of the effects of sex hormones in red-meat producing animals mimic more detailed observations in laboratory animals wherein estrogen treatment lowers adipose tissue lipoprotein lipase (Ramirez, 1981; Wade *et al.*, 1985), lipolysis (Pecquery *et al.*, 1986) and fatty acid synthesis (Hansen *et al.*, 1980). Administration of estrogens to intact bovine males does not further enhance the muscle mass or reduce the fat mass but rather increases fatness (Prior *et al.*, 1983).

Androgen administration can alter body composition in both castrated males and intact females (Roche & Quirke, 1986). The androgen effects on carcass composition, even in intact males may be the result of conversion to estrogens by aromatization (Gray *et al.*, 1979). Combinations of androgens and estrogens or estrogens and progestins have been used to favorably alter body composition (Meissonnier, 1983; Roche & Quirke, 1986; Mersmann, 1987*a*).

Somatotropin

Another major endocrine influence on adiposity is somatotropin or growth hormone. Not only is this anterior pituary hormone associated with increased skeletal muscle growth but when administered exogenously it produces decreased fat accretion (reviewed in Mersmann, 1987*a*). The most complete documentation of growth hormone effects in meat-producing species is in pigs (Machlin, 1972; Chung *et al.*, 1985; Etherton *et al.*, 1987; Sillence & Etherton, 1987). Sizeable reduction in fat accretion is observed in this specics concomitant with increased muscle mass, daily gain and efficiency of feed utilization. Less is known about the effect of exogenous growth hormone in cattle (Trenkle, 1981; Hart & Johnsson, 1986) or sheep (Muir *et al.*, 1983; Johnsson *et al.*, 1985; Butler-Hogg & Johnsson, 1987) but the reduction in fat mass probably occurs. Somatotropin increases nitrogen retention in cattle (Eisemann *et al.*, 1986*a*), elevates plasma free fatty acid concentration and utilization and enhances whole body protein synthesis (Eisemann, 1986*b*). There are somatotropin receptors on adipose tissue cells of rats (Gavin *et al.*, 1982) so that the effect of this hormone to decrease fat mass may be directly on the adipocyte. Growth hormone stimulation of lipolysis in adipose tissue *in vitro* usually requires an extended incubation period but growth hormone will stimulate porcine adipose tissue lipolysis in the presence of theophylline, suggesting that the coupling to cAMP production is not very efficient (Mersmann *et al.*, 1975*b*; Mersmann, 1986*c*).

Somatotropin does not increase plasma free fatty acid concentration when acutely infused into pigs (Mersmann, 1986*c*). The generally longer incubation time necessary for observation of growth hormone effects seems to fit with current thoughts of localized somatomedin production although production of other proteins cannot be excluded. Support for this hypothesis comes from in-vivo studies in which injection of somatotropin into pigs produces an increase in serum somatomedin (IGF-I) but only after about 6 h (Sillence & Etherton, 1987). Growth hormone reduces the insulin-stimulated lipogenic activity of porcine (Walton & Etherton, 1986) and ovine (Vernon, 1982) adipose tissue *in vitro*. Thus, the somatotropin-induced decrease in carcass adiposity may result from a direct effect on adipose tissue, possibly mediated through local somatomedin production, to inhibit insulin-stimulated lipogenesis and concomitantly to stimulate insulin-inhibited lipolysis through counteraction of insulin effects. If somatotropin effects on adipose tissue are mediated by somatomedins *in vivo*, then the autocrine or paracrine effects of somatomedins are the reverse of those observed—somatomedins generally having insulin-like effects on adipose tissue (e.g. Walton *et al.*, 1987).

Plasma growth hormone concentration is increased by growth hormone releasing hormone (Ling *et al.*, 1985; Hall *et al.*, 1986) and inhibited by somatostatin (Martin & Millard, 1986). The growth effects and carcass compositional changes induced by growth hormone might be accomplished by administration of the releasing factor (Moslely *et al.*, 1985; Etherton *et al.*, 1986; Schanbacher, 1986) or by inhibition of the effects of somatostatin, the inhibitor. Immunization of sheep to somatostatin may increase growth rate (Spencer, 1985, 1986). Because many growth hormone effects are purportedly mediated through somatomedins, the administration of these proteinaceous factors could directly increase growth and repartition body composition towards less fat and more muscle mass; when one of the somatomedins (IGF-I) is given to hypophysectomized rats, growth rate is increased (Froesch *et al.*, 1986). Manipulation of the synthesis, degradation or release of any of these central nervous system factors that control plasma concentration of somatotropin or of the effector molecules might provide a useful approach in the future. The pragmatic utilization of these various approaches awaits determination of optimal conditions and dosages coupled with lower production costs and availability of better delivery systems. Genetic technology has solved the problem of obtaining large quantities of these peptide factors although cost effectiveness remains to be established in the marketplace.

More recent reports on somatotropin-treated pigs have established the equivalency of recombinant and pituitary hormone in this species (Evock *et al.*, 1988) and its effectiveness in young pigs (25 kg) coupled with information on the interaction between hormone treatment and dietary energy concentration (Campbell *et al.*, 1988). Several papers from a conference on the topic have described the effects of somatotropin on porcine growth (Pork Industry Conference Proceedings, 1987) and also provide information on the insertion of the somatotropin gene into porcine DNA. A number of studies have established that administration of growth hormone releasing hormone and several analogs cause an increase in serum somatotropin in cattle (Enright *et al.*, 1987; Scarborough *et al.*, 1988), sheep (Kensinger *et al.*, 1987) and pigs (Petitclerc *et al.*, 1987). Symposium articles have described the role of somatomedin i.e. IGF-I (Clemmons *et al.*, 1987) and somatomedin binding proteins (Spencer, 1987) in growth. The serum binding protein(s) for somatomedin (IGF-I) have been measured in porcine (Buonomo *et al.*, 1987), and bovine and ovine species (Hossner *et al.*, 1988); some properties of the bovine and ovine binding proteins are also described. A major breakthrough was recently announced in the popular press: a time release implant for administering somatotropin to pigs for six weeks was described (Anon., 1988).

Insulin

A major endocrine influence on adipose tissue accretion in mammals is insulin (Etherton, 1982; Martin *et al.*, 1984; Weekes, 1986). Generally, insulin stimulates anabolic and inhibits catabolic adipose tissue lipid metabolism pathways (Fig. 3). Diabetic animals have smaller fat depots and usually the lipogenic carbon flux and enzyme activities associated with lipogenesis are reduced. For example, in alloxan-diabetic pigs, glucose incorporation into lipids is dramatically reduced; it can be restored by replacement therapy with insulin (Romsos *et al.*, 1971). Alloxan-diabetic cattle appear to have reduced lipogenesis because of reduced feed intake rather than because of a direct insulin effect on the metabolic pathway (Prior & Smith, 1982; Smith *et al.*, 1983). Fetuses from diabetic pregnant sows have greater plasma insulin concentrations in response to the elevated maternal glucose concentration (Kasser *et al.*, 1982) and have

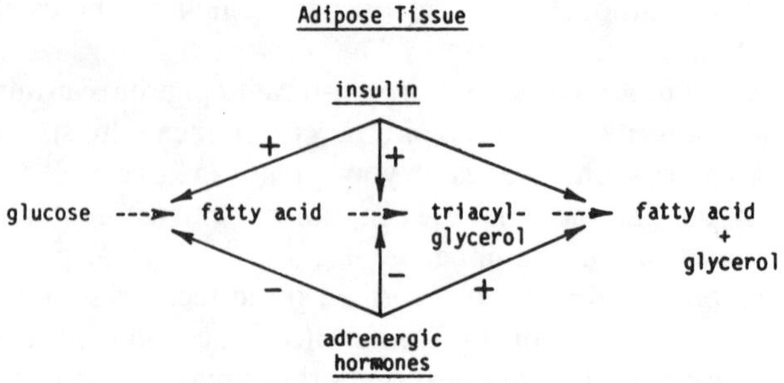

Adipose Tissue

insulin

glucose ---> fatty acid ---> triacylglycerol ---> fatty acid + glycerol

adrenergic hormones

Fig. 3. Control of adipose tissue metabolism by insulin and adrenergic hormones.

elevated adipose tissue fatty acid synthesis rates *in vitro* (Kasser *et al.*, 1981). The fetuses have increased fat deposition. Certainly insulin receptors are present on porcine adipocytes (Etherton & Walker, 1982; Chung *et al.*, 1983; Etherton *et al.*, 1984) and probably are present on bovine adipocytes (Vernon *et al.*, 1985*b*) although these are not always demonstrable (Vasilatos *et al.*, 1983). Unfortunately, exogenous insulin added to acute incubations of bovine, ovine or porcine adipocytes or adipose tissue slices *in vitro* produces only modest or no stimulation of lipogenesis in most laboratories (Vasilatos *et al.*, 1983; Vernon *et al.*, 1985*b*; Mersmann & Hu, 1987). The modest insulin response may result from improper incubation conditions although, to date, conditions have not been found, in most laboratories, to produce consistent or large insulin responses (Mersmann & Hu, 1987; Rule *et al.*, 1987). Insulin added to bovine (Etherton & Evock, 1986), porcine (Walton & Etherton, 1986) or ovine (Vernon, 1982) adipose tissue during culture for several days maintains lipogenic capacity lost during culture without insulin. One laboratory has achieved considerable insulin stimulation of lipogenesis during acute incubation in both bovine (Etherton & Evock, 1986) and porcine (Walton & Etherton, 1986) adipose tissue so that in the future when incubation conditions to produce insulin stimulation are understood and reproducible, studies *in vitro* may be more productive in establishing mechanisms of insulin action on adipocytes in the red-meat producing species.

Insulin inhibits lipolysis even in a tissue that responds only modestly to insulin stimulation of lipogenesis. Lipolysis, stimulated to high rates by an adrenergic hormone, is strongly inhibited by insulin in porcine adipose tissue (Mersmann & Hu, 1987; Rule *et al.*, 1987). This result purportedly indicates the presence of functional insulin receptors on this cell under incubation conditions *in vitro* that are very similar to those used for attempts to demonstrate effects on lipogenesis. Insulin inhibition of lipolysis *in vivo* is suggested because acute insulin infusion into pigs decreases plasma free fatty acid concentration (Mersmann, 1986*c*). Exogenous insulin also inhibits epinephrine-stimulated lipolysis in bovine adipocytes (Yang & Baldwin, 1973*b*; Jones & Marchello, 1983). Regardless of the role of insulin in lipogenesis or lipolysis *in vivo*, it remains to be determined whether this hormone can be manipulated to effect the desired decrease in adipose tissue deposition without major detrimental changes in function of other tissues.

Adrenergic Agonists

The adrenergic hormones and neurotransmitters represented by dopamine, epinephrine and norepinephrine provide another major regulatory system for adipose tissue lipid deposition (Fain & Garcia-Sainz, 1983; Lafontan & Berlan, 1985). Generally, adrenergic hormones or neurotransmitters stimulate catabolic and inhibit anabolic adipose tissue metabolism (Fig. 3). Although the physiological role of these substances appears to be for acute mobilization of energy stores including glycogen and fat, they also can be involved in longer term regulation such as energy mobilization when energy intake is marginal. Thus, the concentration of an adrenergic substance reaching target tissues may be very important to maintain homeostasis during the usual overnight fast in animals raised for meat production, especially in nonruminants. The degree to which these hormones and neurotransmitters stimulate adipose tissue receptors coupled with the degree of responsiveness of the tissue will dictate the extent to which lipogenesis is inhibited and lipolysis concomitantly is stimulated.

Adrenergic receptors are of two types, α- and β- and each is usually divided into subtypes, α_1 and α_2 and, β_1 and β_2 (Table 3). Compounds specific for β_1 and β_2-adrenergic receptors interact with a specific membrane bound receptor coupled to the enzyme adenylate cyclase by a GTP-dependent coupling protein, G_s to increase the intracellular production of cAMP (Fig. 2). The α_2-adrenergic receptor is coupled to adenylate cyclase by a GTP-dependent inhibitory coupling protein, G_i so that stimulation causes a decrease in intracellular cAMP concentration. Thus, stimulation of β_1 or β_2-adrenergic receptors increases cAMP concentration and stimulation of α_2-adrenergic receptors acts antagonistically to decrease cAMP concentration. Unlike several other mammalian species, no α_2-

Table 3
Alpha- and beta-adrenergic function[a]

Receptor	*Function*	*Agonist*	*Antagonist*
α_1	Constrict arteries	Methoxamine	Prazosin
α_2	Inhibit lipolysis	Clonidine	Yohimbine
β_1	Increase heart rate	Dobutamine	Practolol
β_2	Bronchodilate	Terbutaline	Butoxamine

[a]More details of adrenergic function are found in pharmacology and endocrinology texts, e.g. Martin (1985—Table 8-2).

adrenergic agonist-stimulated inhibition of lipolysis is demonstrable in rat or pig adipose tissue (Mersmann, 1984*c*); there have been no attempts to demonstrate α_2-adrenergic inhibition of lipolysis in bovine or ovine adipose tissue. The α_1-adrenergic receptor is not coupled to cAMP metabolism but rather controls phosphatidylinositol metabolism (Fain & Garcia-Sainz, 1983). Beta-adrenergic receptors have been demonstrated by binding techniques on adipose tissue cells (Lafontan & Berlan, 1985) including those from cattle (Jaster & Wegner, 1981) and pigs (Bocklen *et al.*, 1986).

There is an enigma regarding the physiological regulation of adipose tissue metabolism by norepinephrine and epinephrine because both stimulate α- and β-adrenergic receptors. Norepinephrine is usually more potent towards β_1- than β_2-adrenergic receptors compared with epinephrine whereas epinephrine is more potent toward α-adrenergic receptors than norepinephrine. In fact, in rabbit adipose tissue *in vitro*, because of the predominant activity of α_2-adrenergic inhibition, it was thought for many years that epinephrine did not stimulate lipolysis (Lafontan, 1979). In human fat cells (Lafontan *et al.*, 1983), there is a preponderance of α_2- over β-adrenergic receptors so that receptor number does not explain the lipolytic function of infused epinephrine *in vivo* wherein plasma free fatty acids are increased, not decreased. The circulating concentration of norepinephrine is usually greater than that of epinephrine in most species (Buhler *et al.*, 1978) but the physiological function of these two adrenergic agonists to preferentially stimulate α- or β-adrenergic receptors is not known. Specificity *in vivo* appears to be dependent on factors other than the density of receptor subtypes.

Oral administration of analogs of norepinephrine increases muscle mass and decreases fat mass in growing cattle, chickens, pigs and sheep (Table 4). This effect is a repartitioning of dietary components away from

Table 4
Repartitioning effects of β-adrenergic agonists[a]

Species	*Dose (mg/kg diet)*	*Gain (%)*	*Feed (%)*	*Muscle (%)*	*Fat (%)*
Cattle	≤ 1	−10	−10	+13	−25
Chickens	≤ 1	+ 3	+ 1	+ 2	− 6
Pigs	≤ 1	0	− 5	+ 8	− 9
Rats	≤ 2	+20	+10	+16	− 6
Sheep	≤10	+ 2	− 5	+25	−30

[a]Approximate projections from published data amalgamating several studies on each species involving both clenbuterol and cimaterol. Data sources are indicated in several reviews (Ricks *et al*., 1984; Hanrahan *et al*., 1986; Mersmann, 1987*b*).

fat accretion to muscle accretion. Consequently, the compounds have been designated repartitioning agents although it should be noted that the mechanism(s) is unknown and that this effect is not unique to norepinephrine analogs since sex steroids and somatotropin also repartition growth. The observed decrease in fat mass could result from direct adrenergic effects on adipose tissue (Fig. 3) by stimulation of lipolysis (Lafontan & Berlan, 1985) and/or inhibition of fatty acid and triacylglycerol biosynthesis (Saggerson, 1985). Adrenergic agonists stimulate adipose tissue lipid degradation in bovine (Yang & Baldwin, 1973*b*), porcine (Mersmann *et al*., 1974) and ovine (Duquette *et al*., 1984) adipose tissue. Inhibition of fatty acid biosynthesis is not observed in porcine adipose tissue and although triacylglycerol biosynthesis is inhibited by isoproterenol, the inhibition is observed only under specific incubation conditions (Rule *et al*., 1987).

There are several investigations *in vitro* with clenbuterol, one of the norepinephrine analogs that when fed, decreases fat accretion. Incubation of ovine adipose tissue (Thornton *et al*., 1985) or chicken hepotocytes (Campbell & Scanes, 1985) with clenbuterol decreases acetate incorporation into adipose tissue or hepatic lipids, respectively; clenbuterol also stimulates lipolysis in adipose tissue from both species *in vitro*. If these observations *in vitro* prevail *in vivo*, a mechanism for decreased adiposity is revealed. Porcine adipose tissue does not respond to clenbuterol *in vitro* by inhibition of lipogenesis or stimulation of lipolysis (Mersmann, 1987*b*; Rule *et al*., 1987) so that the mechanism *in vivo* may be different in this species. It appears that adrenergic agonist stimulation of lipolysis in pigs shows extreme structural specificity (Mersmann, 1984*b*) so it is not surprising that a specific norepinephrine analog is ineffective.

Infusion of adrenergic hormones increases plasma free fatty acid concentrations in cattle (Blum *et al.*, 1982), pigs (Mersmann, 1987*b*), and sheep (Bassett, 1970). Plasma free fatty acid concentration is increased in sheep fed cimaterol (another norepinephrine analog that decreases fat accretion when fed) (Beermann *et al.*, 1985) or cattle fed clenbuterol (Eisemann *et al.*, 1988). In sheep fed cimaterol, plasma concentrations of insulin are decreased and thyroxine plus somatotropin are increased (Beermann *et al.*, 1987). Thus, some of the repartitioning effects might result from these plasma hormone changes, possibly with decreased insulin primarily affecting lipolysis and somatotropin primarily stimulating muscle accretion. Although the adrenergically active substances may act directly on adipose tissue, they may also have indirect effects. For example, clenbuterol increases plasma free fatty acid concentration when administered acutely to pigs even though no effects are observed on adipose tissue lipolysis *in vitro* (Mersmann, 1987*b*). The effect *in vivo* could result from release of endogenous adrenergic hormones, by conversion to an active metabolite by nonadipose tissues, by modulation of metabolic substrates, inhibitors or hormones, or by a change in blood flow. Blood flow increased in sheep (Beerman *et al.*, 1986) and cattle (Eisemann *et al.*, 1988) administered norepinephrine analogs. Adipose tissue blood flow has not yet been specifically measured so it is not known if a change in blood flow causes the increased lipolysis. In those species studied, systemic adrenergic stimulation generally decreases blood flow to adipose tissue (Fredholm, 1985). Preliminary results suggest that infused clenbuterol and isoproterenol acutely increase adipose blood flow in pigs (H. J. Mersmann, unpublished observations).

Several recent review articles have considered the effects of β-adrenergic agonists in growing meat-producing animals and have included references to two new norepinephrine analogs, ractopamine and L-644-969 (Beermann, 1987; Convey *et al.*, 1987; Mersmann, 1987*c*; Smith, 1987). Additional papers have been published regarding cimaterol effects on skeletal muscle in sheep (Kim *et al.*, 1987), on ovine adipose tissue metabolism (Hu *et al.*, 1988), on growth and body composition in pigs (Cromwell *et al.*, 1988), and in mice selected for rapid gain (Eisen *et al.*, 1988). Effects of clenbuterol on ovine (Coleman *et al.*, 1988) and bovine (Miller *et al.*, 1988) growth and adipose tissue metabolism and on ovine growth and muscle protein synthesis and degradation (Bohorov *et al.*, 1987; MacRae *et al.*, 1988) have also been reported. A review of the effects of β-adrenergic agonists on all aspects of animal growth and body composition has also appeared (Hanrahan, 1987). Further suggestions of how β-adrenergic agonists increase muscle and decrease fat accretion have been discussed by Mersmann (1987*c*, 1989). Reeds *et al.* (1988) have indicated that the decrease in fat accretion in rats fed clenbuterol can be antagonized by β-adrenergic antagonists but the increase in muscle accretion cannot be antagonized; this suggests different mechanisms of action for clenbuterol in these two tissues in rats.

Regardless of the mechanisms for a given norepinephrine analog in a given species, this class of compound opens up a large number of possibilities of pharmacological manipulation of fat and muscle growth in growing cattle, pigs and sheep.

FUTURE OPPORTUNITIES

Endocrinology

Major developments will result as details of endocrine function emerge. Rather than simply removing an endocrine gland from an animal to minimize hormone concentration or adding an exogenous hormonal material to maximize hormone concentration, endocrine manipulation in the future will be more subtle. In many cases, endocrine manipulation need only occur at a defined time period in the ontogeny of the animal to affect endocrine function for the entire life of the animal. Use of such techniques during an endocrinc 'window of opportunity' will allow major changes in animal growth.

An example of such manipulation is in regard to sex determination in fetal and neonatal animals. Production of androgens must occur at critical times for the genetically determined male to develop as a functional or behavioral male; if this does not occur many of the existing female characteristics will remain (Martin, 1985). Because prepubertal growth characteristics are more favorable in male than in female meat-producing animals, endocrine manipulation of females to have more male-like growth characteristics would be beneficial in the females not kept for replacement of the breeding herd. Injection of androgens into pregnant ewes at a critical period for sexual development of the fetus produces female progeny with growth characteristics similar to the ram rather than the genetically determined female; male progeny are not affected (Klindt *et al.*, 1987; Jenkins *et al.*, 1988).

Endocrine manipulation of the fetus can change fat deposition in the newborn growing animal. In pigs, fetal hypophysectomy yields a smaller number of adipocytes accompanied by increased adipocyte hypertrophy presumably because somatotropin and probably because other trophic hormones such as thyrotropin are removed (Hausman *et al.*, 1987; Ramsay *et al.*, 1987*a,b*). Genetically obese pigs have reduced somatotropin in the fetus as well as in the postpartum animal (reviewed in Mersmann, 1990) so apparently this hormone deficiency is important in regulating fat deposition. Other than sex hormone manipulation, there is no fetal endocrine change yet available to produce lean growing animals but the accumulating evidence indicates that the possibility will arise when the specific endocrine substance(s) and the specific 'window of opportunity' for manipulation are discovered.

Another approach to changing endocrine function in animals is immunization against a specific hormone. This technique may be particularly valuable when used at a specific time in the life cycle of the animal. For example, it may be useful to allow male animals to grow for a considerable period to take advantage of the favorable growth characteristics of the male and then to immunologically castrate them to remove some of the unfavorable characteristics as they enter puberty. Immunological castration can be accomplished by immunization against luteinizing hormone (LH), the releasing hormone for LH, or testosterone (Robertson *et al.*, 1982; Schanbacher, 1984; Falvo *et al.*, 1986). Another example is immunization of female animals against estrogenic hormones. Heifers immunized against estradiol tend to grow faster and more efficiently than intact heifers; there may be differences in the biological mechanism depending on whether immunization produces a low titer (Wise & Ferrell, 1984) or a high titer (Crouse *et al.*, 1987; St. John *et al.*, 1987).

Immunization of growing animals against somatostatin, which inhibits the release of somatotropin, could improve growth characteristics by increasing the release of pituitary somatotropin. This would be expected to produce leaner, more muscular animals as observed with injection of exogenous somatotropin. Immunization in sheep has been shown to improve growth rates with no change in body composition in some experiments (Spencer *et al.*, 1983*a,b*; Spencer, 1985, 1986). However, increased growth rate was not always observed (e.g. Bass *et al.*, 1987).

There is evidence that prolactin improves animal growth rates and carcass characteristics (Ohlson *et al.*, 1981; Klindt *et al.*, 1985). Other evidence for these effects comes from studies involving manipulation of the photoperiod whereby prolactin concentration is changed (Forbes, 1982; Tucker *et al.*, 1984). However, there is considerable inconsistency in the results from studies which have examined prolactin stimulation of growth. The extreme complexity of the relationships between plasma endocrine concentration and the growth of various tissues (Klindt *et al.*, 1985) suggests that the understanding of the subtleness of endocrine manipulation of growth rates and compartmentalization of growth within individual tissues is at an embryonic state.

Genetics

In addition to the classical selection techniques for reducing body fat, the future should allow selection for specific endocrine patterns or tissue enzyme concentrations. Selection processes that more directly target the metabolic or endocrine factors responsible for adiposity will be more successful and should yield more rapid achievement of the goals than classical techniques. The heritability estimates for the NADPH-generating dehydrogenases in porcine adipose tissue are high so that selection against these activities could yield rapid achievement of reduced body fat (Strutz & Rogdakis, 1979). Unfortunately, it is not clear which enzyme(s) or endocrine substances should be targeted because so little is known about bioregulation of the accretion of fat. The knowledge base regarding fat accretion in cattle, pigs and sheep is broad but not sophisticated. Furthermore, most of what is known relates to tissues *in vitro* and this may not be relevant to the complexity of the interactions *in vivo*.

Modulation of an enzyme or hormone that is specific or at least relatively specific for control of adipose tissue hypertrophy during postnatal growth would allow direct manipulation of adipose tissue accretion. Insulin would be an excellent candidate because it stimulates anabolic and inhibits catabolic paths of adipose tissue lipid metabolism; however, insulin has profound effects on many tissues and obviously creation of diabetes. Endocrines or enzymes will have to be found that, although no more specific for adipose tissue, will at least not have major negative effects on the animal as observed with insulin. Likewise, manipulation of the control of adipocyte hyperplasia could change fat accretion dramatically. However, little is known regarding the regulation of hyperplasia, especially in cattle, pigs or sheep *in vivo*.

Genes may be implanted in mammalian cells so that in the future additional copies of a gene may yield the desired bioregulation (Anderson,

1986). Implantation of the gene for somatotropin has been achieved in several species but at the present time, the only mammalian species exhibiting increased growth is the mouse (Evans & Hollaender, 1986; Wagner & Jochle, 1986). Considering the complexity of the chronology of gene function in mammalian organisms, it is not surprising that although genes can be transferred, the desired phenotypic phenomena are not readily expressed. In time, given knowledge about the control of mammalian gene function, such techniques may yield desirable effects on animal growth that now can be produced only by administration of exogenous substances. At some time in the future, when regulation of gene function is understood, the technology to turn a given gene on and off at will in one specific tissue will emerge. This technology will allow subtle control and manipulation of animal metabolism but it may not become a reality for many years. Possibly the most immediate uses of genetic engineering techniques in animal production are the provision of proteinaceous materials such as somatotropin, growth hormone releasing hormone, or somatomedin for use in promoting animal growth and of vaccines for use in animal health (Petters, 1986).

Immunology

Immunization of animals against adipose tissue cells could result in decreased fat accretion in the growing animal by destroying a proportion of the adipocytes contributing to the depot mass. Antibodies to rat adipocyte plasma membrane cause destruction of rat adipocytes *in vitro* and when these antibodies are injected into rats, carcass fat concentration is decreased (Flint *et al.*, 1986). Refinement of this approach in terms of dosage of antibody, specificity of antibody and timing of immunization could lead to a viable method of decreasing adiposity in cattle, pigs and sheep. Polyclonal antibodies to ovine (Nassar & Hu, 1988) and monoclonal antibodies to porcine (Killefer *et al.*, 1988; Wright & Hausman, 1988) adipose tissue membrane proteins could be used to identify precursor cells, study hyperplasia and reduce actual fat deposition.

Pharmacology

The use of anorectic agents to reduce appetite in circumstances of excessive energy intake is an approach that would result in reduced fat deposition as is observed in the control of some types of obesity (Curtis-Prior, 1983). Anorectic compounds may have practical impact if an agent is found that does not leave residues and the use of which in the animal can be titrated so that although it decreases adiposity it does not also decrease muscle deposition.

Other pharmacological agents might be useful in activating the cascade for stimulation of lipolysis. Analogs of cAMP might be active over long periods or inhibitors of cAMP-phosphodiesterase might effectively increase the amount of cAMP present by inhibiting its destruction. Such compounds are available but have not been tailored structurally or in application for specific use in growing cattle, pigs or sheep. Although there is less information, direct activation of cAMP-dependent protein kinase or hormone-sensitive lipase could also be used to manipulate fat deposition.

Pharmacological intervention in metabolic pathways by stimulation or inhibition of individual enzymes becomes viable as the specific role of the enzyme in the regulation of fat deposition becomes known. Analogs of substrates, products or allosteric effectors might all provide means of regulating a pathway so that reduced deposition will occur (Curtis-Prior, 1983).

Summary

There are numerous approaches to changing metabolic activity in adipose tissue of growing animals so that fat deposition is reduced. Some of these have been demonstrated to be effective *in vivo* but for essentially none has the mechanism of action been demonstrated. In many cases the only impediment to achieving new approaches is the imagination of the investigator. The ultimate mechanism may not be that assumed in theory, instead the desired end result produced by serendipity could readily yield major breakthroughs. Species specificity of endocrine function and tissue metabolism will undoubtedly temper the applicability of some discoveries.

REFERENCES[†]

Allen, C. E. (1976). Cellularity of adipose tissue in meat animals. *Federation Proc.*, **35**, 2302–7.

Anderson, D. B. & Kauffman, R. G. (1973). Cellular and enzymatic changes in porcine adipose tissue during growth. *J. Lipid Res.*, **14**, 160–8.

Anderson, W. F. (1986). Genetic engineering of animals. In *Genetic Engineering of Animals*, ed. J. W. Evans & A. Hollaender. Plenum, New York, pp. 7–14.

[†]Literature for this chapter was considered through May 1987, with a partial update through September 1988.

Anon. (1982). Marketing according to value: three foreign systems. *Pig American*, **7,** 28–32.
Anon. (1988). Time-released PST. *Pork,* **8,** 8.
Baile, C. A., Della-Fera, M. A. & McLaughlin, C. L. (1983). Hormones and feed intake. *Proc. Nutr. Soc.*, **42,** 113–27.
Bakke, H. (1975). Serum levels of non-esterified fatty acids and glucose in lines of pigs selected for rate of gain and thickness of backfat. *Acta Agr. Scand.*, **25,** 113–16.
Bass, J. J., Gluckman, P. D., Fairclough, R. J., Peterson, A. J., Davis, S. R. & Carter, W. D. (1987). Effect of nutrition and immunization against somatostatin on growth and insulin-like growth factors in sheep. *J. Endocr.*, **112,** 27–31.
Bassett, J. M. (1970). Metabolic effects of catecholamines in sheep. *Aust. J. Biol. Sci.*, **23,** 903–14.
Bauman, D. E. (1976). Intermediary metabolism of adipose tissue. *Federation Proc.*, **35,** 2308–13.
Beermann, D. H. (1987). Effects of beta adrenergic agonists on endocrine influence and cellular aspects of muscle growth. *Recip. Meat Conf. Proc.*, **40,** 57–63.
Beermann, D. H., Campion, D. R. & Dalrymple, R. H. (1985). Mechanisms responsible for partitioning tissue growth in meat animals. *Recip. Meat Conf. Proc.*, **38,** 105–14.
Beermann, D. H., Butler, W. R., Fishell, V. K., Bergman , E. N. & McCann, J. P. (1986). Preliminary observations on the effects of cimaterol on heart rate, blood flow, plasma insulin concentration and net glucose uptake in the hind-quarters of growing lambs. *J. Anim. Sci.*, **63,** Suppl. 1, 225 (Abst. No. 160).
Beermann, D. H., Butler, W. R., Hogue, D. E., Fishell, V. K., Dalrymple, R. H., Ricks, C. A. & Scanes, C. G. (1987). Cimaterol-induced muscle hypertrophy and altered endocrine status in lambs. *J. Anim. Sci.*, **65,** 1514–24.
Berg, R. T. & Walters, L. E. (1983). The meat animal: changes and challenges. *J. Anim. Sci.*, **57,** Suppl. 2, 133–46.
Bertrand, H. A., Stacy, C., Masoro, E. J., Byung, P. Y., Murata, I. & Maeda, H. (1984). Plasticity of fat cell number. *J. Nutr.*, **114,** 127–31.
Blum, J. W., Froehli, D. & Kunz, P. (1982). Effects of catecholamines on plasma free fatty acids in fed and fasted cattle. *Endocrinology,* **110,** 452–6.
Bocklen, E., Flad, S., Muller, E. & von Faber, H. (1986). Comparative determination of beta-adrenergic receptors in muscle, heart and backfat of Pietrain and Large White pigs. *Anim. Prod.*, **43,** 335–40.
Bohorov, O., Buttery, P. J., Correia, J. H. R. D. & Soar, J. B. (1987). The effect of the β-2-adrenergic agonist clenbuterol or implantation with oestradiol plus trenbolone acetate on protein metabolism in wether lambs. *Brit. J. Nutr.*, **57,** 99–107.
Brockman, R. P. & Laarveld, B. (1986). Hormonal regulation of metabolism in ruminants; A review. *Livestock Prod. Sci.*, **14,** 313–34.
Buhler, H. U., Da Prada, M., Haefely, W. & Picotti, G. B. (1978). Plasma adrenaline, noradrenaline and dopamine in man and different animal species. *J. Physiol.*, **276,** 311–20.

Buonomo, F. C., Lauterio, T. J., Baile, C. A. & Campion, D. R. (1987). Determination of insulin-like growth factor 1 (IGF1) and IGF binding protein levels in swine. *Domest. Anim. Endocrinol.*, **4,** 23–31.
Butler-Hogg, B. W. & Johnsson, I. D. (1987). Bovine growth hormone in lambs: effects on carcass composition and tissue distribution in crossbred females. *Anim. Prod.*, **44,** 117–24.
Buttery, P. J. (1985). Exogenously applied growth promotents for use in ruminants—present and future applications. *Reviews in Rural Science,* **6,** 141–9.
Byers, F. M. (1982). Nutritional factors affecting growth of muscle and adipose tissue in ruminants. *Federation Proc.*, **41,** 2562–6.
Campbell, R. M. & Scanes, C. G. (1985). Adrenergic control of lipogenesis and lipolysis in the chicken *in vitro*. *Comp. Biochem. Physiol.*, **82C,** 137–42.
Campbell, R. G., Steele, N. C., Caperna, T. J., McMurtry, J. P., Solomon, M. B. & Mitchell, A. D. (1988). Interrelationships between energy intake and endogenous porcine growth hormone administration on the performance, body composition and protein and energy metabolism of growing pigs weighing 25 to 55 kilograms live weight. *J. Anim. Sci.*,**66,** 1643–55.
Campion, D. R. & Crouse, J. D. (1975). Predictive value of USDA beef quality grade factors for cooked meat palatability. *J. Food Sci.*, **40,** 1225–8.
Chung, C. S., Meserole, V. K. & Etherton, T. D. (1983). Temporal nature of insulin binding and insulin-stimulated glucose metabolism in isolated swine adipocytes. *J. Anim. Sci.*, **56,** 58–63.
Chung, C. S., Etherton, T. D. & Wiggins, J. P. (1985). Stimulation of swine growth by porcine growth hormone. *J. Anim. Sci.*, **60,** 118–30.
Cianzio, D. S., Topel, D. G., Whitehurst, G. B., Beitz, D. C. & Self, H. L. (1985). Adipose tissue growth and cellularity: changes in bovine adipocyte size and number. *J. Anim. Sci.*, **60,** 970–6.
Clemmons, D. R., Dehoff, M., McCusker, R., Elgin, R. & Busby, W. (1987). The role of insulin-like growth factor I in the regulation of growth. *J. Anim. Sci.*, **65,** Suppl. 2, 168–79.
Coleman, M. E., Ekeren, P. A. & Smith, S. B. (1988). Lipid synthesis and adipocyte growth in adipose tissue from sheep chronically fed a beta-adrenergic agent. *J. Anim. Sci.*, **66,** 372–8.
Conover, T. E. (1987). Does citrate transport supply both acetyl groups and NADPH for cytoplasmic fatty acid synthesis? *TIBS,* **12,** 88–9.
Convey, E. M., Rickes, E., Yang, Y. T., McElligott, M. A. & Olson, G. (1987). Effects of the beta-adrenergic agonist L-644,969 on growth performance, carcass merit and meat quality. *Recip. Meat Conf. Proc.*, **40,** 47–55.
Cromwell, G. L., Kemp, J. D., Stahly, T. S. & Dalrymple, R. H. (1988). Effects of dietary level and withdrawal time on the efficacy of cimaterol as a growth repartitioning agent in finishing swine. *J. Anim. Sci.*, **66,** 2193–9.
Crouse, J. D., Smith, G. M. & Mandigo, R. W. (1978). Relationship of selected beef carcass traits with meat palatability. *J. Food Sci.*, **43,** 152–7.
Crouse, J. D., Schanbacher, B. D., Cross, H. R., Seideman, S. C. & Smith, S. B. (1987). Growth and carcass traits of heifers as affected by hormonal treatment. *J. Anim. Sci.*, **64,** 1434–40.
Curtis-Prior, P. B. (1983). *Biochemical Pharmacology of Obesity*. Elsevier, Amsterdam.

Della-Fera, M. A. & Baile, C. A. (1984). Control of feed intake in sheep. *J. Anim. Sci.*, **59,** 1362–8.

De Wilde, R. O. & Lauwers, H. (1984). The effect of parenteral use of estradiol, progesterone, testosterone and trenbolone on growth and carcass composition in pigs. *J. Anim. Sci.*, **59,** 1501–9.

DiGirolamo, M. & Mendlinger, S. (1971). Role of fat cell size and number in enlargement of epididymal fat pads in three species. *Am. J. Physiol.*, **221,** 859–64.

DiMarco, N. M., Whitehurst, G. B. & Beitz, D. C. (1986). Evaluation of prostaglandin E_2, as a regulator of lipolysis in bovine adipose tissue. *J. Anim. Sci.*, **62,** 363–9.

Duquette, P. F., Scanes, C. G. & Muir, L. A. (1984). Effects of ovine growth hormone and other anterior pituitary hormones on lipolysis of rat and ovine adipose tissue *in vitro*. *J. Anim. Sci.*, **58,** 1191–7.

Eisemann, J. H., Tyrrell, H. F., Hammond, A. C., Reynolds, P. J., Bauman, D. E., Haaland, G. L., McMurtry, J. P. & Varga, G. A. (1986*a*). Effect of bovine growth hormone administration on metabolism of growing Hereford heifers: Dietary digestibility, energy and nitrogen balance. *J. Nutr.*, **116,** 157–63.

Eisemann, J. H., Hammond, A. C., Bauman, D. E., Reynolds, P. J., McCutcheon, S. N., Tyrrell, H. F. & Haaland, G. L. (1986*b*). Effect of bovine growth hormone administration on metabolism of growing Hereford heifers: Protein and lipid metabolism and plasma concentrations of metabolites and hormones. *J. Nutr.*, **116,** 2504–15.

Eisemann, J. H., Huntington, G. B. & Ferrell, C. L. (1988). Effects of dietary clenbuterol on metabolism of the hindquarters in steers. *J. Anim. Sci.*, **66,** 342–53.

Eisen, E. J., Croom, W. J. Jr & Helton, S. W. (1988). Differential response to the β-adrenergic agonist cimaterol in mice selected for rapid gain and unselected controls. *J. Anim. Sci.*, **66,** 361–71.

Enright, W. J., Zinn, S. A., Chapin, L. T. & Tucker, H. A. (1987). Growth hormone response of bull calves to growth hormone-releasing factor (42504). *Proc. Soc. Exp. Biol. Med.*, **184,** 483–8.

Enser, M. (1973). Clearing-factor lipase in muscle and adipose tissue of pigs. *Biochem. J.*, **136,** 381–5.

Etherton, T. D. (1982). The role of insulin-receptor interactions in regulation of nutrient utilization by skeletal muscle and adipose tissue: a review. *J. Anim. Sci.*, **54,** 58–67.

Etherton, T. D. & Allen, C. E. (1980). Effects of age and adipocyte size on glucose and palmitate metabolism and oxidation in pigs. *J. Anim. Sci.*, **50,** 1073–84.

Etherton, T. D. & Evock, C. M. (1986). Stimulation of lipogenesis in bovine adipose tissue by insulin and insulin-line growth factor. *J. Anim. Sci.*, **62,** 357–62.

Etherton, T. D. & Kensinger, R. S. (1984). Endocrine regulation of fetal and postnatal meat animal growth. *J. Anim. Sci.*, **59,** 511–28.

Etherton, T. D. & Walker, O. A. (1982). Characterization of insulin binding to isolated swine adipocytes. *Endocrinology*, **110,** 1720–4.

Etherton, T. D. & Walton, P. E. (1986). Hormonal and metabolic regulation of lipid metabolism in domestic livestock. *J. Anim. Sci.*, **63,** Suppl. 2, 76–88.

Etherton, T. D., Bauman, D. E. & Romans, J. R. (1977). Lipolysis in subcutaneous and perirenal adipose tissue from sheep and dairy steers. *J. Anim. Sci.*, **44,** 1100–6.

Etherton, T. D., Wangsness, P. J., Hammers, V. M. & Ziegler, J. H. (1982). Effect of dietary restriction on carcass composition and adipocyte cellularity of swine with different propensities for obesity. *J. Nutr.*, **112,** 2314–23.

Etherton, T. D., Chung, C. S. & Wiggins, J. P. (1984). Receptor-dependent and independent degradation of insulin by isolated swine adipocytes at 37°C. *J. Anim. Sci.*, **59,** 366–75.

Etherton, T. D., Wiggins, J. P., Chung, C. S., Evock, C. M., Rebhun, J. F. & Walton, P. E. (1986). Stimulation of pig growth performance by porcine growth hormone and growth hormone-releasing factor. *J. Anim. Sci.*, **63,** 1389–99.

Etherton, T. D., Wiggins, J. P., Evock, C. M., Chung, C. S., Rebhun, J. F., Walton, P. E. & Steele, N. C. (1987). Stimulation of pig growth performance by porcine growth hormone: determination of the dose-response relationship. *J. Anim. Sci.*, **64,** 433–43.

Evans, J. W. & Hollaender, A. (Eds) (1986). *Genetic Engineering of Animals.* Plenum, New York.

Evock, C. M., Etherton, T. D., Chung, C. S. & Ivy, R. E. (1988). Pituitary porcine growth hormone (pGH) and a recombinant pGH analog stimulate pig growth performance in a similar manner. *J. Anim. Sci.*, **66,** 1928–41.

Fain, J. N. (1980). Hormonal regulation of lipid mobilization from adipose tissue. In *Biochemical Actions of Hormones*, Vol. 7, ed. Gerald Litwack. Academic Press, Inc., New York, pp. 119–204.

Fain, J. N. & Garcia-Sainz, J. A. (1983). Adrenergic regulation of adipocyte metabolism. *J. Lipid Res.*, **24,** 945–66.

Falvo, R. E., Chandrashekar, V., Arthur, R. D., Kuenstler, A. R., Hasson, T., Awoniyi, C. & Schanbacher, B. D. (1986). Effect of active immunization against LHRH or LH in boars: reproductive consequences and performance traits. *J. Anim. Sci.*, **63,** 986–94.

Flint, D. J., Coggrave, H., Futter, C. E., Gardner, M. J. & Clarke, T. J. (1986). Stimulatory and cytotoxic effects of an antiserum to adipocyte plasma membranes on adipose tissue metabolism *in vitro* and *in vivo. Int. J. Obesity,* **10,** 69–76.

Forbes, J. M. (1982). Effects of lighting pattern on growth, lactation and food intake of sheep, cattle and deer. *Livestock Prod. Sci.*, **9,** 361–74.

Fredholm, B. B. (1985). Nervous control of circulation and metabolism in white adipose tissue. In *New Perspectives in Adipose Tissue: Structure, Function and Development,* eds A. Cryer & R. L. R. Van. Butterworths, London, pp. 45–64.

Froesch, E. R., Schmid, C., Zangger, I., Schoenle, E., Eigenmann, E. & Zapf, J. (1986). Effects of IGF/somatomedins on growth and differentiation of muscle and bone. *J. Anim. Sci.*, **63,** Suppl. 2, 57–75.

Galbraith, H. & Topps, J. H. (1981). Effect of hormones on the growth and body composition of animals. *Nutrition Abstr. and Rev.—Series B,* **51,** 521–40.

Gavin, J. R., Saltman, R. J. & Tollefsen, S. E. (1982). Growth hormone receptors in isolated rat adipocytes. *Endocrinology,* **110,** 637–43.

Gray, J. M., Nunez, A. A., Siegel, L. I. & Wade, G. N. (1979). Effects of testosterone on body weight and adipose tissue: role of aromatization. *Physiol. Behav.,* **23,** 465–9.

Gray, J. M., Dudley, S. D. & Wade, G. N. (1981). In vivo cell nuclear binding of 17β [^{3}H]estradiol in rat adipose tissues. *Am. J. Physiol.,* **240,** E43–46.

Greenwood, M. R. C. & Hirsch, J. (1974). Postnatal development of adipocyte cellularity in the normal rat. *J. Lipid Res.,* **15,** 474–83.

Gurr, M. I., Kirtland, J., Phillip, M. & Robinson, M. P. (1977). The consequences of early overnutrition for fat cell size and number: the pig as an experimental model for human obesity. *Int. J. Obesity,* **1,** 151–70.

Hall, T. R., Harvey, S. & Scanes, C. G. (1986). Control of growth hormone secretion in the vertebrates: a comparative survey. *Comp. Biochem. Physiol.,* **84A,** 231–53.

Hanrahan, J. P. (Ed.) (1987). *Beta-Agonists and Their Effects on Animal Growth and Carcass Quality.* Elsevier, London.

Hanrahan, J. P., Quirke, J. F., Bomann, W., Allen, P., McEwan, J. C., Fitzsimons, J. M., Kotzian, J. & Roche, J. F. (1986). β-Agonists and their effects on growth and carcass quality. In *Recent Advances in Animal Nutrition,* eds W. Haresign & D. J. A. Cole. Butterworths, London, pp. 125–38.

Hansen, F. M., Fahmy, N. & Nielsen, J. H. (1980). The influence of sexual hormones on lipogenesis and lipolysis in rat fat cells. *Acta Endocrinol.,* **95,** 566–70.

Hanson, R. W. & Ballard, F. J. (1967). The relative significance of acetate and glucose as precursors for lipid synthesis in liver and adipose tissue from ruminants. *Biochem, J.,* **105,** 529–36.

Hart, I. C. & Johnsson, I. D. (1986). Growth hormone and growth in meat producing animals. In *Control and Manipulation of Animal Growth,* eds P. J. Buttery, D. B. Lindsay & N. B. Haynes. Butterworths, London, pp. 135–59.

Haugebak, C. D., Hedrick, H. B. & Asplund, J. M. (1974*a*). Relationship between extramuscular adipose tissue lipoprotein lipase activity and intramuscular lipid deposition in fattening lambs. *J. Anim. Sci.,* **39,** 1026–31.

Haugebak, C. D., Hedrick, H. B. & Asplund, J. M. (1974*b*). Adipose tissue accumulation and cellularity in growing and fattening lambs. *J. Anim. Sci.,* **39,** 1016–25.

Hausman, G. J. (1982). Histochemically detectable lipoprotein lipase activity in adipose tissue of pigs and normal and decapitated pig fetuses. *Acta Anat.,* **114,** 281–90.

Hausman, G. J. (1985). The comparative anatomy of adipose tissue. In *New Perspectives in Adipose Tissue: Structure, Function and Development*, eds A. Cryer & R. L. R. Van. Butterworths, London, pp. 1–22.

Hausman, G. J. & Kauffman, R. G. (1986*a*). Mitotic activity in fetal and early postnatal porcine adipose tissue. *J. Anim. Sci.,* **63,** 659–73.

Hausman, G. J. & Kauffman, R. G. (1986*b*). The histology of developing porcine adipose tissue. *J. Anim. Sci.,* **63,** 642–58.

Hausman, G. J. & Martin, R. J. (Eds) (1987). *Biology of the Adipocyte: Research Approaches.* Van Nostrand Reinhold, New York.

Hausman, G. J. & Richardson, R. L. (1982). Histochemical and ultrastructural analysis of developing adipocytes in the fetal pig. *Acta Anat.*, **114,** 228–47.
Hausman, G. J., Campion, D. R. & Martin, R. J. (1980). Search for the adipocyte precursor cell and factors that promote its differentiation. *J. Lipid Res.*, **21,** 657–70.
Hausman, G. J., Hentges, E. J. & Thomas, G. B. (1987). Differentiation of adipose tissue and muscle in hypophysectomized pig fetuses. *J. Anim. Sci.*, **64,** 1255–61.
Henry, Y. (1977). Developpement morphologique et metabolique du tissu adipeux chez le porc: influence de la selection, de l'alimentation et du mode d'elevage. *Ann. Biol. anim. Bioch. Biophys.*, **17,** 923–52.
Henry, Y. (1985). Dietary factors involved in feed intake regulation in growing pigs: A review. *Livestock Prod. Sci.*, **12,** 339–54.
Hermans, P. G. C. (1973). The development of adipose tissue in swine foetuses (A morphological study). *Tijdschr. Diergeneesk,* **98,** 662–7.
Hetzer, H. O. & Harvey, W. R. (1967). Selection for high and low fatness in swine. *J. Anim. Sci.*, **26,** 1244–51.
Hietanen, E. & Greenwood, M. R. C. (1977). A comparison of lipoprotein lipase activity and adipocyte differentiation in growing male rats. *J. Lipid Res.*, **18,** 480–9.
Hood, R. L. & Allen, C. E. (1977). Cellularity of porcine adipose tissue: effects of growth and adiposity. *J. Lipid Res.*, **18,** 275–84.
Hood, R. L. & Thornton, R. F. (1979). The cellularity of ovine adipose tissue. *Aust. J. Agric. Res.*, **30,** 153–61.
Hossner, K. L., Link, G. A. & Yemm, R. S. (1988). Comparison of insulin-like growth factor I serum binding proteins in sheep and cattle: Species differences in size and endogenous binding capacities. *J. Anim. Sci.*, **66,** 1401–8.
Houpt, K. A., Houpt, T. R. & Pond, W. G. (1979). The pig as a model for the study of obesity and of control of food intake: a review. *Yale J. Biol. Med.*, **52,** 307–29.
Houpt, T. R. (1984). Controls of feeding in pigs. *J. Anim. Sci.*, **59,** 1345–53.
Hu, C. Y., Novakofski, J. & Mersmann, H. J. (1987). Hormonal control of porcine adipose tissue fatty acid release and cyclic AMP concentration. *J. Anim. Sci.*, **64,** 1031–7.
Hu, C. Y., Suryawan, A., Forsberg, N. E., Dalrymple, R. H. & Ricks, C. A. (1988). Effect of cimaterol on sheep adipose tissue lipid metabolism. *J. Anim. Sci.*, **66,** 1393–400.
Imoto, S. & Namioka, S. (1983). Acetate–glucose relationship in growing pigs. *J. Anim. Sci.*, **56,** 867–75.
Ingle, D. L., Bauman, D. E. & Garrigus, U. S. (1972). Lipogenesis in the ruminant: in vitro study of tissue sites, carbon source and reducing equivalent generation for fatty acid synthesis. *J. Nutr.*, **102,** 609–16.
Jaster, E. H. & Wegner, T. N. (1981). Beta-adrenergic receptor involvement in lipolysis of dairy cattle subcutaneous adipose tissue during dry and lactating state. *J. Dairy Sci.*, **64,** 1655–63.
Jenkins, T. G., Ford, J. J. & Klindt, J. (1988). Postweaning growth, feed efficiency and chemical composition of sheep as affected by prenatal and postnatal testosterone. *J. Anim. Sci.*, **66,** 1179–85.

Johnsson, I. D., Hart, I. C. & Butler-Hogg, B. W. (1985). The effects of exogenous bovine growth hormone and bromocriptine on growth, body development, fleece weight and plasma concentrations of growth hormone, insulin and prolactin in female lambs. *Anim. Prod.*, **41,** 207–17.

Jones, S. J. & Marchello, J. A. (1983). Lipolysis in subcutaneous adipose tissue from cattle varying in frame size and length of time on a finishing diet. *J. Anim. Sci.*, **57,** 343–8.

Jungas, R. L. (1968). Fatty acid synthesis in adipose tissue incubated in tritiated water. *Biochemistry,* **7,** 3708–15.

Kasser, T. R., Martin, R. J. & Allen, C. E. (1981). Effect of gestational alloxan diabetes and fasting on fetal lipogenesis and lipid deposition in pigs. *Biol. Neonate,* **40,** 105–12.

Kasser, T. R., Gahagan, J. H. & Martin, R. J. (1982). Fetal hormones and neonatal survival in response to altered maternal serum glucose and free fatty acid concentrations in pigs. *J. Anim. Sci.*, **55,** 1351–9.

Kensinger, R. S., McMunn, L. M., Stover, R. K., Schricker, B. R., Maccecchini, M. L., Harpster, H. W. & Kavanaugh, J. F. (1987). Plasma somatotropin response to exogenous growth hormone releasing factor in lambs. *J. Anim. Sci.*, **64,** 1002–9.

Khachadurian, A. K., Adrouni, B. & Yacoubian, H. (1966). Metabolism of adipose tissue in the fat tail of the sheep *in vivo*. *J. Lipid Res.*, **7,** 427–36.

Killefer, J., Hu, C. Y. & Banowetz, G. M. (1988). Identification of proteins unique to porcine adipocyte plasma membranes. *FASEB J.*, **2,** A1414 (Abst. No. 6500).

Kim, Y. S., Lee, Y. B. & Dalrymple, R. H. (1987). Effect of the repartitioning agent cimaterol on growth, carcass and skeletal muscle characteristics in lambs. *J. Anim. Sci.*, **65,** 1392–9.

Kirtland, J. & Gurr, M. I. (1980). Fat cell synthesis in pigs assessed after administration of tritiated thymidine *in vivo*. *J. Agric. Sci., Camb.*, **95,** 325–31.

Klindt, J., Jenkins, T. G. & Leymaster, K. A. (1985). Relationships between some estimates of growth hormone and prolactin secretion and rates of accretion of constituents of body gain in rams. *Anim. Prod.*, **41,** 103–11.

Klindt, J., Jenkins, T. G. & Ford, J. J. (1987). Prenatal androgen exposure and growth and secretion of growth hormone and prolactin in ewes postweaning. *Proc. Soc. Exp. Biol. Med.*, **185,** 201–5.

Kris-Etherton, P. M. & Etherton, T. D. (1982). The role of lipoproteins in lipid metabolism of meat animals. *J. Anim. Sci.*, **55,** 804–17.

Lafontan, M. (1979). Inhibition of epinephrine-induced lipolysis in isolated white adipocytes of aging rabbits by increased alpha-adrenergic responsiveness. *J. Lipid Res.*, **20,** 208–16.

Lafontan, M. & Berlan, M. (1985). Plasma membrane properties and receptors in white adipose tissue. In *New Perspectives in Adipose Tissue: Structure, Function and Development,* eds A. Cryer & R. L. R. Van. Butterworths, London, pp. 145–82.

Lafontan, M., Berlan, M. & Villeneuve, A. (1983). Preponderance of α_2- over β_1-adrenergic receptor sites in human fat cells is not predictive of the lipolytic effect of physiological catecholamines. *J. Lipid Res.*, **24,** 429–40.

Leat, W. M. F. & Cox, R. W. (1980). Fundamental aspects of adipose tissue growth. In *Growth in Animals*, ed. T. L. J. Lawrence. Butterworths, London, pp. 137–74.

Ling, N., Zeytin, F., Bohlen, P., Esch, F., Brazeau, P., Wehrenberg, W. B., Baird, A. & Guillemin, R. (1985). Growth hormone releasing factors. *Ann. Rev. Biochem.*, **54,** 403–23.

Lister, D. (1976). Effects of nutrition and genetics on the composition of the body. *Proc. Nutr. Soc.*, **35,** 351–6.

Lister, D. (1980). Metabolism and growth. *Reprod. Nutr. Develop.*, **20,** 225–33.

Lu, F. C. & Rendel, J. (Eds) (1976). *Anabolic Agents in Animal Production.* Georg Thieme, Stuttgart.

McNamara, J. P. & Hillers, J. K. (1986). Adaptations in lipid metabolism of bovine adipose tissue in lactogenesis and lactation. *J. Lipid Res.*, **27,** 150–7.

McNamara, J. P. & Martin, R. J. (1982). Muscle and adipose tissue lipoprotein lipase in fetal and neonatal swine as affected by genetic selection for high or low backfat. *J. Anim. Sci.*, **55,** 1057–61.

McNamara, J. P., Dehoff, M. H., Collier, R. J. & Bazer, F. W. (1985). Adipose tissue fatty acid metabolism during pregnancy in swine. *J. Anim. Sci.*, **61,** 410–15.

Machlin, L. J. (1972). Effect of porcine growth hormone on growth and carcass composition of the pig. *J. Anim. Sci.*, **35,** 794–800.

MacRae, J. C., Skene, P. A., Connell, A., Buchan, V. & Lobley, G. E. (1988). The action of the β-agonist clenbuterol on protein and energy metabolism in fattening wether lambs. *Brit. J. Nutr.*, **59,** 457–65.

Martin, C. R. (1985). *Endocrine Physiology.* Oxford University Press, New York, pp. 499–570.

Martin, J. B. & Millard, W. J. (1986). Brain regulation of growth hormone secretion. *J. Anim. Sci.*, **63,** Suppl. 2, 11–26.

Martin, R. J., Gobble, J. L., Hartsock, T. H., Graves, H. B. & Ziegler, J. H. (1972). Characterization of an obsese syndrome in the pig. *Proc. Soc. Exp. Biol. Med.*, **143,** 198–203.

Martin, R. J., Ramsay, T. G. & Harris, R. B. S. (1984). Central role of insulin in growth and development. *Domes. Anim. Endocrinol.*, **1,** 89–104.

Meissonnier, E. (Ed.) (1983). *Anabolics in Animal Production.* Office International des Epizooties, Paris.

Mersmann, H. J. (1984*a*). Effect of sex on lipogenic activity in swine adipose tissue. *J. Anim. Sci.*, **58,** 600–4.

Mersmann, H. J. (1984*b*). Specificity of β-adrenergic control of lipolysis in swine adipose tissue. *Comp. Biochem. Physiol.*, **77C,** 39–42.

Mersmann, H. J. (1984*c*). Absence of α-adrenergic inhibition of lipolysis in swine adipose tissue. *Comp. Biochem. Physiol.*, **79C,** 165–70.

Mersmann, H. J. (1986*a*). Lipid metabolism in swine. In *Swine in Cardiovascular Research,* Vol. 1, eds H. C. Stanton & H. J. Mersmann. CRC Press, Boca Raton, Florida, pp. 75–104.

Mersmann, H. J. (1986*b*). Comparison of plasma free-fatty-acid and blood-glycerol concentrations with measurement of lipolysis in porcine adipose tissue *in vitro*. *J. Anim. Sci.*, **63,** 757–69.

Mersmann, H. J. (1986*c*). Acute effects of metabolic hormones in swine. *Comp. Biochem. Physiol.*, **83A,** 653–60.

Mersmann, H. J. (1987*a*). Nutritional and endocrinological influences on the composition of animal growth. In *Progress in Food and Nutrition Science,* Vol. 11, ed. R. F. Chandra. Pergamon Press, New York, pp. 175–201.

Mersmann, H. J. (1987*b*). Acute metabolic effects of adrenergic agents in swine. *Am. J. Physiol.*, **252,** E85–95.

Mersmann, H. J. (1987*c*). Primer on beta adrenergic agonists and their effect on the biology of swine. In *The Repartitioning Revolution: Impact of Somatotropin and Beta Adrenergic Agonists on Future Pork Production.* Univ. of Illinois Pork Industry Conference, pp. 19–45.

Mersmann, H. J. (1989). Potential mechanisms for repartitioning of growth by β-adrenergic agonists. In *Animal Growth Regulation*, eds D. R. Campion, R. J. Martin & G. J. Hausman. Plenum Publishing Corp., New York, pp. 337–57.

Mersmann, H. J. (1990). Characteristics of obese and lean swine. In *Swine Nutrition,* eds E. R. Miller, D. E. Uillrey & A. J. Lewis. Butterworths, London. (In press.)

Mersmann, H. J., Houk, J. M., Phinney, G., Underwood, M. C. & Brown, L. J. (1973). Lipogenesis by *in vitro* liver and adipose preparations from neonatal swine. *Am. J. Physiol.*, **224,** 1123–9.

Mersmann, H. J., Brown, L. J., Underwood, M. C. & Stanton, H. C. (1974). Catecholamine-induced lipolysis in swine. *Comp. Biochem. Physiol.*, **47B,** 263–70.

Mersmann, H. J., Goodman, J. R. & Brown, L. J. (1975*a*). Development of swine adipose tissue: morphology and chemical composition. *J. Lipid Res.*, **16,** 269–79.

Mersmann, H. J., Phinney, G., Brown, L. J. & Arakelian, M. C. (1975*b*). Factors influencing the lipolytic response in swine (*Sus domesticus*) adipose tissue. *Gen. Pharm.*, **6,** 193–9.

Mersmann, H. J. & MacNeil, M. D. (1985). Relationship of plasma lipid concentration to fat deposition in pigs. *J. Anim. Sci.*, **61,** 122–8.

Mersmann, H. J. & MacNeil, M. D. (1986). Variables in estimation of adipocyte size and number with a particle counter. *J. Anim. Sci.*, **62,** 980–91.

Mersmann, H. J. & Hu, C. Y. (1987). Factors affecting measurements of glucose metabolism and lipolytic rates in porcine adipose tissue slices *in vitro*. *J. Anim. Sci.*, **64,** 148–64.

Metz, S. H. M., Lopes-Cardozo, M. & Van den Bergh, G. (1974). Inhibition of lipolysis in bovine adipose tissue by butyrate and β-hydroxybutyrate. *FEBS Letters,* **47,** 19–22.

Miller, M. F., Garcia, D. K., Coleman, M. E., Ekeren, P. A., Lunt, D. K., Wagner, K. A., Procknor, M., Welsh, T. H. Jr & Smith, S. B. (1988). Adipose tissue, longissimus muscle and anterior pituitary growth and function in clenbuterol-fed heifers. *J. Anim. Sci.*, **66,** 12–20.

Miller, W. H., Faust, I. M. & Hirsch, J. (1984). Demonstration of *de novo* production of adipocytes in adult rats by biochemical and radioautographic techniques. *J. Lipid Res.*, **25,** 336–47.

Moseley, W. M., Krabill, L. F., Friedman, A. R. & Olsen, R. F. (1985). Administration of synthetic human pancreatic growth hormone-releasing factor

for five days sustains raised serum concentrations of growth hormone in steers. *J. Endocr.*, **104,** 433–9.
Muir, L. A. (1985). Mode of action of exogenous substances on animal growth—an overview. *J. Anim. Sci.*, **61,** Suppl. 2, 154–80.
Muir, L. A., Wien, S., Duquette, P. F., Rickes, E. L. & Cordes, E. H. (1983). Effects of exogenous growth hormone and diethylstilbestrol on growth and carcass composition of growing lambs. *J. Anim. Sci.*, **56,** 1315–23.
Nassar, A. R. & Hu, C. Y. (1988). Development of antibodies specific to ovine adipocyte plasma membranes. *FASEB J.* **2,** A1106 (Abst. No. 4705).
Nilsson-Ehle, P., Garfinkel, A. S. & Schotz, M. C. (1976). Intra- and extracellular forms of lipoprotein lipase in adipose tissue. *Biochim. Biophys. Acta,* **431,** 147–56.
Ohlson, D. L., Spicer, L. J. & Davis, S. L. (1981). Use of active immunization against prolactin to study the influence of prolactin on growth and reproduction in the ram. *J. Anim. Sci.*, **52,** 1350–9.
Pecquery, R., Leneveu, M.-C. & Giudicelli, Y. (1986). Estradiol treatment decreases the lipolytic responses of hamster white adipocytes through a reduction in the activity of the adenylate cyclase catalytic subunit. *Endocrinology,* **118,** 2210–16.
Petitclerc, D., Pelletier, G., Lapierre, H., Gaudreau, P., Couture, Y., Dubreuil, P., Morisset, J. & Brazeau, P. (1987). Dose response of two synthetic human growth hormone-releasing factors on growth hormone release in heifers and pigs. *J. Anim. Sci.*, **65,** 996–1005.
Petters, R. M. (1988). Recombinant DNA, gene transfer and the future of animal agriculture. *J. Anim. Sci.*, **62,** 1759–68.
Pike, B. V. & Roberts, C. J. (1980). The metabolic activity of bovine adipocytes before and after parturition. *Res. Vet. Sci.*, **29,** 108–10.
Pork Industry Conference Proceedings (1987). *The Repartitioning Revolution: Impact of Somatotropin and Beta Adrenergic Agonists on Future Pork Production.* University of Illinois.
Pothoven, M. A., Beitz, D. C. & Thornton, J. H. (1975). Lipogenesis and lipolysis in adipose tissue of *ad libitum* and restricted-fed beef cattle during growth. *J. Anim. Sci.*, **40,** 957–62.
Preston, R. L. (1975). Biological responses to estrogen additives in meat producing cattle and lambs. *J. Anim. Sci.*, **41,** 1414–30.
Prior, R. L. & Jacobson, J. J. (1979). Effects of fasting and refeeding and intravenous glucose infusion on in vitro lipogenesis in bovine adipose tissue. *J. Nutr.*, **109,** 1279–84.
Prior, R. L. & Smith, S. B. (1982). Hormonal effects on partitioning of nutrients for tissue growth: role of insulin. *Federation Proc.*, **40,** 2545–9.
Prior, R. L., Kohlmeier, R. H., Cundiff, L. V., Dikeman, M. E. & Crouse, J. D. (1977). Influence of dietary energy and protein on growth and carcass composition in different biological types of cattle. *J. Anim. Sci.*, **45,** 132–46.
Prior, R. L., Smith, S. B. & Jacobson, J. J. (1981). Metabolic pathways involved in lipogenesis from lactate and acetate in bovine adipose tissue: effects of metabolic inhibitors. *Arch. Biochem. Biophys.*, **211,** 202–10.
Prior, R. L., Smith, S. B., Schanbacher, B. D. & Mersmann, H. J. (1983). Lipid metabolism in finishing bulls and steers implanted with oestradiol-17-β

dipropionate. *Anim. Prod.*, **37,** 81–8.
Ramirez, Israel (1981). Estradiol-induced changes in lipoprotein lipase, eating, and body weight in rats. *Am. J. Physiol.*, **240,** E533–8.
Ramsay, T. G., Hausman, G. J. & Martin, R. J. (1987*a*). Pre-adipocyte proliferation and differentiation in response to hormone supplementation of decapitated fetal pig sera. *J. Anim. Sci.*, **64,** 735–44.
Ramsay, T. G., Hausman, G. J. & Martin, R. J. (1987*b*). Central endocrine regulation of the development of hormone responses in porcine fetal adipose tissue. *J. Anim. Sci.*, **64,** 745–51.
Reeds, P. J., Hay, S. M., Dorward, P. M. & Palmer, R. M. (1988). The effect of β-agonists and antagonists on muscle growth and body composition of young rats. *Comp. Biochem. Physiol.*, **89C,** 337–41.
Ricks, C. A., Baker, P. K. & Dalrymple, R. H. (1984). Use of repartitioning agents to improve performance and body composition of meat animals. *Recip. Meat Conf. Proc.*, **37,** 5–11.
Robelin, J. (1981). Cellularity of bovine adipose tissues: developmental changes from 15 to 65 percent mature weight. *J. Lipid Res.*, **22,** 452–7.
Robertson, I. S., Fraser, H. M., Innes, G. M. & Jones, A. S. (1982). Effect of immunological castration on sexual and production characteristics in male cattle. *Vet. Rec.*, **111,** 529–31.
Robertson, J. P., Faulkner, A. & Vernon, R. G. (1981). L-Lactate as a source of carbon for fatty acid synthesis in adult and foetal sheep. *Biochim. Biophys. Acta*, **665,** 511–18.
Roche, J. F. & O'Callaghan, D. (1984). *Manipulation of Growth in Farm Animals.* Martinus Nijhoff, Boston.
Roche, J. F. & Quirke, J. F. (1986). The effects of steroid hormones and zenobiotics on growth of farm animals. In *Control and Manipulation of Animal Growth,* eds P. J. Buttery, D. B. Lindsay & N. B. Haynes. Butterworths, London, pp. 39–51.
Romsos, D. R., Leveille, G. A. & Allee, G. L. (1971). In vitro lipogenesis in adipose tissue from alloxan–diabetic pigs. *Comp. Biochem. Physiol.*, **40A,** 569–78.
Rosebrough, R. W. & Steele, N. C. (1986). Effect of dietary protein status on glucose utilization by the pentose cycle in pig adipose tissue. *Nutr. Res.*, **6,** 275–85.
Rule, D. C., Smith, S. B. & Mersmann, H. J. (1987). Effects of adrenergic agonists and insulin on porcine adipose tissue lipid metabolism *in vitro*. *J. Anim. Sci.*, **65,** 136–49.
Rule, D. C., Smith, S. B. & Mersmann, H. J. (1988*a*). Glycerolipid biosynthesis in porcine adipose tissue in vitro. I. Assay conditions for homogenates. *J. Anim. Sci.*, **66,** 1656–64.
Rule, D. C., Smith, S. B. & Mersmann, H. J. (1988*b*). Glycerolipid biosynthesis in porcine adipose tissue *in vitro*. II. Synthesis by various types of cellular preparations. *J. Anim. Sci.*, **66,** 1665–75.
Saggerson, E. D. (1985). Hormonal regulation of biosynthetic activities in white adipose tissue. In *New Perspectives in Adipose Tissue: Structure, Function and Development,* eds A. Cryer & R. L. R. Van. Butterworths, London, pp. 87–120.

Savell, J. W., Branson, R. E., Cross, H. R., Stiffler, D. M., Wise, J. W., Griffin, D. B. &, Smith, G. C. (1987). National consumer retail beef study: palatability evaluations of beef loin steaks that differed in marbling. *J. Food Sci.*, **52,** 517–32.

Scarborough, R., Gulyas, J., Schally, A. V. & Reeves, J. J. (1988). Analogs of growth hormone-releasing hormone induced release of growth hormone in the bovine. *J. Anim. Sci.*, **66,** 1386–92.

Schanbacher, B. D. (1984). Manipulation of endogenous and exogenous hormones for red meat production. *J. Anim. Sci.*, **59,** 1621–30.

Schanbacher, B. D. (1986). Growth hormone releasing factor (GRF): physiological and immunological studies. In *Control and Manipulation of Animal Growth,* eds P. J. Buttery, D. B. Lindsay & N. B. Haynes. Butterworths, London, pp. 259–77.

Seideman, S. C., Cross, H. R., Oltjen, R. R. & Schanbacher, B. D. (1982). Utilization of the intact male for red meat production: a review. *J. Anim. Sci.*, **55,** 826–40.

Sidhu, K. S., Emery, R. S., Parr, A. F. & Merkel, R. A. (1973). Fat mobilizing lipase in relation to fatness in lambs. *J. Anim. Sci.*, **36,** 658–62.

Sillence, M. N. & Etherton, T. D. (1987). Determination of the temporal relationship between porcine growth hormone, serum IGF-1 and cortisol concentrations in pigs. *J. Anim. Sci.*, **64,** 1019–23.

Smith, S. B. (1983). Contribution of the pentose cycle to lipogenesis in bovine adipose tissue. *Arch. Biochem. Biophys.*, **221,** 46–56.

Smith, S. B. (1984). Evidence that phosphofructokinase limits glucose utilization in bovine adipose tissue. *J. Anim. Sci.*, **58,** 1198–204.

Smith, S. B. (1987). Effects of β-adrenergic agonists on cellular metabolism. *Recip. Meat Conf. Proc.*, **40,** 65–74.

Smith, S. B. & Crouse, J. D. (1984). Relative contributions of acetate, lactate and glucose to lipogenesis in bovine intramuscular and subcutaneous adipose tissue. *J. Nutr.*, **114,** 792–800.

Smith, S. B. & Prior, R. L. (1984). Pentose cycle flux and fatty acid synthesis in bovine adipose tissue slices incubated with 6-aminonicotinamide. *Proc. Soc. Exp. Biol. Med.*, **175,** 98–105.

Smith, S. B. & Prior, R. L. (1986). Comparisons of lipogenesis and glucose metabolism between ovine and bovine adipose tissues. *J. Nutr.*, **116,** 1279–86.

Smith, S. B., Prior, R. L. & Mersmann, H.J. (1983). Interrelationships between insulin and lipid metabolism in normal and alloxan-diabetic cattle. *J. Nutr.*, **113,** 1002–15.

Smith, G. M., Crouse, J. D., Mandigo, R. W. & Neer, K. L. (1977). Influence of feeding regime and biological type on growth, composition and palatability of steers. *J. Anim. Sci.*, **45,** 236–53.

Smith, S. B., Prior, R. L., Ferrell, C. L. & Mersmann, H. J. (1984). Interrelationships among diet, age, fat deposition and lipid metabolism in growing steers. *J. Nutr.*, **114,** 153–62.

Spencer, E. M. (1987). Role of somatomedin C/insulin-like growth factor I binding proteins in regulation of growth. *J. Anim. Sci.*, **65,** Suppl. 2, 180–5.

Spencer, G. S. G. (1985). Hormonal systems regulating growth: a review. *Livestock Prod. Sci.*, **12,** 31–46.

Spencer, G. S. G. (1986). Hormonal manipulation of animal production by immunoneutralization. In *Control and Manipulation of Animal Growth,* eds P. J. Buttery, D. B. Lindsay & N. B. Haynes. Butterworths, London, pp. 279–91.

Spencer, G. S. G., Garssen, G. J. & Hart, I. C. (1983*a*). A novel approach to growth promotion using auto-immunisation against somatostatin. I. Effects on growth and hormone levels in lambs. *Livestock Prod. Sci.,* **10,** 25–37.

Spencer, G. S. G., Garssen, G. L. & Bergstrom, P. L. (1983*b*). A novel approach to growth promotion using auto-immunisation against somatostatin. II. Effects on appetite, carcass composition and food utilisation in lambs. *Livestock Prod. Sci.,* **10,** 469–77.

St. John, L. C., Ekeren, P. A., Crouse, J. D., Schanbacher, B. D. & Smith, S. B. (1987). Lipogenesis in adipose tissue from ovariectomized and intact heifers immunized against estradoil and(or) implanted with trenbolone acetate. *J. Anim. Sci.,* **64,** 1428–33.

Standal, N., Vold, E., Trygstad, O. & Foss, I. (1973). Lipid mobilization in pigs selected for leanness or fatness. *Anim. Prod.,* **16,** 37–42.

Steffen, D. G., Brown, L. J. & Mersmann, H. J. (1978). Ontogenic development of swine adipose tissue lipases. *Comp. Biochem. Physiol.,* **59B,** 195–8.

Steffen, D. G., Phinney, G., Brown, L. J. & Mersmann, H. J. (1979). Ontogeny of glycerolipid biosynthetic enzymes in swine liver and adipose tissue. *J. Lipid Res.,* **20,** 246–53.

Strutz, C. & Rogdakis, E. (1979). Phenotypic and genetic parameters of NADPH-generating enzymes in porcine adipose tissue. *Z. Tierzuchtg. Zuchtgsbiol.,* **96,** 170–85.

Thornton, R. F., Tume, R. K., Payne, G., Larsen, T. W., Johnson, G. W. & Hohenhaus, M. A. (1985). The influence of the β_2-adrenergic agonist, clenbuterol, on lipid metabolism and carcass composition of sheep. *Proc. New Zealand Soc. Anim. Prod.,* **45,** 97–101.

Trenkle, A. H. (1976). The anabolic effect of estrogens on nitrogen metabolism of growing and finishing cattle and sheep. In *Anabolic Agents in Animal Production,* eds F. C. Lu & J. Rendel. Georg Thieme, Stuttgart, pp. 79–88.

Trenkle, A. H. (1981). Endocrine regulation of energy metabolism in ruminants. *Federation Proc.,* **40,** 2536–41.

Tucker, H. A. & Merkel, R. A. (1987). Applications of hormones in the metabolic regulation of growth and lactation in ruminants. *Federation Proc.,* **46,** 300–6.

Tucker, H. A., Petitclerc, D. & Zinn, S. A. (1984). The influence of photoperiod on body weight gain, body consumption, nutrient intake and hormone secretion. *J. Anim. Sci.,* **59,** 1610–20.

Vasilatos, R., Etherton, T. D. & Wangsness, P. J. (1983). Preparation of isolated bovine adipocytes: validation of use for studies characterizing insulin sensitivity and binding. *Endocrinology,* **112,** 1667–73.

Vernon, R. G. (1981). Lipid metabolism in the adipose tissue of ruminant animals. In *Lipid Metabolism in Ruminant Animals,* ed. W. W. Christie. Pergamon, Oxford, pp. 279–362.

Vernon, R. G. (1982). Effects of growth hormone on fatty acid synthesis in sheep adipose tissue. *Int. J. Biochem.,* **14,** 255–8.

Vernon, R. G. (1986). The growth and metabolism of adipocytes. In *Control and Manipulation of Animal Growth*, eds P. J. Buttery, D. B. Lindsay & N. B. Haynes. Butterworths, London, pp. 67–84.

Vernon, R. G. & Clegg, R. A. (1985). The metabolism of white adipose tissue *in vivo* and *in vitro*. In *New Perspectives in Adipose Tissue: Structure, Function and Development,* eds A. Cryer & R. L. R. Van. Butterworths, London, pp. 65–86.

Vernon, R. G. & Taylor, E. (1986). Acetyl-CoA carboxylase of sheep adipose tissue: problems of the assay and adaption during fetal development. *J. Anim. Sci.,* **63,** 1119–25.

Vernon, R. G., Finley, E. & Taylor, E. (1985*a*). Fatty acid synthesis from amino acids in sheep adipose tissue. *Comp. Biochem. Physiol.,* **82B,** 133–6.

Vernon, R. G., Finley, E., Taylor, E. & Flint, D. J. (1985*b*). Insulin binding and action on bovine adipocytes. *Endocrinology,* **116,** 1195–9.

Vodovar, N., Desnoyers, F. & Etienne, M. (1977). Etude morphologique du tissu adipeux de couverture au stade de sa formation chez le foetus de porc. *Ann. Biol. anim. Bioch. Biophys.,* **17,** 775–86.

Wade, G. N., Gray, J. M. & Bartness, T. J. (1985). Gonadal influences on adiposity. *Int. J. Obesity,* **9,** Suppl. 1, 83–92.

Wagner, T. E. & Jochle, W. (1986). Recombinant gene transfer in animals: the potential for improving growth in livestock. In *Control and Manipulation of Animal Growth,* eds P. J. Buttery, D. B. Lindsay & N. B. Haynes. Butterworths, London, pp. 293–313.

Walton, P. E. & Etherton, T. D. (1986). Stimulation of lipogenesis by insulin in swine adipose tissue: antagonism by porcine growth hormone. *J. Anim. Sci.,* **62,** 1584–95.

Walton, P. E., Etherton, T. D. & Chung, C. S. (1987). Exogenous pituitary and recombinant growth hormones induce insulin and insulin-like growth factor 1 resistance in pig adipose tissue. *Domest. Anim. Endocrinol.,* **4,** 183–9.

Weekes, T. E. C. (1983). The hormonal control of fat metabolism in animals. *Proc. Nutr. Soc.,* **42,** 129–36.

Weekes, T. E. C. (1986). Insulin and growth. In *Control and Manipulation of Animal Growth,* eds P. J. Buttery, D. B. Lindsay & N. B. Haynes. Butterworths, London, pp. 187–206.

Whitehurst, G. B., Beitz, D. C., Cianzio, D. & Topel, D. G. (1981). Fatty acid synthesis from lactate in growing cattle. *J. Nutr.,* **111,** 1454–61.

Wise, T. & Ferrell, C. (1984). Effects of immunization of heifers against estradiol on growth, reproductive traits, and carcass characteristics. *Proc. Soc. Exp. Biol. Med.,* **176,** 243–8.

Wood, J. D., Gregory, N. G., Hall, G. M. & Lister, D. (1977). Fat mobilization in Pietrain and Large White pigs. *Brit. J. Nutr.,* **37,** 167–86.

Wood, J. D., Enser, M. B. & Restall, D. J. (1978). The cellularity of backfat in growing pigs and its relationship with carcass composition. *Anim. Prod.,* **27,** 1–10.

Wright, J.T. & Hausman, G. J. (1988). Ontogenetic expression of two cell surface antigens associated with adipogenesis and angiogenesis in fetal porcine tissue. *FASEB J.,* **2,** A1414 (Abst. No. 6499).

Yang, Y. T. & Baldwin, R. L. (1973*a*). Preparation and metabolism of isolated cells from bovine adipose tissue. *J. Dairy Sci.*, **56,** 350–65.
Yang, Y. T. & Baldwin, R. L., (1973*b*). Lipolysis in isolated cow adipose cells. *J. Dairy Sci.*, **56,** 366–74.
Yang, Y. T., White, L. S. & Muir, L. A. (1982*a*). Lactate metabolism and cytosolic NADH reducing equivalents in ovine adipocytes. *Int. J. Biochem.*, **14,** 335–40.
Yang, Y. T., White, L. S. & Muir, L. A. (1982*b*). Glucose metabolism and effect of acetate in ovine adipocytes. *J. Anim. Sci.*, **55,** 313–20.
York, D. A. (1985). The role of hormone status in the development of excess adiposity in animal models of obesity. In *New Perspectives in Adipose Tissue: Structure, Function and Development.* eds A. Cryer & R. L. R. Van. Butterworths, London, pp. 407–45.

Chapter 5

Practical Methods of Controlling Fatness in Farm Animals

J. J. BASS, B. W. BUTLER-HOGG & A. H. KIRTON
MAFTech,
Ruakura Agricultural Centre,
Hamilton, New Zealand

INTRODUCTION

Practical methods for reducing fat in meat-producing animals are of two basic types, short term methods which have immediate effects on stock and long term selection for markets of the future. This chapter, which has concentrated on ruminants, discusses existing practical and potential methods farmers can utilise to manipulate carcass fatness. The traditional farming methods for changing fatness such as breed selection, castration, altering live weight and nutrition, as well as the more recent short term methods utilising growth promoters have been described. However the most important factor in determining whether farmers will use existing or new technology for manipulating carcass fatness will always be economic. The returns for producing leaner stock must be financially worthwhile for the farmer and the meat industry.

MATURITY

In 1972 it was proposed that when different sheep breeds were slaughtered at the same degree of maturity (proportion of mature size) their carcass compositions would be similar (McClelland & Russel, 1972).

Further testing of this hypothesis (McClelland *et al.*, 1976) led to the conclusion that most breed and sex differences in composition observed at the same degree of maturity largely disappeared when the composition was expressed as a percentage of carcass weight. However small breed differences may still exist: Australian researchers (Butterfield *et al.*, 1983) have reported that a large mature sized strain of Merinos was slightly fatter at maturity than a smaller strain. The relationships between maturity and fatness apply equally to cattle, sheep and pigs (Kempster *et al.*, 1982*a*). When compared at the same live/carcass weight, smaller mature sized animals will be fatter and when compared at the same age, the differences will be in the same direction, but smaller.

In practical terms, although the weights of all tissues increase with increasing carcass weight, fat increases at a faster rate than muscle or bone (Kempster *et al.*, 1982*a*; Kirton, 1982; Black, 1983). As a result, the proportion of fat increases with increasing carcass weight and the proportions of muscle and bone decrease. Therefore, one method for controlling the level of fatness in farm animals is to slaughter at the appropriate live/carcass weight (level of maturity) for the particular species, breed and sex to produce the level of fatness appropriate for the target market. The alternative strategy which stands or falls on economic criteria is to produce heavier, fatter animals and trim the meat back to the level of fatness required by the marketplace.

SEX

Management practices relative to the sex of farm animals provide little scope for altering the fatness of female carcasses which comprise half the population of domestically farmed animals (the use of growth promoters and hormone-like compounds will be discussed elsewhere in this chapter). The main possibilities involve accepting that fatness increases as carcass weight increases within any female group and therefore involve slaughtering at lighter weights or alternatively trimming off the excess fat once it has been produced. However, the practice of castration (orchidectomy) which in some countries is commonly applied to the male half of the population under more intensive farming systems for management reasons (Robertson, 1966; Kiley, 1976) results in increased fat deposition in the castrate.

Female cattle (Berg & Butterfield, 1976) and sheep (Kirton *et al.*, 1982) are on average fatter than castrate males of the same breed from the same environment, and the latter are fatter than entire males when the comparisons are based on animals of similar size/weight. On this basis, ewe lamb carcasses may contain 2–4% of carcass weight more chemical fat than wether carcasses, which in turn contain 2–4% more fat than ram carcasses (Kirton, 1982; Kirton *et al.*, 1982). Ram–ewe differences of a similar order (3–8%) have been reported by other workers when carcass fat content was measured by dissection (Fourie *et al.*, 1970; Butler-Hogg *et al.*, 1984). Differences tend to be larger at heavier weights. Most reviews compare castrates with entires (Turton, 1969; Field, 1971; Seideman *et al.*, 1982) and a few experiments have also included females in the same comparisons (Walker, 1950; Everitt & Jury, 1966; Everitt & Evans, 1970; Fredeen *et al.*, 1971; Kirton *et al.*, 1982). Although sometimes claimed that the difference between rams and ewes of the same weight is merely a reflection of potential differences in ultimate mature size (McClelland *et al.*, 1976) with compositions expected to be similar at those mature sizes, this explanation does not explain the greater relative fatness of wethers when they reach maturity at a similar weight to rams (Butterfield *et al.*, 1984).

Pigs differ from sheep and cattle in that castrated males (barrows) are fatter than the female (gilt) (Evans & Kempster, 1979), but are similar in that entire males (boars) are on average leanest (Wood *et al.*, 1979; Wood & Riley, 1982).

With current consumer opposition to high fat diets and with meat identified as an important source of dietary fat, low fat animal products should be produced; in this context, unmodified or modified entire males (cryptorchids with testes intra-abdominally or short scrotum with testes retained against the abdominal wall) best meet this requirement with the additional advantages of faster growth rates and greater feed conversion efficiencies than castrates. The main problem preventing the adoption of meat production from entire male animals comes from trade discrimination against them in some countries (Robertson, 1966; Kiley, 1976; Seideman *et al.*, 1982).

The meat from older bulls and rams may be tougher than that from equivalent aged females or castrates (Field, 1971; Seideman *et al.*, 1982). Although this applies to a lesser extent in younger animals (Field, 1971; Seideman *et al.*, 1982; Dransfield *et al.*, 1984), if present in the latter, it may be due to poor post-mortem chilling treatments (Tatum, 1981; Dransfield *et al.*, 1984) rather than inferior qualities of the raw material. In Europe, bull beef can be a very satisfactory product (Dransfield *et al.*, 1984).

Although a US review notes some experiments have shown that heavy ram lambs are unsatisfactory in terms of palatability (Seideman *et al.*, 1982), an earlier review (Field, 1971) showed little difference between ram and wether lambs. Although problems have been reported from some concentrate-fed, heavyweight US ram lambs, more recent results on pasture-fed rams up to two years old have shown no palatability problems (Kirton *et al.*, 1983).

In contrast to the situation for sheep and cattle, a well defined 'boar odour' can be detected in the meat (fat) of some male pigs (Seideman *et al.*, 1982). However, consumer reaction in several countries (e.g. UK; Smith *et al.*, 1983) indicates that although detectable by taste panels and also chemically (Seideman *et al.*, 1982) it may not be of serious consumer concern.

We therefore conclude that the leaving of male animals entire, or where there are management benefits, the adoption of the short scrotum or the cryptorchid procedures, will result in the production of faster growing animals with less fat carcasses for the male half of the population.

GENOTYPE

Although between and within breed differences in carcass fatness on a common age or weight basis are well established and the appropriate genetic parameters have been determined which indicate the progress likely from selection, almost all the published and abstracted work has been undertaken on temperate breeds (Hodges, 1987). Only 25% of experimental work from developing countries is abstracted by international abstracting services and much of their domestic animal population is run under conditions where currently recommended animal breeding systems would be difficult to apply even if the genetic information was available.

Sheep

Comparisons have mainly involved breeds of British origin (Table 1). These trials have established that the larger sized meat breeds produce on average lambs that are less fat compared at any carcass weight than smaller sized breeds such as the Southdown and Clun. In all trials where they have been included, the Texel has been the sire breed producing progeny with least fat and most lean, a finding confirmed in a trial

Table 1
Relative fat trim (FT) or carcass weight at constant fat trim for different breeds or crosses of sheep (rankings should be made within columns)

Reference:	*(1)[a]*	*(1)[b]*	*(1)[c]*	*(2)[d]*	*(3)[e]*	*(4)[f]*
Measurement:	*% fat*	*Wt. fat*	*% fat*	*% fat*	*% fat*	*Carcass wt*
Comparative basis:	*Carcass wt*	*Live wt*	*Live wt*	*Carcass wt*	*Carcass wt*	*% FT*
Breed/Cross						
Texel	22·7	2·3	21·5	—	—	19·5
Suffolk	27·4	2·5	25·0	32·9	29·6	19·8
Oxford	26·6	2·2	24·6	—	—	20·4
Colbred	—	—	—	—	29·9	—
Poll Dorset/Dorset Horn	25·8	—	—	33·8	—	—
Dorset Down	29·8	2·6	27·9	33·1	—	17·5
Hampshire	27·6	—	—	33·1	31·6	17·5
Lincoln	26·2	—	—	—	—	—
Border Leicester	—	—	—	33·3	—	19·9
Cheviot	—	—	—	34·4	—	19·0
Romney	—	—	—	34·3	—	—
Clun	—	—	—	—	34·1	—
Southdown	—	3·2	—	38·5	—	16·4

(1) Wolf & Smith (1983).
(2) Kirton *et al.* (1974).
(3) Wood *et al.* (1980).
(4) Read (1982).
[a] Data from O'Ferrell & Timon (1977) as summarised in (1). Percent trimmed fat estimated from best neck. Crossbred progeny from listed sire breeds.
[b] Dissected tissue weight (kg) in half side. Probably crossbred progeny.
[c] Percent dissected fat reported by Wolff *et al.* (1980) for lambs slaughtered at the same live weight summarised in (1). Crossbred progeny from listed sire breeds.
[d] Percent chemical carcass fat at 20 kg carcass weight. Crossbred progeny (except Romney) from listed sire breeds.
[e] Percent dissected fat at the same carcass weight from purebred progeny.
[f] Carcass weight at the same estimated subcutaneous fat (FT) percent. On this basis the Texel still had more than 1% more lean than all other breeds. Crossbred progeny from listed sire breeds.

(Cameron & Drury, 1985) which included two French breeds, the Chamoise and the Charollais. The Chamoise cross lambs were at the fatter end of the scale from the Texel and the Charollais cross lambs were intermediate.

According to photographs, indigenous sheep of Nigeria (Adu & Ngere, 1979) and Sri Lanka (Ravindran *et al.*, 1983) are similar to those from many other countries with less intensive farming systems and show some resemblance to goats with the suggestion that they may have a lower fat content than the breeds of British origin. One Nigerian trial (Adu & Ngere, 1979) reported that dwarf sheep had 7·4% carcass fat; a very low figure in comparison with the normally quoted figure of over 20% fat found in British breed carcasses. Another African trial (Amegee, 1984) reported 3·7% fat in the retail cuts of Vogan sheep. However, in the absence of comparative data, we do not know how well the fat content matches the appearance of these animals, and if it does, whether the low fat content is environmentally or genetically controlled.

Fat-tailed sheep (e.g. Karakul, Naeini, Afshari and others) are known to deposit up to 19% of their carcass weight in the tail which has traditionally been used as a source of cooking fat (Farid *et al.*, 1983; Nik-Khah, 1984). The fat stored in the tail may contribute to the lower level of fat found elsewhere in the various cuts (Farid *et al.*, 1983) compared to more traditional sheep carcasses. However, comparative trials would be required to determine if the lower level of fat reported from these fat-tailed sheep is genetically or environmentally determined.

Although there are differences in fatness between breeds/crosses allowing for selection of the appropriate breed to give the required fatness at some target carcass weight, there is also considerable spread in carcass fatness at any carcass weight between different animals of the same breed or cross. While some of the within breed variation is environmental, it is well established that there is considerable genetic control over the differences in fatness between animals of the same breed.

A summary of some of the heritability values published for different measurements of fatness in sheep is given in Table 2. All estimates show that the tendency to lay down fat in sheep carcasses is moderately heritable indicating that the fat content of sheep carcasses can be reduced by selection (Bennett & Clarke, 1984).

Cattle

Bos taurus breed comparisons of carcass fatness have usually involved British and European beef breeds crossed with beef or dairy (usually Friesian and Jersey) cows (Table 3). Such comparisons have tended to place the Jersey (small mature size) and British beef breeds at the high end of

Table 2
Heritability estimates of measures of fatness in sheep carcasses

Fatness measurement	*Reference*							
	W(1)	*W(2)*	*A(2)*	*A(3)*	*A(4)*	*W(4)*	*W(5)*	*(6)*
Back fat depth	0·51	0·21	—	0·31	0.23	—	0·32	0·37
	—	0·37	0·27	—	—	—	—	—
	—	0·40	0·28	—	—	—	—	—
	—	0·51	—	—	—	—	—	—
% ether-extract	—	—	0·50	0·34	0·32[a]	—	—	—
Fat trim weight	—	—	—	—	0·25[b]	0·50[b]	—	—
% dissectible fat	0·54	0·37	—	—	0·36	—	—	0·43–0·48[c]
Other fat depths[d]	—	—	—	0·21–0·29	0·19	—	—	0·28–0·32

(1) Botkin *et al.* (1969).
(2) Wolf & Smith (1983).
Summarises much earlier work giving the appropriate references.
(3) Bennett *et al.* (1987).
(4) Parratt *et al.* (1987).
(5) Thorsteinsson & Bjornsson (1982).
(6) Cotterill & Roberts (1976).
W = At constant weight.
A = At constant age.
[a] First year's results only.
[b] Dissected fat weight.
[c] Two sample joints were dissected.
[d] A number of subcutaneous fat depths taken at different specified points on the carcass. For (3), six different depths involved.

the fatness scale when compared on a carcass weight basis and the European breeds (Chianina, Charollais, Simmental, Maine-Anjou, Limousin) at the lower end of the scale. The Blonde d'Aquitaine cross has been recorded as the least fat or similar to the Chianina beef animal in two trials (Berg *et al.*, 1978; Bass *et al.*, 1981). Trials involving the zebu (*Bos indicus*), a type adapted to tropical environments, report that their carcasses fall at the lower end of the fatness scale (Cole *et al.*, 1964; Koch *et al.*, 1982*b*; Baker *et al.*, 1984*a*).

More recently, the possibility of producing beef from the water buffalo (*Bos bubalus bubalis*) which comprises just over 10% of the global cattle population has been considered (Cockerill, 1980). In an indirect comparison, buffalo bulls were reported to have less carcass fat than Brahman cross, Hereford and Shorthorn steers (Charles & Johnson,

Table 3

Relative fat trim (FT) or carcass weight at constant fat trim for different breeds or crosses of cattle (ranking should be made within columns)

Reference:	*(1)*		*(2)*		*(3)*		*(4)*	*(5)*	
Measurement:	*% fat trim*		*% fat trim*	*Carcass wt*	*Fat trim (kg)*[d]		*Carcass wt*	*Carcass wt*	
Comparative basis:	*Age*[a]	*Carcass wt*[b]	*Age*	*FT*[c]	*Age*	*Carcass wt*	*FT*[e]	*FT*[f]	*FT*[g]
Breed/cross									
Jersey ×	—	—	22·1	244	12·3	13·4	—	—	—
Hereford	19·0	20·4	—	—	—	—	—	—	—
Hereford ×	—	—	—	—	—	—	245	214	224
Hereford × Angus	20·9	21·6	21·7	265	9·5	10·6	—	—	—
Angus	21·5	22·6	—	—	7·8	10·4	—	—	—
Angus ×	—	—	—	—	—	—	221	205	206
South Devon ×	—	—	20·0	290	7·6	9·0	279	237	251
Friesian ×	—	—	—	—	10·0	9·9	266[h]	—	—
Red Poll ×	20·2	21·7	21·0	264	—	—	—	—	—
Brown Swiss ×	16·8	16·1	17·6	328	—	—	—	—	—
Maine-Anjou ×	15·7	14·3	16·5	357	9·4	8·2	—	—	—
Simmental ×	—	—	15·6	359	10·4	8·8	286	258	272
Limousin ×	—	—	15·1	357	8·9	9·0	—	247	—
Charolais ×	—	—	15·2	376	7·3	6·8	317	268	275
Chianina ×	12·3	11·4	13·0	413	—	—	—	—	—

(1) Koch *et al.* (1979) — For crossbred animals, results averaged over Angus and Hereford dams.
(2) Koch *et al.* (1982*b*) — As above.
(3) Bass *et al.* (1981) — Crossbreds produced from Angus dams.
(4) Kempster *et al.* (1982*b*) — Crossbreds produced from Friesian dams.
(5) Kempster *et al.* :1982*c*) — Crossbreds produced from Hereford × Friesian or Blue–Grey dams.
[a] Same age. [b] Same carcass weight. [c] At 19% fat trim. [d] From left hindquarter.
[e] At 7·8% estimated subcutaneous fat on carcass from animals raised on a 24 month system.
[f] At 7·8% estimated subcutaneous fat from animals raised under a winter fattening system.
[g] At 8·2% estimated subcutaneous fat on animals raised in a summer fattening system.
[h] Purebred Friesians.

Table 4
Heritability estimates of measures of fatness in cattle carcasses

Fatness measurement	*Reference*[a]								
	1	*2*	*3*	*4*	*5*	*7*	*8*	*Average*	*NZ*[b]—
Fat trim weight	—	—	0·46	0·50	0·39	0·94	—	0·57	—
Kidney fat weight	—	—	—	0·72	—	—	—	0·72	—
Fat thickness	0·24	0·43	0·50	0·43	0·57	0·68	0·50	0·48	0·26
Marbling	0·17	0·62	0·31	0·73	0·31	0·34	0·56	0·42	—

[a] Reference 1–8 given as footnotes to Table 3 in the paper of Koch *et al.* (1982*a*).
[b] New Zealand reference–Johnsson *et al.* (1986).

1972). An international summary reported that generally the buffalo carcass had a low fat content (Rao & Nagarcenkar, 1977). While better direct comparative composition data between the beef species is desirable, given that the buffalo thrives in a tropical environment whereas *Bos taurus* thrives in a temperature environment, such a comparison may be academic.

A summary of heritability values for fat measurements in beef carcasses is given in Table 4. Because of the expense involved in the collection of complete composition data (separable, dissectible or chemical fat) in beef carcasses compared with sheep, such data is not readily available for cattle. The information in Table 4 shows that, as for sheep, fat measurements in cattle are moderately heritable indicating that fat levels in cattle can be reduced by a within breed selection programme.

Pigs

Little comparative breed data are available for pig carcass composition. Probing live animals for backfat thickness (mm) ranked breeds in the order Landrace (15·1), Duroc (14·5), Yorkshire (14·0) and Hampshire (12·9) at 90 kg live weight (Kennedy *et al.*, 1985). A carcass study (Smith & Pearson, 1986) ranked 64–65 kg pigs in the order Landrace (27·8% fat), Large White (23·7%) and Duroc (21·9%). The other coloured British breeds are fatter than the Large White (Yorkshire), Landrace and Welsh breeds while the Belgian Pietrain is reported as a well muscled, low fat breed (Bichard, 1968). However, the above sampling demonstrates some of the known breed variation available, but covers a low proportion of the 300 breeds or local varieties of pigs found on a worldwide basis (Ollivier & Sellier, 1982).

A review of pig genetics (Ollivier & Sellier, 1982) has reported heritabilities for percentage carcass fat and backfat thickness generally ranging from 0·5 to 1·0 with a recent report based on very large numbers reporting heritabilities of 0·4 to 0·6 for backfat thickness in four breeds of pig (Smith & Pearson, 1986). Heritabilities of this order were realised in an 8 to 10 generation US selection programme for and against backfat thickness in the Yorkshires and Duroc breeds (Hetzer & Harvey, 1967). National programmes to reduce the fat content of carcasses have been in operation for a much longer time for pigs in contrast to the situation for sheep and cattle. Effective selection against backfat has contributed much to the improvement in this characteristic in Danish pigs (Fredeen, 1958).

NUTRITION

Carcass fatness can be readily manipulated by nutrition in pigs (Campbell & Dunkin, 1983; Campbell *et al.*, 1983; Zhang *et al.*, 1986) and in poultry (Fisher, 1984) because they are monogastrics having high voluntary feed intakes and the pattern of digestive physiology allows a wide variation in energy input to reach the sites of tissue deposition. Under conditions of very controlled feeding, carcass fatness can be manipulated in ruminants (Orskov *et al.*, 1983; Butler-Hogg & Johnsson, 1986; see Tables 5, 6, 7 and 8) but under practical feeding situations this is more difficult because of the fibrous nature of feeds allowing a limited range of energy input and the equalizing nature of the fermentation process (Jagusch & Rattray, 1979; Kirton *et al.*, 1981). In spite of differences in the control of inputs in pig/poultry systems versus cattle/sheep systems, and the differences in their digestive physiology, there are examples of nutritional manipulation of carcass composition in cattle (Price *et al.*, 1980; Sully & Morgan, 1982; O'Donovan, 1984) and sheep (Kellaway, 1973; O'Donovan, 1984) which indicate that at the tissue level the mechanisms involved in the partitioning of nutrients is similar for cattle, sheep, pigs and poultry. The opportunity and potential to use direct nutritional manipulation as a practical option diminishes as we move from the controlled input (in terms of both quantity and composition of diet) systems of pig and poultry production and cattle feedlot systems to rangeland cattle and sheep production systems in the temperate and the semi-arid zones of the world. In the exten-

Table 5
The influence of energy intake on carcass fatness in sheep

Sex	*Energy content of diet (MJ ME/kg)*	*Feeding level*	*Carcass weight (kg)*	*Growth rate (g/d)*	*Sub-cutaneous fat % dissected*	*Total fat % dissected*	*Chemical fat (%)*	*Fat depth (mm)*	*No. of animals*	*Reference*
Wether (castrate male)	11·6	*ad lib.*	18·0[a]	171		26·1[b]	33·0	8·9	8	(1)
	7·4	*ad lib.*	15·9	106		21·5	27·7	5·5	8	(1)
	White clover	On offer	16·6	227		19·0	26·1	5·6	8	(1)
	Phalaris	On offer	16·0	198		18·0	24·3	4·8	8	(1)
Male, Wether, Female	11·3	*ad lib.*	16·7[c]	200	17·3	29·8	0·7	5·0	12	(2)
	7·5	*ad lib.*	15·3	151	14·6	7·1	28·2	4·3	12	(2)
Male		*ad lib.*	19·3	455	10·2			4·5	10	(3)
		70% *ad lib.*	16·8	400	8·3			2·6	10	(3)
Female		*ad lib.*	19·3	361	11·6			4·7	10	(3)
		70% *ad lib.*	17·3	336	11·1			4·1	10	(3)
Wether	Pasture	High	13·5	187			3·2		25	(4)
		Low	13·7	130			3·2		14	
	Pasture	High	20·2	High				7·0	—[e]	(5)
		Medium	20·0	Medium[d]				4·5		
		Low	21·4	Low[d]				5·2		

(1) Soeparno & Lloyd Davies (1982).
(2) Ahmad & Lloyd Davies (1986).
(3) Jones *et al.* (1983).
(4) Kirton (1976).
(5) Jagusch & Rattray (1979).

[a] All killed at 33·4 kg live weight.
[b] Leg dissection only.
[c] Target final live weight 35 kg.
[d] Both medium and low received a check in their growth.
[e] Numbers of animals not given.

Table 6
The influence of energy intake on carcass fatness in cattle

Sex	Energy content	Feeding level	Carcass weight (kg)	Growth rate (g/d)	Subcutaneous fat % dissected	Total fat % dissected	No. of Animals	Reference
Angus bull	8·4 MJ ME/kg DM (estimated)	*ad lib.*	183		5·5	21·3	16	(1)
		70% *ad lib.*	183		3·7	15·0	7	
Angus steer		*ad lib.*	183		6·2	25·7	11	
		70% *ad lib.*	183		5·2	20·3	8	
Angus heifer		*ad lib.*	183		7·4	29·7	16	
		70% *ad lib.*	183		6·9	26·8	8	
Holstein bull		*ad lib.*	183		3·5	16·6	19	
		70% *ad lib.*	183		2·5	12·1	11	
Holstein steer		*ad lib.*	183		4·2	19·4	12	
		70% *ad lib.*	183		2·8	14·1	8	
Holstein heifer		*ad lib.*	183		5·35	23·5	16	
		70% *ad lib.*	183		4·9	18·8	9	
Angus steer		*ad lib.*	219	800	9·3	25·8	9	(2)
		Restricted	219	400	7·2	21·8	9	
Danish Red bull	[a]	*ad lib.*	250	1170	16·4		42	(3)
		85% *ad lib.*	250	1160	14·9		42	
		75% *ad lib.*	250	1010	12·1		42	
		55% *ad lib.*	250	600	10·2		42	

(1) Fortin *et al.* (1980).
(2) Murray *et al.* (1974).
(3) Anderson (1975).
[a] Scale-fed according to Scandinavian feeding units scale, based on weight.

Table 7

The influence of energy intake on carcass fatness in pigs

Sex	Energy content	Feeding level	Carcass weight (kg)	Growth rate (g/d)	Sub-cutaneous fat % dissected	Total fat % dissected	Chemical fat (%)	Fat depth (mm)	No. of animals	Reference
Entire male							26·5	16·9	3	
		34·2 MJ/d		900						
	14·5 MJ DE/kg	29·4 MJ/d		779			25·7	16·8	3	
		24·9 MJ/d		679			24·3	16·2	3	
		20·3 MJ/d		557			22·7	14·2	3	
		14·5 MJ/d		419			18·1	13·1	3	1[a]
Female		34·2 MJ/d		832			30·6	20·3	3	
		29·4 MJ/d		780			26·3	16·9	3	
		24·9 MJ/d		699			25·3	16·6	3	
		20·3 MJ/d		540			23·5	14·9	3	
		14·5 MJ/d		409			18·4	13·3	3	
Castrate male	11·4[b]	100%[c]	63·5	470		22·2		15·7	12	
	12·9	87·5%	64·2	495		20·2		13·0	12	2[d]
	15·0	75·0%	66·2	542		22·5		15·1	12	
	18·3	62·5%	66·7	567		24·5		16·9	12	
Entire male	14·9 MJ DE/kg	*ad lib.*	54·7	820	24·5[f]			23·4	3	
		Medium	54·9	733	21·4			21·1	3	
		Low	55·3	630	20·6			18·8	3	3[e]
Castrate male		*ad lib.*	54·7	796	27·0	27·4			3	
		Medium	54·9	710	24·8	22·3			3	
		Low	55·3	598	23.8	21·0			3	

(1) Campbell *et al.* (1983).
(2) Lawrence (1977).
(3) Campbell & King (1982).

[a] All slaughtered at 45 kg live weight. *Ad lib.* intake was 34·2 MJ/d for both sexes.
[b] MJ DE/kg DM.
[c] Relative feeding level.
[d] Slaughtered at 87 kg live weight.
[e] Slaughtered at 70 kg live weight.
[f] Percent dissectible fat in the ham.

Table 8
The influence of energy intake on carcass fatness in poultry

Sex	*Energy content (MJ ME/kg DM)*	*Feeding level*	*Body wt (BW kg)*	*Growth rate (g/d)*	*Fat pad wt (% BW)*	*Total body chemical fast (%)*	*No. of birds*	*Reference*
Male	13·4	*ad lib.*	1·93	38		12·7	10	(1)
	10·1	*ad lib.*	1·40	27		5·7	12	
Male	14·1[a]	*ad lib.*	1·77[b]		2·19		10	(2)
	13·4	*ad lib.*	1·75		1·89		10	
	12·7	*ad lib.*	1·69		1·84		10	
Female	14·1	*ad lib.*	1·44		2·60		10	
	13·4	*ad lib.*	1·42		2·21		10	
	12·7	*ad lib.*	1·41		2·14		10	
Male and female	10·9	*ad lib.*	1·64		1·46	9·8	24	(3)
	11·7	*ad lib.*	1·69		1·72	10·5	24	
	12·5	*ad lib.*	1·68		2·08	12·4	24	
	13·4	*ad lib.*	1·74		2·13	12·1	24	
	14·2	*ad lib.*	1·76		2·39	13·4	24	
	15·1	*ad lib.*	1·78		2·70	14·5	24	

(1) Fisher (1984).
(2) Kubena *et al.* (1974).
(3) Jackson *et al.* (1982).
[a] Starter ration 0–4 weeks, 13·8 MJ ME/kg DM.
[b] All killed at 8 weeks of age.

sive situation, where the quality and quantity of feed varies in a reasonably cyclical fashion it is the management systems that have evolved which may be considered as having an indirect effect on carcass fatness, through the effect of weight on carcass composition.

The feed consumed by an animal may be considered to be made up of protein, energy and specific limiting nutrients, usually amino acids and minerals. These components may be derived from a variety of sources, and differ in digestibility, metabolisable energy (ME) content, protein content and form, which in turn will influence the amount of undegraded digestible protein (UDP) available at the intestine for absorption. The relative proportions of both, and their interactions will then influence the effective protein/energy (P/E) ratio of the diet.

Both singly and in combination, the various components of diets can influence carcass fatness, to varying extents. The main components are energy and protein, although it is recognised that there are complex interactions between these two (MacRae & Lobley, 1986) which suggest that such a simple approach should be taken with caution. Nevertheless, much of the data in the literature considers either energy or protein, and its effect on fatness, particularly in the case of cattle and sheep. Studies of the interactions between protein and energy, and the effect of changing P/E ratios of diets have mainly been done with pigs and poultry (Griffiths *et al.*, 1977; Campbell & King, 1982).

A growth model, based on preruminant milk-fed lambs (Black, 1974), has been developed which explains a large part of the apparent confusion over the effects of nutrition on body and carcass composition. It is suggested that this model can be extended in a general sense to cover ruminant animals. In cattle infused abomasally the responses observed were in line with predictions based on the model reinforcing the view that at the tissue level similar mechanisms of energy/protein partitioning are operating in cattle and sheep and possibly also in monogastrics. At the simplest level the model indicates that protein retention (PR) is dependent upon protein intake (PI) in diets that are protein deficient. As PI increases, PR increases linearly until the point is reached when PI is no longer limiting the animals protein requirements and hence ability to respond. This relationship is independent of energy intake. As energy intake level is increased the point where PR is no longer influenced by PI also increases (Fig. 1). When considered as the P/E ratio, results from pig studies (e.g. Campbell *et al.*, 1983) show that as the P/E ratio declines (i.e. the protein concentration in the diet is reduced) so carcass fatness increases in a curvilinear fashion, and the actual level of fatness at any given P/E ratio will be affected by the total energy intake level (Fig. 2). For

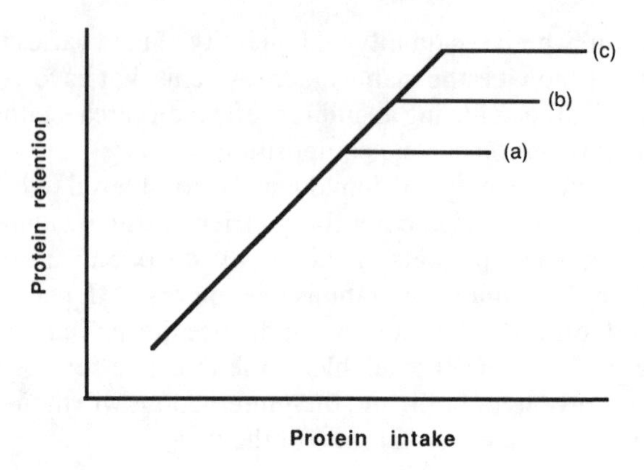

Fig. 1. The influence of protein intake on protein retention: protein retention (PR) increases linearly with increasing protein intake (PI), until PI is no longer limiting. With PI in excess of requirements PR plateaus, at a level dependent upon energy intake (EI). As EI increases (a–c) so the plateau level of PR increases.

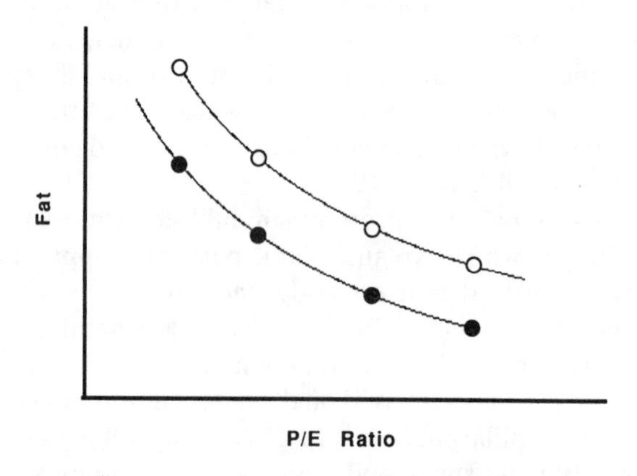

Fig. 2. The influence of the protein/energy (P/E) ratio on body fatness: as the P/E ratio declines, body fatness at any specific weight increases, in a curvilinear fashion. At any given P/E ratio, animals fed a high energy diet (○ — ○) will be fatter than those on a low energy diet (● — ●).

example, the work of Campbell *et al.* (1983) showed that when the protein content of an experimental diet was reduced from 20% to 10%, the rate of fat deposition in 40 kg pigs fed a low energy intake increased from approximately 135 g/day to 215 g/day, while the increase in fat gain in pigs fed a high energy intake increased from 235 g/day to 305 g/day.

In this section the effects of nutrition on carcass fatness will be considered under the following headings but with the expressed reservations of this approach clearly in mind:

1. Growth paths.
2. Energy (E).
3. Protein (P).
4. Energy density.
5. P/E ratios.
6. Limiting nutrients.

Growth Paths

The growth path followed by an animal during its lifetime is a reflection of its total nutrition, and this may vary considerably in both quantity and quality over time. Such variation in nutrition may be imposed directly or indirectly by man or nature with the result that growth rate (kg live weight gain/day) varies over time. As a consequence, animals may follow diverse growth patterns to arrive at a specific live weight at a particular age.

In the early days of intensive pig production carcass fatness was manipulated by restricted feeding of a given diet in the later stages of growth, in the belief that this was the most effective method of controlling the late developing fat tissue (e.g. McMeekan, 1940). Subsequent studies and re-analyses of some of this original work (Davies, 1983) discounted this notion, and attention is now focusing on the influence of protein, energy and their interactions on composition, particularly in pigs and poultry (e.g. Auckland & Fulton, 1973; Kubena *et al.*, 1974; Morgan, 1980).

Figures 3(a)–(f) are examples of the types of growth paths followed by farm animals to achieve specific carcass weights. It is emphasised that carcass weight is the criterion on which the comparisons are made. As progression is made from 3(a) to 3(f) so the level of control and type of production system changes from highly controlled, intensive (pigs, fed cattle, poultry systems) to uncontrolled, extensive grazing where animals must survive the vagaries of changes in feed quality and quantity with the seasons.

However, most of the following discussion focuses on cattle and sheep as paths are due to variation in feed availability rather than specific nutrients.

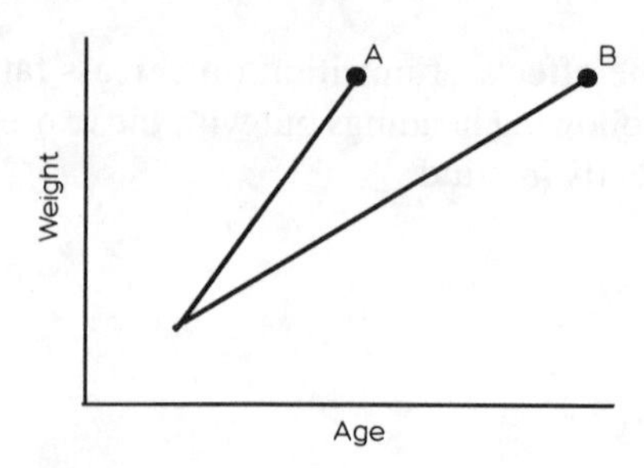

Fig. 3(a). High growth rate (A) versus lower growth due to feed restriction (B): animals at A will generally be fatter than at B.

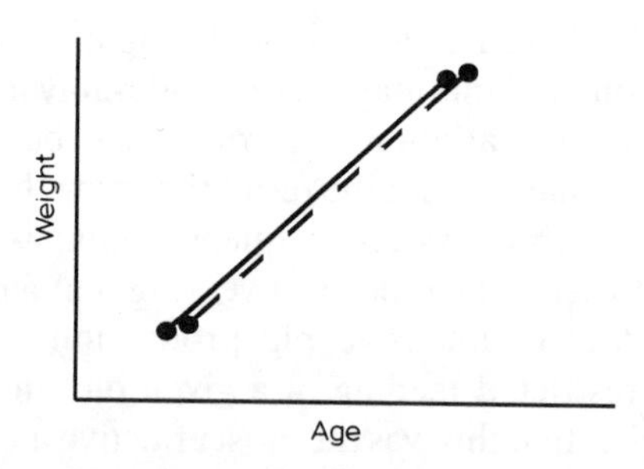

Fig. 3(b). Parallel growth paths but different diets: animals fed on pasture (● —— ●) contained less fat than grain-fed (● - - - ●) animals at the same weight.

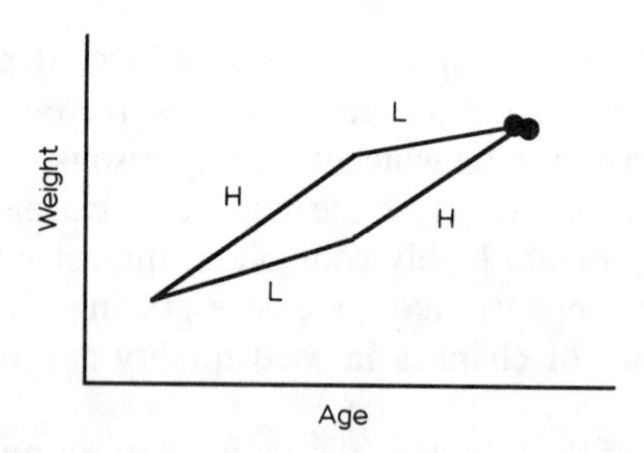

Fig. 3(c). Periods of high (H) and low (L) growth: at the same age and weight, animals from the LH path were fatter than those from the HL path.

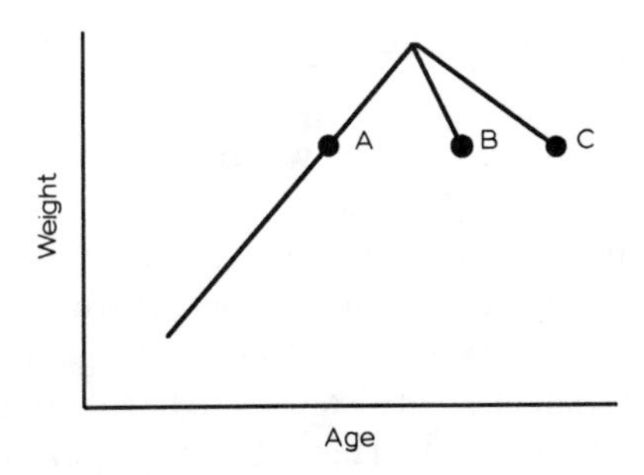

Fig. 3(d). Weight loss following normal growth: animals which lost weight rapidly (B) or slowly (C) to reach the same weight as A were fatter (B) and leaner (C) than A.

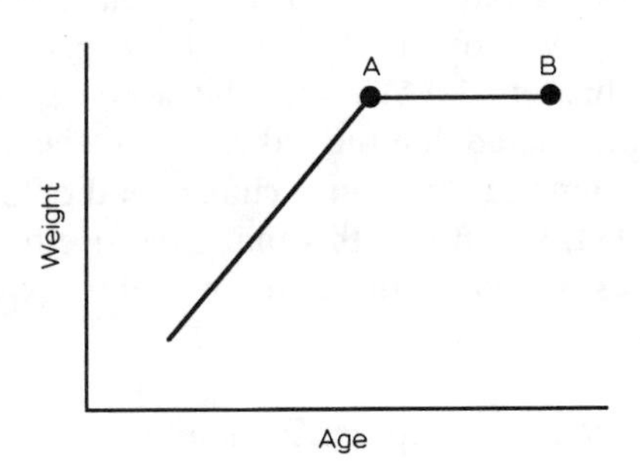

Fig.3(e). Period of weight maintenance: unrestricted growth animals (A) at the same weight as maintenance animals (B) will be leaner than B.

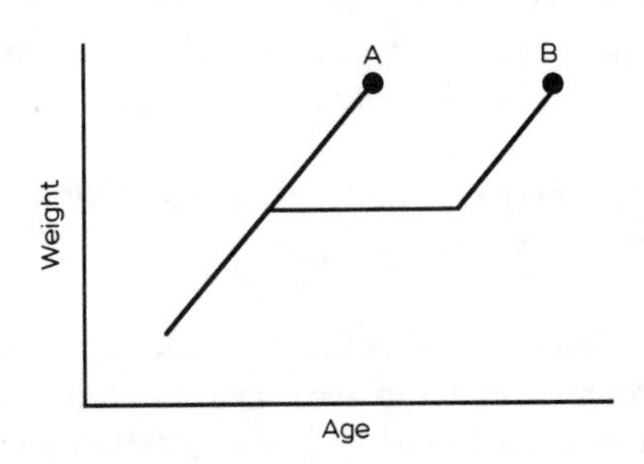

Fig. 3(f). Unrestricted growth compared with weight maintenance and subsequent recovery of weight: unrestricted animals (A), compared with others (B) may be fatter, no different or leaner, depending upon the time period and weight gained during recovery.

High Versus Low Growth Rate (Fig. 3(a))

In general, the faster growing animal will be fatter than its slower grown counterpart. This is because at any stage of growth there is some maximal rate of muscle growth which appears to be related to age as well as protein intake (e.g. Butler-Hogg & Johnsson, 1986). Additional energy intake above maintenance is then partitioned towards fat gain. The faster growing animal (cattle and sheep) will also tend to have proportionately more of its carcass fat subcutaneously. This could be important where grading systems use a measure of subcutaneous fatness to determine carcass value. The fatter, faster-grown animals are more likely to be downgraded under these circumstances than if carcass weight alone is the major or sole criterion of value.

However, there are always exceptions to the rule, and there are examples in cattle (Sully & Morgan, 1982) and sheep (Searle *et al.*, 1982; Butler-Hogg & Johnsson, 1986) where the slower grown animals were the fatter. It was postulated that the reduction in feed intake imposed on the slow growing animals resulted in a change in the P/E ratio of the diet (Margan *et al.*, 1982) and it was this that gave rise to the carcass composition differences, similar to those that would be expected in pigs and poultry.

Parallel Growth Paths, but Different Diets (Fig. 3(b))

An example is grass versus grain feeding, where cattle were fed restricted amounts of grain to achieve the same growth rates as their *ad libitum* grass-fed counterparts (Sully & Morgan, 1978). The grass-fed steers tended to have slightly lower fat depths, less subcutaneous fat and significantly less total fat in the side (24 vs 16%).

Periods of High (H) or Low (L) Growth Rate so that Animals Achieve the Same Weight at the Same Age (Fig. 3(c))

The HL pathway is typical of that found in the pig industry, whereas the opposite might occur under extensive grazing systems. Under experimental conditions it has been found with sheep (Butler-Hogg & Johnsson, 1986) that animals which followed a HL growth path had leaner carcasses than those which followed the LH path. The effect on subcutaneous carcass fat was sufficient to change the grade of the HL animals compared with the LH ones. Ledin (1983) also found reduced carcass fatness in lambs from a LH path compared with a H path, both groups being killed at the same weight.

Weight Loss Following Normal Growth (Fig. 3(d))

The practical application of this growth path has been tried, but with limited success and repeatability. Part of the variability may be caused by the differential effect of the rate of weight loss. As proposed in the model of Black (1974) and verified experimentally in sheep (Butler-Hogg, 1984; O'Donovan, 1984) a high rate of loss results in fatter carcasses than in animals at that same weight which have not suffered a period of loss, while a slow rate of loss gives leaner carcasses. As a practical option to reduce carcass fatness, the use of planned weight loss needs further investigation. The general consensus is that under extensive grazing systems based on pasture, such as in New Zealand, it is difficult to manipulate carcass composition by such means. The subject area of weight loss and compensatory gain has been reviewed (O'Donovan, 1984).

A Period of Weight Maintenance (Fig. 3(e))

The weight maintenance may have been achieved by a steady reduction in the amount of available feed, or it may have occurred because there was a progressive decline in the quality of available feed. In general, animals subjected to a period of maintenance tend to have fatter carcasses than expected for that weight (Winter, 1971; Foot & Tulloh, 1977). However, if the period of maintenance is also associated with weaning, then, at least for the first few weeks postweaning, body fat is being extensively mobilised, and so the carcass is leaner than expected (Searle & Griffiths, 1976).

Weight Maintenance Followed by ad libitum *Feeding with Possible Compensatory Growth* (Fig. 3(f))

Depending upon the amount of weight gained following the maintenance period, sheep carcasses may be fatter, leaner (O'Donovan, 1984) or not different from continuously grown controls (Murray & Slezacek, 1976). Under *ad libitum* feeding conditions, the carcass and body composition of an animal will tend towards the genetically defined composition for that weight and stage of maturity. The longer the recovery period the greater is the likelihood that the group which had suffered some nutritional setback will be no different from their unaffected counterparts.

Table 9
The influence of protein intake on carcass fatness in sheep

Sex	*Protein content (g crude protein/ kg DM)*	*Feeding level (kg DM/day)*	*Growth rate (g/d)*	*Ether extract (% carcass dry matter)*	*No. of animals*	*Reference*
Male and female	100	0·916	305	63·7	4	(1)[a]
		0·818	250	63·6	4	
		0·669	128	60·8	4	
	125	0·897	249	58·6	4	
		0·791	210	60·0	4	
		0·646	149	60·1	4	
	150	0·918	260	52·0	4	
		0·776	216	56·1	4	
		0·661	134	58·0	4	
	175	0·911	287	53·4	4	
		0·800	212	53·5	4	
		0·659	146	58·8	4	
	200	0·897	266	54·1	4	
		0·783	217	53·8	4	
		0·677	148	58·6	4	
Male	110	*ad lib.*	191	60·4	5	(2)[b]
	157	*ad lib.*	270	53·7	5	
	194	*ad lib.*	330	50·2	5	
Female	110	*ad lib.*	171	63·5	5	
	157	*ad lib.*	225	58·8	5	
	194	*ad lib.*	301	55·3	5	

(1) Andrews & Orskov (1970).
(2) Orskov *et al.* (1971).
[a] All killed at 40 kg live weight.
[b] Estimated carcass weight 15 kg.

Energy

If protein intake is not limiting the potential of the animals to respond to changes in energy intake, the general rule is that the greater the energy intake (and hence usually the faster the growth rate) the fatter will be the carcass at any weight (see Tables 5, 6, 9 and 10; e.g. Jones *et al.*, 1983 in sheep; Murray *et al.*, 1974; Patterson *et al.*, 1985, in cattle; Campbell & King, 1982; Campbell *et al.*, 1983, in pigs and Griffiths *et al.*, 1977, in poultry). In sheep and cattle the effects of varying intake are greatest on the subcutaneous fat depot (Murray & Slezacek, 1976; Murray *et al.*, 1974). However, it should be kept in mind that there can also be an interaction between energy intake and energy density of the diet which may produce unexpected results (Sully & Morgan, 1978).

Table 10
The influence of protein intake[a] on carcass fatness in bulls

Protein content (g CP/kg DM)	*Feeding level (kg DM/day)*	*Growth rate (g/d)*	*Fat depth (mm)*	*No. of Animals*
88	7·50	1080	6·7	40
100	7·77	1190	8·4	39
113	7·89	1260	8·3	39
127	7·57	1270	7·9	39

Lemenager *et al.* (1981).
[a] Fed for 140 days, protein supplement for the last 70 days.

Protein

Manipulating fatness by changing energy intake is limited in scope (up to 5% units difference in fatness at a given weight) when compared with the changes in composition that can be induced by varying protein intake (up to 15% units difference; Black, 1974). This has been well demonstrated in pigs (Davies, 1983; Campbell *et al.*, 1984) and in milk-fed lambs (Black, 1974), but is more difficult to achieve in ruminants, except perhaps where initial protein intakes are extremely low (see Tables 7, 8, 11 and 12). Nutritional regimes imposed on pigs specifically take advantage of the effect of varying protein intake at different stages of growth. As the pig increases in weight, in general, the level of protein in the diet

Table 11
The influence of protein intake on carcass fatness in pigs

Sex	Protein content (g CP/kg)	Feeding level	Carcass weight (kg)	Growth rate (g/d)	Total dissected fat in ham (%)	Chemical fat (%)	Fat depth (mm)	No. of Animals	Reference
Entire male	170	Low	55·3	630	20·6		18·8	3	(1)
	210		54·6	673	20·8		16·0	3	
	231		55·2	650	20·3		16·7	3	
	170	Medium	54·9	733	21·4		21·1	3	
	210		55·3	776	21·7		18·3	3	
	231		55·0	685	22·6		21·0	3	
	170	High	54·7	820	24·5		23·4	3	
	210		55·5	820	23·8		20·0	3	
	231		55·0	772	23·6		23·2	3	
Entire male and female	151	[a]				20·9[c]	13·8	16	(2)[b]
	252					15·3	10·3	16	

(1) Campbell & King (1982).
(2) Campbell & Dunkin (1983).
[a] Scale feeding based on weight.
[b] All killed at 45 kg live weight.
[c] Percent chemical fat in the whole body.

Table 12
The influence of protein intake on carcass fatness in poultry

Sex	Protein content (g CP/kg DM)	Feeding level	Body weight (BW) (kg)	Growth rate (g/d)	Fat pad (% BW)	Total body chemical fat (%)	No. of birds	Reference
Male and female	16	*ad lib.*	1·63[a]	33[b]	3·12[a]	15·5	12	(1)
	20	*ad lib.*	1·73	35	2·55	13·7	12	
	24	*ad lib.*	1·77	36	1·92	11·9	12	
	28	*ad lib.*	1·74	35	1·67	10·9	12	
	32	*ad lib.*	1·73	35	1·73	10·7	12	
	36	*ad lib.*	1·74	35	1·49	10·3	12	
Male	18		1·23[c]		1·95[a]		10	(2)
	23		1·77		2·11		10	
	28		1·75		2·00		10	
	33		1·66		—		10	
Male	15.8	3·50[d]		52[e]	3·19[f]		32	(3)
	18·6	3·35		52	2·61		32	
	21·4	3·30		51	2·22		32	

(1) Jackson *et al.* (1982).
(2) March *et al.* (1984).
(3) Griffiths *et al.* (1977).
[a] At 49 days of age.
[b] From 1–49 days of age.
[c] Estimated 7 week weight, calculated as mean of 6 and 8 week live weights.
[d] Total feed consumed over the period 4–8 weeks of age.
[e] From 4–8 weeks of age.
[f] At 8 weeks of age.

is progressively reduced. The end result is a pig carcass considerably lower in fatness than one at the same weight which came from a pig fed a high protein diet throughout life (Campbell & King, 1982; Campbell *et al.*, 1984).

Energy Density

The energy density (megajoules (MJ) of metabolisable energy (ME) per kilogram of dry matter (DM)) in the diet can have an effect on fatness which appears to be independent of total energy intake. In poultry, the proportion of fat in the gain increased by 6·7 g/kg gain per MJ/kg increase in ME (Fisher, 1984). Studies with cattle and sheep have shown that animals fed high energy-dense diets were fatter than those fed diets of lower energy density even though they had grown more slowly (Soeparno & Lloyd Davies, 1982). In other cases (Purchas & Lloyd Davies, 1974; Bond *et al.*, 1982; Jones *et al.*, 1985*a* with cattle; Ahmad & Lloyd Davies, 1986 with sheep) the high energy density diets resulted in faster growth and fatter carcasses, similar to results in poultry and pigs. The apparent anomaly of slower growth and fatter carcasses leads on to a consideration of the effect on composition of the P/E ratio of the diet.

Protein/Energy Ratio

The proportion of protein in the diet compared with total dietary energy, is referred to as the protein/energy (P/E) ratio. Although a particular feed may have an apparent P/E ratio based on proximal analysis of the diet, it is the true P/E ratio present at the duodenum which can exert an influence on the chemical composition of live weight gain (and hence fatness). The effect of changing P/E ratio has been well demonstrated in pigs (Campbell & Dunkin, 1983; Zhang *et al.*, 1986) and poultry (Donaldson *et al.*, 1956; Bartov *et al.*, 1974) but the effect can also occur in sheep (Searle *et al.*, 1982) and cattle (Orskov *et al.*, 1983).

The effect of changing P/E ratio on fatness has been described earlier (see Fig. 2). Using the P/E ratio of the diet to control carcass fatness has been an option used by both the pig and poultry industries because animals are monogastric and the composition of the diet as well as the level of feeding are capable of being controlled.

As the proportion of protein in the diet increases, the proportion of fat in the gain decreases, until a minimum value is reached. Further changes in the protein content then cause an increase in the proportion of fat in the gain. The actual protein level at which this occurs varies with the level of feeding (Campbell *et al.*, 1985). As the feeding level goes up, so the protein content of the diet giving minimal fat gain declines. In practice, particularly with pigs, this means that the protein content of the diet can be reduced as the animals grow and consume greater quantities of feed.

Limiting Nutrients

If a specific nutrient e.g. an amino acid, is limiting in a diet, it may influence the response to increased intake and hence affect carcass fatness. It is in that context that brief mention is made of limiting nutrients, rather than as a means of controlling fatness. A great deal of work has been done to determine the sequence of limiting amino acids in pig and poultry diets. Generally lysine availability is of most concern, followed by methionine. In sheep, the availability of methionine, cysteine and histidine may be important. Under rangeland conditions, where sheep and cattle are consuming mature, lowly digestible pastures for up to six months of the year inadequate dietary nitrogen intake is limiting (Hunting & Siebert, 1980). The provision of a source of non-protein nitrogen, such as a urea lick, or in some cases a protein meal, can produce improvements in weight gain. Hennessy *et al.* (1983) found that giving a protein meal to young cattle consuming poor quality hay resulted in positive weight gains via its effect on food intake and hencc availability of glucose. However, in many cases, supplementation merely prevents, or at least diminishes the rate of body weight loss.

Nutritional Control

In the practical situation the potential to influence carcass fatness in sheep and cattle through nutrition is limited. However, as the degree of control over both stock and pasture increases, so the potential for manipulation also increases. But, unless big changes in composition occur (as through protein nutrition) the changes in carcass fatness achieved through nutritional manipulation may not be sufficiently large to reward the producer for his efforts. In the semi-intensive lamb, the cattle finishing systems of the UK, or the feedlot systems of the US, there is some scope in affecting carcass fatness by manipulating the dietary P/E ratio. This can be done by the inclusion of sources of protein which are high in UDP in the silage component of the diet (MacRae & Lobley, 1980).

Controlling the energy supply, either directly or through manipulation of stocking rate, is probably the major way in which cattle and sheep producers might attempt to control carcass fatness. However, with current grading systems generally rewarding fatter carcasses the incentives to apply extra management skill and effort are not high. In contrast, within the pig industry, where clearly defined and accepted grading systems operate, the nutritional manipulation of carcass fatness is both feasible and rewarding. At the very practical level it is reassuring to see that the manipulation of composition via protein rather than energy is the more effective way, as predicted by the growth model (Black, 1974).

GROWTH PROMOTERS

Growth promoters for domestic meat-producing animals have been developed to provide a cost effective way of manipulating growth and carcass composition in order to fulfil the increasing world need for lean meat. Growth promoters in general improve live weight gain, feed efficiency and lean growth of meat-producing animals (Heitzman, 1979). The effect of growth promoters on carcass fat is much more variable than that on live weight.

Growth promoters have been classified into three main groups according to their mode of action: antibiotics, ionophores and anabolic agents (Van Der Wal & Berende, 1983).

The first two types of growth promotants listed have no significant effect on carcass composition. Antibiotics, such as terramycin, have little effect on carcass fatness. The ionophores, such as monensin, increase propionic acid production in the rumen and improve the efficiency of feed conversion in ruminants (Davis & Erhart, 1976) and stimulate live weight gain but do not appear to have an effect on carcass composition of steers (Potter *et al.*, 1976). The anabolic agents may offer a practical way of increasing carcass weight and decreasing carcass fatness and these compounds will be dealt with in greater detail in this section because many of them are or have been used to reduce the fatness of meat animals.

Commercial Anabolic Agents

Commercially available anabolic agents can be divided into two basic types—natural hormones and xenobiotics; combinations of these products are also often used commercially (Table 13).

Table 13
Commercial anabolic agents

Type	*Agent*	*Commercial*
1. Natural hormones	Oestradiol	Compudose
	Testosterone	
	Progesterone	
2. Xenobiotics	Zeranol	Ralgro
	Trenbolone acetate	TBA
	Melengestrol acetate	MGA
3. Combination products	Oestradiol+progesterone	Synovex S
	Oestradiol+testosterone	Synovex H
	Trenbolone acetate+Zeranol	
	Trenbolone acetate+Oestradiol	

The stilbenes, which includes diethylstilbestrol (DES), are no longer registered for use in meat-producing animals in the EEC and North America, so will not be discussed in detail.

The natural anabolic agents, in particular, have found wide approval by the scientific community (Henderson, 1983), because of their low, often undetectable residue levels in meat and their health safety record (Hoffmann, 1981). The recent (1986) EEC decision to ban all growth promotants appears to be related to political pressure rather than scientific evidence and therefore natural and xenobiotic anabolic agents still used in North America, Australasia and other parts of the world will be discussed.

Mode of Action of Anabolics

Oestrogens

The stimulatory effects of oestrogens on lean growth are thought to be achieved mainly through an increase in somatotropins or growth hormone (GH) (Davies *et al.*, 1977) and not a change in GH clearance rate from the plasma (Gopinath & Kilts, 1984). GH is considered one of the main hormones controlling fat deposition by decreasing lipogenesis whilst not affecting lipolysis in long term GH treatment (Lewis, 1987). The increased GH levels in oestradiol-treated animals may be a result of increased hypothalamic growth hormone releasing hormone (GHRH) or the modulation of GRF pituitary receptors (Trenkle, 1983).

Oestradiol not only increases plasma GH but also increases the number of GH receptors in the liver (Breier *et al.*, 1986) and as a consequence increases plasma concentrations of insulin-like growth factor-1, a hormone which is thought to stimulate muscle and bone growth and decrease fat deposition (Lewis, 1987). These responses may not be direct effects of oestradiol, but result from increased GH levels. However, in wether lambs, the effect on growth of DES, a synthetic oestrogen, appeared before an increase in GH was observed (Muir *et al.*, 1983). In primates and humans, physiological doses of oestrogens stimulate GH and somatomedin C (Copeland *et al.*, 1984) whereas pharmacological doses depress somatomedin C in acromegalics. This indicates that oestrogens can affect the growth promoting hormones independently of GH. Oestrogens have also been shown to bind to androgen and oestrogen muscle receptors, suggesting that oestrogens may have a direct role in stimulating growth (Meyer & Rapp, 1985). However, DES does not bind to muscles and appears to act indirectly on lean growth.

Zeranol, an oestrogenic xenobiotic, initially increases GH in sheep, but long term infusions produced no significant effects on GH (Olsen *et al.*, 1977).

Plasma insulin concentrations have frequently been shown to be increased by oestrogen treatment, which may result from the diabetogenic effects of elevated GH concentrations (Trenkle, 1983). The rise in insulin may contribute to an increase in lean growth, if, as has been shown by Lewis (1987), elevated GH concentrations suppress the lipogenic response of adipose tissue to insulin.

Androgens

The major mechanisms by which androgens stimulate growth and as a consequence decrease carcass fat is by direct action on muscle fibres (Wainman & Shipounoff, 1941). Specific androgen receptors have been demonstrated on rat striated muscles (Juny & Baulieu, 1972), and the receptor numbers decrease with castration and are stimulated by testosterone. Associated with this receptor response to testosterone is a marked increase in muscle fibre diameter. Protein synthesis in cultured muscle cells is enhanced by testosterone and blocked by anti-androgens, such as cyproterone acetate (Juny & Baulieu, 1972). However, the direct action of androgens on cultured muscle cells fails to explain all the growth promoting effects of androgens. There may be other direct as well as indirect mechanisms by which lean growth is stimulated by androgens. One of these may involve the GH-IGF-I axis, which is stimulated by oestrogens, since testosterone propionate stimulates plasma GH concentrations (Powers & Florini, 1975) and rams have higher IGF-I plasma concentrations than comparable wethers and ewe lambs (Bass, personal observation).

The mode of action of TBA (trenbolone acetate), a synthetic androgen, which is considered to be 3–5 times more androgenic than testosterone (Neumann, 1975), is uncertain, but it does bind to muscle receptors in a manner similar to testosterone (Bouffault & Willemart, 1983). TBA, unlike testosterone propionate, fails to stimulate GH and insulin.

Progestagens

The mechanism of action of progesterone is poorly understood. Progesterone may interact with muscle androgen receptor sites (Bouffault & Willemart, 1983) and so could directly affect muscle growth. However, melengestrol acetate (MGA), a synthetic, orally active progestagen, may act through a different mechanism in heifers. In heifers MGA may block ovulation and corpus luteum development and so allow the development of immature pre-ovulatory ovarian follicles, which possibly lead to enhanced oestrogen production. This may account for the reported increased lean growth in heifers. However, MGA has been shown to decrease cortisol and cortiscosterone concentrations which may increase daily gain in cattle (Purchas *et al.*, 1971).

Animal Response to Anabolics

Oestrogens

In general commercially available oestrogenic growth promoters increase protein and decrease fat deposition (Table 14). This is particularly so in steers. In bulls there is an increase in lean meat and a decrease in bone weight but no effect on fat trim (Table 15).

Ralgro has been reported (Staigmiller *et al.*, 1978) to produce increased live weight gain in heifers, but studies on pasture (Bass *et al.*, 1984) found that Ralgro only stimulated the growth of spayed heifers and had little effect on the carcass composition of spayed or entire heifers (Table 16).

Table 14
Fat deposition of control and anabolic treated calves, steers and pigs

Animals	*Treatment*	*Fat deposition*	
		(g)	*(%)*
Male calves	Control	26·6	100
	120 mg DES	21·6	81
Steers	Control	15·7	100
	20 mg oestrogen	10·0	64
	20 mg oestrogen+200 mg TBA	14·5	93
	20 mg oestrogen+400 mg TBA	9·1	58
Castrated male pigs	Control	47·5	100
	20 mg oestrogen+140 mg TBA	35·0	82

From Van Der Wal & Berende (1983).

Table 15
Composition of Friesian bulls at 15–16 months after Ralgro treatment (36 mg/70 days) Bass *et al.* (1984)

	Control	*3-month treatment*	*6-month treatment*
Carcass weight (kg)	223·2	225·6	229·2
Left side composition (kg) (adjusted for carcass wt)			
Meat	79·4	81·1	80·4**
Bone	24·3	23·2	23·4**
Trimmed fat	7·9	7·5	7·9

**$P < 0{\cdot}01$

The carcass composition response to oestradiol is affected slightly by nutrition (Table 17), with the well nourished low stocked (3/ha) Compudose treated steers having 3·1% more trimmed lean meat and 2·6% less bone than untreated steers. There was no significant effect on trimmed fat percentage at the lower stocking rate (3/ha) after Compudose treatment, whereas longer treatment with oestradiol at a stocking rate of 5/ha not only increased the percentage of meat but also decreased bone percentage, as well as decreasing trimmed fat by 2% (Table 17).

Studies on the effects of oestrogens in sheep have involved DES or hexoestrol up to 1970 and subsequently zeranol (Ralgro). Although

Table 16
Composition of heifer carcasses treated with Ralgro (36 mg)

	Entire	*Ralgro and entire*	*Spayed*	*Ralgro and spayed*
Carcass weight (kg)	143·3	150·5	135·0	150·1*
Left hind quarter composition (kg) (adjusted for hot carcass weight)				
Meat	48·5	47·7	47·3	47·3
Bone	12·9	13·1	13·5	12·5
Trimmed fat	8·5	8·1	8·4	8·5

*$P < 0{\cdot}05$.

Table 17
Carcass composition of steers treated with Compudose (oestradiol-17β) grazed at two stocking rates (3 or 5/ha)

	3/ha		*5/ha*	
	Control	*Compudose*	*Control*	*Compudose*
No.	25	25	25	25
Carcass weight (kg)	239	260***	229	247***
Carcass Composition (%)				
Meat	61·8	64·9***	65·2	68·0***
Bone	20.8	18·2***	17·8	17·0*
Trimmed fat	16·6	16·3	17·0	15·0*

*$P < 0{\cdot}05$.
***$P < 0{\cdot}001$
Treatment: Compudose 200, Elanco Products, Indianapolis, USA; two treatments at 200 day intervals.
3/ha slaughtered at 14 months; 5/ha slaughtered at 26 months (Bass *et al.*, 1989).

Ralgro has been shown to stimulate growth of feedlot lambs (Wilson *et al.*, 1972), there has been a number of trials in which the growth effect has been minimal (Bass *et al.*, 1984). Pasture-fed lambs have shown no consistent growth response to Ralgro and the effect on carcass composition was also minimal (Bass *et al.*, 1984).

The effect of oestradiol-17β on sheep growth and composition is limited. In a recent trial oestradiol given at different doses to pasture-fed lambs decreased carcass fat in both wethers and ewes (Table 18). The oestradiol treatment caused a greater decrease in fatness in the ewe lambs, than in the wethers, when slaughtered at both 14 and 23 weeks of age. These results do not appear to be in agreement with the survey of the effects of steroids reported by Van Der Wal & Berende (1983), which indicates that male animals respond better than females to oestrogen treatment. These differences may be associated with differences in age and sexual maturity of the lambs reported on.

Oestrogen treatment had little effect on percentage chemical fat in the carcass of wether lambs at different levels of nutrition (Table 19). The weight of subcutaneous fat dissected from the hindquarter was decreased by oestrogen at the highest nutritional level, but there was no similar effect at low levels of nutrition. The results from this trial indicate that there is a significant oestradiol/nutritional interaction on subcutaneous fat weight and that the reduction of carcass fatness is more likely to be achieved in well fed sheep.

In pigs oestrogens have little effect on weight gain but tend to decrease the thickness of subcutaneous backfat in castrated boars (Heitman & Clegg, 1957) and have the greatest effect in boars (Plimpton & Teague, 1972).

Poultry, unlike ruminants, respond to oestrogen treatment by an increase in fat deposition rather than a decrease and this has been used commercially when a fatter carcass is required (Ryley *et al.*, 1970).

Androgens

The main anabolic androgen used commercially is the xenobiotic TBA. TBA in general increases weight gain, carcass weight and food conversion efficiency and tends to decrease fat deposition in the carcass. The carcass composition response tends to be less than that achieved with TBA plus oestrogenic growth promoters (Bouffault & Willemart, 1983).

The greatest effect of TBA on carcass fatness is found in cows (Table 20) and the combination of TBA plus an oestrogen is used commercially. The effect of TBA on carcass fatness is reduced when animals are on restricted feeding regimes (Table 20), which may be related as much to maturity as nutrition.

Combinations of Anabolic Agents

The most common commercially used androgen in combination growth

Table 18
Composition of lambs at 14 and 23 weeks of age after treatment with oestradiol-17β

	Age (weeks)	*Control*	*Treatment dose Dose cm*						
			0·5	*0·75*	*1·0*	*2·0*	*3·0*	*4·0*	*RSD*
Wethers									
Hot carcass (kg)	14	10·0	10·0	10·3	10·1	10·9	10·7	—	0·89
	23	13·8	14·3	13·8	14·0	14·4	14·1	—	1·33
% ether extract	14	16·6	15·8	15·4	13·9	14·7	13·2	—	1·66
	23	22·8	22·2	20·9	20·4	21·2	19·6	—	3·05
% water	14	60·0	60·8	61·1	62·2	61·7	62·8	—	1·47
	23	54·7	54·8	55·1	56·2	54·7	56·4	—	2·42
% protein	14	18·1	18·3	18·5	18·6	18·4	18.7	—	0·51
	23	17·0	17·4	18·3	17·7	18·2	18·1	—	0·82
Ewes									
Hot carcass (kg)	14	9·6	10·4	10·6	10·3	10·0	9·9	10·5	0·85
	23	13·8	14·0	14·1	14·1	14·4	13·3	14·0	1·32
% ether extract	14	18·2	16·8	16·7	14·2	14·5	14·3	13·8	2·68
	23	25·1	23·4	21·5	18·6	19·4	18·2	18·7	3·28
% water	14	58·6	59·8	59·6	61·5	61·5	61·7	62·0	2·11
	23	53·0	54·0	55·0	56·9	56·7	57·4	56·3	2·56
% protein	14	17·9	18·2	18·4	18·6	18·7	18·4	18·8	0·72
	23	16·5	17·1	18·0	18·6	18·3	18·5	19·1	0·85

Percentage composition data adjusted for hot carcass weight (Bass *et al.*, 1987).
Dose cm—length of silastic implant.

Table 19

Effect of oestradiol on carcass composition of lambs at different levels of nutrition (lucerne pellets)

	Ad lib.		*1·5% of body wt/day*		*Effect*	
	Control	*Oestradiol*	*Control*	*Oestradiol*	*Nutrition*	*Oestradiol*
Hot carcass weight (kg)	28·9	31·2	18·0	18·1	***	+
Carcass composition (%)						
Ether extract	38·3	32·9	32·4	32·7	–	+
Water	44·3	47·8	49·2	48·6	–	–
Protein	14·0	15·2	14·3	14·2	–	–
Hind quarter (g)						
Subcutaneous fat	762	498	828	849	–	*
Intermuscular fat	305	322	224	234	–	–
Muscle	2098	2304	2045	1973	–	–
Bone	535	581	535	578	–	*

*$P < 0·05$.
***$P < 0·001$.
Treatment—Silastic, oestradiol-17β implants weighing 94 mg and retained for 12 weeks.
Carcass composition percent adjusted to same carcass and hind quarter weight (Bass *et al.*, Pers. comm., 1986)
Hind quarter components adjusted to same hind quarter weight.

Table 20
Effect of TBA (300 mg) on carcass composition of cull cows

Dietary conditions:	*Free access to pasture*		*Limited access to pasture*	
Treatment:	*Control*	*TBA*	*Control*	*TBA*
No.	8	8	8	8
Daily weight gain (kg)	0·75	1·06	0·56	0·74*
Kidney fat index	2·37	1·77	2·02	1·96
11th rib fat index	31·4**	23·7	25·5	24·0

*$P < 0{\cdot}05$.
**$P < 0{\cdot}01$.

Kidney fat index: $\frac{\text{Perirenal and precrural fat}}{\text{Weight cold carcass}} \times 100$

11th rib fat index: $\frac{\text{Fat of ribs 10 and 11}}{\text{Weight of bone of 11th rib}} \times 100$

After Bouffault & Willemart (1983).

promoters is TBA which has been used in conjunction with zeranol and oestradiol. These combination products improve feed conversion efficiency and carcass weight with little effect on carcass fat in veal calves, steers and bulls (Table 21).

Oestradiol-17β (5·6 mg) plus TBA (40 mg) improved live weight gain of treated lambs but had no effect on carcass weight (Grandadam *et al.*, 1975). Treated castrated male pigs were significantly heavier with improved feed conversion efficiency, and backfat thickness was significantly lower than in the controls (Table 22).

In more detailed comparative studies on lambs (Grandadam *et al.*, 1975) none of the growth promoters alone or in combination altered the proportions of water, crude protein or fat in the carcass. The lack of response in this trial as well as some others may be the result of an inappropriate dose level or , nutritional treatment, etc.

Another commercial combination product is Synovex S used in steers (oestradiol/progesterone) and Synovex H in heifers (oestradiol/testosterone). In one study, Synovex S increased growth rates in steers but did not affect the ether extract content of the carcass (Rumsey, 1982) and a similar result was found with Synovex H in heifers (Reid, 1983). Similar

Table 21
Effect of Zeranol and TBA on carcass weight and composition

Treatment			*Controls*	*Treatment*
				Zeranol 36 mg+TBA 140 mg day 60
Male	*veal calves*	No.	136	136
	Cold carcass	kg	110·3	119·8
	% carcasses in three visual categories	Lean	1·4	0·7
		Covered	97·2	97·2
		Fat	1·4	2·1
				Zeranol 36 mg+TBA 140 mg day 53
Female	*veal calves*	No.	190	193
	Cold carcass	kg	97·5	105·7
	% carcasses in three visual categories	Lean	3	2·2
		Covered	92·6	97·8
		Fat	4·4	—
				Zeranol 36 mg+TBA 40 mg
Young bulls		No.	53	53
	Carcass weight	kg	369·5	380·2
Steers		No.	7	7
	Carcass weight	kg	365·5	376·8

From review by Bouffault & Willemart (1983).

Table 22
Effect of 124 days of oestradiol-17β (20 mg) plus TBA (140 mg) on castrated male pigs with a start weight of 26 kg

	Control	*Oestradiol+TBA*
No.	14	14
Carcass weight (kg)	70·6 ± 0·75	75·2 ± 0·49***
Backfat thickness (mm)[a]	28·5 ± 1·17	24·3 ± 0·47**

**$P < 0·01$.
***$P < 0·001$.
[a]Backfat thickness at level of sacrum and 3rd rib.
From Grandadam *et al.* (1975).

Table 23
Effect of Synovex L on pasture-fed wether lambs

		Dose			
		Control	*75 mg*	*100 mg*	*Overall treatment effect*
Hot carcass (kg)	(1)	13·4	13·9	14·0	*
	(2)	21·8	24·3	23·5	
Fat depth 13th rib[a]	(1)	4·6	4·6	5·4	NS
	(2)	4·8	6·0	4·5	

(1) Age at slaughter—14 weeks.
(2) Age at slaughter—23 weeks.
[a]Fat depth adjusted for hot carcass weight (Bass *et al.* pers. comm., 1986).
*$P < 0{\cdot}05$.
NS = not significant.

results on pasture-fed wether lambs have also been found (Table 23). Synovex increases carcass weight in steers, heifers and wether lambs but seems to have little effect on carcass composition.

Developmental Growth Promoters

Growth Hormone and the Somatotropic Axis

Somatotropin or growth hormone (GH) is one of the most important growth controlling hormones, which decreases fat deposition and increases muscle and bone growth (Fig. 4). GH secretion from the anterior pituitary is under the dual control of a stimulatory growth hormone releasing hormone (GHRH) and an inhibitory peptide, somatostatin (SRIF). These two factors are released from nerve endings in the median eminence into the hypophyseal-portal circulation and interact with the somatotrophs of the anterior pituitary. Both GHRH and SRIF neurons are under the influence of other neurons. GH secretion is also regulated by other feedback mechanisms which probably include the insulin-like growth factors.

The first step in GH action is the binding of the hormone to a cell membrane receptor. The cell surface binding sites for GH are heterogeneous with at least two affinity states for the somatogenic receptor in the liver. The binding capacity for GH of the higher affinity site in the liver is correlated with growth rate in ruminants and this has been related to IGF-I

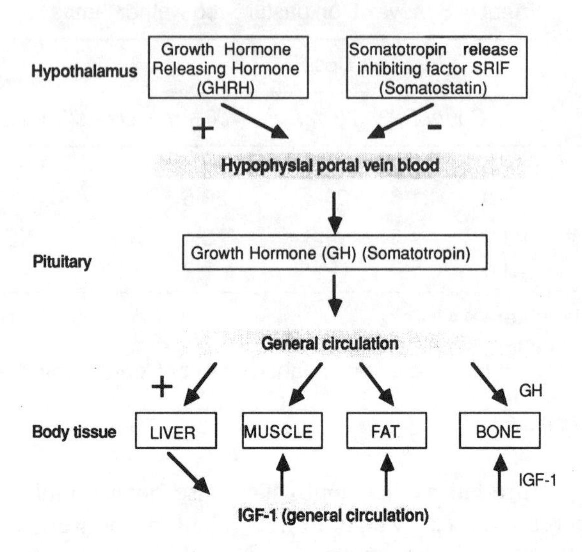

Fig. 4. The somatotrophic axis. The hypothalamic hormones stimulate (GHRH) or inhibit (SRIF) the release of GH from the anterior pituitary gland. The GH circulates through the bloodstream to stimulate tissue growth directly, or indirectly by generating IGF-1 in peripheral tissues.

production. Nutrition is a dominant influence on GH receptors in cattle livers. In ruminants the number and affinity state of the GH receptors varies with nutritional intake. The high affinity GH receptor is not demonstrable at maintenance intake. Oestradiol, a growth promotant in ruminants, has a major effect on the capacity of the GH receptor. The GH receptor is therefore under active endocrine and nutritional regulation. In the ruminant such regulation determines the state of the somatotropic axis and controls the effect of circulating GH on growth and lipolysis.

GH may act directly on tissues such as fat or its effects may be mediated by other hormones such as IGF-I and IGF-II. A major GH-dependent growth factor is IGF-I, which is considered to be an essential component of GH-controlled growth. IGF-I has direct effects on muscle, adipose tissue, and chondrocytes. The GH-stimulated IGF-I is thought to be mainly produced by the liver which is considered the main source of circulating IGF-I. GH also controls IGF-I synthesised locally in a number of other tissues. The relative importance to growth of GH-controlled circulating or locally produced IGF-I is as yet unclear, although locally produced IGF-I can stimulate cell multiplication of differentiating chondrocytes.

Growth hormone has recently been produced by recombinant DNA technology and looks destined to have a major effect on dairy farming by increasing milk production in cows where high energy diets are fed (Fonk *et al.*, 1983). The same product could also have an effect on meat production if long acting delivery systems are developed and they are cost effective for producers using both intensive and extensive farming systems.

Most of the growth trials using either extracted or recombinant GH have been carried out on sheep or pigs. The sheep trials using recombinant GH have shown GH has a limited and variable effect on carcass weight, but consistently depresses fat deposition in the carcass (Table 24).

Other sheep trials have found a similar decrease in fat but a small but significant increase in carcass weight after GH treatment. In other species i.e. pigs (Machlin, 1972; Campbell *et al.*, 1988) and cattle (Moseley *et al.*, 1982), GH treatment has increased growth and decreased carcass fat. When recombinant human growth hormone was administered to pigs there was no effect on carcass composition (Baile *et al.*, 1983) indicating a degree of species specificity for the actions of GH. The major problem associated with GH is the delivery system because of the relatively large daily doses required and the instability of GH in the body. This problem is being addressed by a number of the major pharmaceutical companies,

Table 24
Effect of somatotropin (GH) on carcass weight and shoulder composition of 22-week old lambs (Johnsson *et al.*, 1987)

		Somatotropin mg/kg/d			*Treatment*
	0 (control)	*0·025*	*0·1*	*0·25*	
Hot carcass (kg)	19·5	20·1	18·6	18·1	*
Subcutaneous fat (g)	388	412	350	280	**
Intermuscular fat (g)	700	657	559	534	**
Muscle (kg)	1·38	1·50	1·51	1·54	

*$P < 0.05$.
**$P < 0.01$.

and already long acting delivery systems active for two to three weeks are being tested on meat-producing animals.

The regulation of GH action by pharmacological or immunological means appears a practical route to manipulate growth and carcass composition of ruminants for immediate market requirements, and long term industry needs may be met by early genetic selection for some aspect of the somatotropic axis status.

The GH axis is one of the major growth-controlling axes, and is capable of being manipulated in a number of ways. The commercialisation of these methods will depend to an important extent on consumer attitudes.

Growth Hormone Releasing Hormone (GHRH)

GHRH is a peptide produced by the hypothalamus which stimulates GH secretion as already described (Fig. 4). Results from trials on steers indicate that GHRH can increase lean growth in cattle (Plouzek & Trenkle, 1986) and when used in association with thyroid releasing hormone (TRH) there can be an additive effect on cattle growth.

The lengthy and costly process of producing GHRH limits its present usefulness as an anabolic agent in agriculture. These commercial limitations may be overcome by developing patent synthetic analogues and slow release formulations for use as implants.

Somatostatin (SRIF) Immunisation

Immunisation against somatostatin, a hormone which inhibits GH (Fig. 4), was reported by Spencer (1986) to significantly increase the growth rate of a primitive breed of sheep. This immunological procedure has been repeated by a number of workers (Varner *et al.*, 1980; Bass *et al.*, 1987) although the improvement in weight gain associated with SRIF immunisation reported by Spencer (1986) has not been consistently repeated.

The majority of trials also failed to show a change in carcass composition, although a decrease in perinephric fat weight has been reported in lambs (Table 25).

Although the growth responses of farm animals to SRIF immunisation have been variable in published and unpublished commercial trials, the possibility exists that because the immunological procedures are not standardised it is these which are the cause of the variable responses.

Table 25
Effect of immunisation of lambs against somatostatin

		Hot carcass (kg)	*Perinephric fat (kg)*	*Carcass fat (kg)*
Pasture-fed	Control	22·8	0·92	36·0
	Immunised	21·8	0·67	33·0
Lucerne pellet-fed	Control	30·5	0·93	37·8
	Immunised	32·3	0·82	40·4
	Significance of immunisation	***	NS	NS

Lambs immunised 5 times at 4-weekly intervals (Bass *et al.*, 1986).
$+P < 0{\cdot}1$.
$^{***}P < {\cdot}0{\cdot}001$.

Somatomedins (Insulin-like Growth Factors, IGF)

The anabolic effects of GH are now considered to be mediated, at least in part, by IGFs (Gluckman & Bass, 1987). IGF-I when administered to rats shows ability to restore growth in hypophysectomised and diabetic rats and to stimulate growth in normal rats (Schoenle *et al.*, 1982). However the doses of IGF-I used have been very high; perhaps associated with the circulating IGF-I bound to specific IGF plasma-binding proteins, which are involved in the transport of the IGFs to the tissues. There is also the possibility that IGFs will only affect cells previously potentiated by GH and therefore both hormones are required to achieve a major change in carcass fatness. These problems will have to be resolved before the recombinant IGFs can be used to routinely stimulate lean growth in meat-producing animals.

β-Agonists

The catecholamines, adrenaline and noradrenaline (also called epinephrine and norepinephrine) are produced by the adrenal medulla and the sympathetic branches of the autonomic nervous system. Under stress the hormones are released and stimulate catabolism which mobilises energy. Synthetic compounds with catecholamine-like activity, called β-adrenergic agonists or simply β-agonists have been shown to stimulate lipolysis by binding to the β-receptors in adipose tissue. This repartitions nutrients from fat to muscle by promoting lipolysis and reducing lipogenesis in adipose cells, while at the same time inhibiting protein degradation in muscles.

In domestic animals β-agonists have been shown to decrease fat of poultry (Dalrymple *et al.*, 1983), sheep (Baker *et al.*, 1984*b*), cattle (Ricks *et al.*, 1984) and swine (Jones *et al.*, 1985*b*).

Cimaterol, a β-agonist, developed by the American Cyanamid Co., has been tested on lambs. Carcass fat was reduced and yield of lean meat improved (Table 26).

Table 26
Effects of Cimaterol on lambs (Wolff *et al.*, 1987)

	Control	*Cimaterol (mg/d)*		
		0·9	*1·8*	*Standard error of difference*
Hot carcass (kg)	18·7	20·2	20·8	0·3
Carcass fat gain (g/d)	31	20	15	6
Carcass protein gain (g/d)	10	19	25	3

The catecholamine receptors have been sequenced and the genes coding for them are available for the production of transgenic animals, which may allow the amplification of the catecholamine system without the administration of synthetic β-agonists.

Transgenic Animals

Gene transfer technology may provide agriculture with new products as well as enhancing the efficiency of production of existing products. A substantial number of injected genes have now been expressed in transgenic animals (Wagner, 1987), the growth hormone transgenic mice having elevated GH levels and a dramatic increase in live weight gain. This technology may not only enhance lean production but opens the possibility of protein pharmaceuticals being produced in livestock. The present imprecise gene injection procedures require considerable refinement before the desired production traits are controlled without unexpected side effects. The commercial production of lean carcasses by genetic engineering techniques seems to be some way away in the future.

CONCLUSIONS

This chapter, which has mainly concentrated on ruminants, has described and discussed many of the ways that can be used to reduce carcass fat in meat producing animals. The practical, economic and political implications of the methods described have to be assessed by farmers and meat industries before a particular method can be selected.

A farmer must make immediate decisions on how to manipulate the fat of existing stock and long term decisions on how to make sure that the fatness requirements of future markets can be met. Important factors in these decisions are that an acceptable and reliable method for evaluating fatness in live animals and carcasses is used throughout the industry and that the price paid for the stock reflects a premium for the product required by the market. Without these basics, there is little hope of farmers being able to implement long term policies, which will be financially rewarding and provide the type of meat required by consumers.

The procedures discussed in this chapter which produce significant short term reductions in carcass fat are the traditional methods of breed selection, non-castration of males, and slaughter to light weights which have been and are still acceptable by the animal-conscious city dweller. The use of growth promoters which reduced carcass fatness dramatically in some cases is now being threatened by the 1986 EEC Directive, but more importantly there is real concern by consumers in many countries about any chemical treatments associated with food or animals. The naturally occurring anabolics such as oestradiol, progesterone and testosterone are considered to be safer than the xenobiotics, but even these are now under pressure from consumer lobbies and others. The protein hormones, such as GH may provide a more 'natural' way of reducing fat as they are digestible but this still does not overcome the consumer objection to chemical manipulation of farm animals.

A method acceptable to the consumer for reducing fat is the manipulation of nutrition. Carcass fatness is readily manipulated by altering the nutrition of monogastrics, although it is possible to carry out similar changes in ruminants albeit only under very carefully controlled conditions. Practical farming, especially when stock are extensively grazed, precludes the use of special nutritional regimes to routinely reduce carcass fat of ruminants. The manipulation of growth paths of ruminants has some effect on fatness but the response is often variable. This area requires more detailed investigation if it is to become a practical farming option for extensive as well as intensive farming systems. Farmers can also slaughter at the appropriate live/carcass weight to achieve a particular level of fatness. This option is dependent on the ratio of economic return from carcass weight versus carcass fatness. As carcass weight is often the main factor in payment, reducing carcass fat by slaughtering at lighter weights is often not an economic option and is often not acceptable as a national meat industry strategy, because of the reduction in meat production.

The long term strategies which can be used to reduce carcass fatness are the manipulation of genotype by changing breed type and selection within breed for leanness. Selection schemes for lean breeding stock are a natural and attractive long term option but they require a high level of organisation in the farming industry to enable the implementation of national breeding schemes. These may be difficult to apply in many countries where herd sizes are small and centralised organisation not available. This is particularly so if sophisticated early biochemical traits of leanness such as NADPH/IGF levels etc., are selected for. Centralised testing requires an associated national AI system along with the selection programme so that the lean genotype that has been identified can be rapidly disseminated.

The production of new genotypes by genetic engineering offers tremendous potential for meat industries, but as with all technology this will be restricted to farming industries which disseminate the new genes effectively and make it financially rewarding for genetic engineers and farmers.

One of the main reasons for the success of pig industries in several countries in reducing carcass fat was the realisation that consumers demanded lean meat and as a consequence premiums were paid, providing the incentive for the development of selection systems. Unless the beef and sheep industries also heavily penalise overfat stock, there will be little incentive for farmers to undertake major programmes for reducing fatness of lamb or beef.

REFERENCES

Adu, I. F. & Ngere, L. O. (1979). The indigenous sheep of Nigeria. *Wld Rev. Anim. Prod.*, **15**(3), 51–62.

Ahmad, N. A. & Lloyd Davies, H. (1986). Effect of sex and dietary energy concentration on feed conversion ratio, growth and carcass characteristics in Merino × Border Leicester lambs. *Proc. Aust. Soc. Anim. Prod.*, **16**, 119.

Amegee, Y. (1984). The Vogan sheep (West African dwarf × West African long-legged) in Togo. 2. Carcass quality of unfinished lambs. *Anim. Breed Abstr.*, **53,** Abstract 201.

Anderson, H. R. (1975). The influence of slaughter weight and level of feeding on growth rate, feed conversion and carcass composition of bulls. *Livestock. Prod. Sci.*, **2,** 341–55.

Andrews, R. P. & Orskov, E. R. (1970). The nutrition of the early weaned lamb. II. The effect of dietary protein concentration, feeding level and sex on body composition at two liveweights. *J. Agric. Sci., Camb.*, **75,** 19–26.

Auckland, J. N. & Fulton, R. B. (1973). Effect of feeding restricted amounts of a medium and a high protein diet during the finishing period on growth, fat deposition and feed efficiency of male and female broilers. *J. Sci. Food Agric.*, **24,** 709–17.

Baile, C. A., Della-Fera, M. A. & McLaughlin, C. L. (1983). Performance and carcass quality of swine injected daily with bacterially-synthesized human growth hormone. *Growth,* **47,** 225–36.

Baker, J. F., Long, C. R. & Cartwright, T. C. (1984*a*). Characterization of cattle of a five breed diallel. V. Breed and heterosis effects on carcass merit. *J. Anim. Sci.*, **59,** 922–33.

Baker, P. K., Dalrymple, R. H., Ingle, D. L. & Ricks, C. A. (1984*b*). Use of a β-adrenergic agonist to alter muscle and fat deposition in lambs. *J. Anim. Sci.*, **59,** 1256.

Bartov, I. (1979). Nutritional factors affecting quantity and quality of carcass fat in chickens. *Fed. Proc.*, **38,** 2627–30.

Bartov, I., Bornstein, S. & Lipstein, B. (1974). Effect of calorie to protein ratio on the degree of fatness in broilers fed on practical diets. *Brit. Poult. Sci.*, **15,** 107–17.

Bass, J. J., Carter, A. H., Johnson, D. L., Baker, R. L. & Jones, K. R. (1981). Sire-breed comparison of carcass composition of steers from Angus dams. *J. Agric. Sci.*, **97,** 515–22.

Bass, J. J., Jagusch, K. T., Jones, K. R., Reardon, T. F. & Day, A. M. (1984). Effect of Ralgro on growth, body composition and behaviour of lambs, heifers and bulls. *Proc. NZ Soc. Anim. Prod.*, **44,** 211–13.

Bass, J. J., Gluckman, P. D., Fairclough, R. J., Peterson, A. J., Davis, S. R. & Carter, W. D. (1987). Effect of nutrition and immunization against somatostatin on growth and insulin-like growth factors in sheep. *J. Endocrin.*, **112,** 27–31.

Bass, J. J., Carter, W. D., Duganzich, D. M., Kirton, A. H., Breier, B. H. & Gluckman, P. D. (1989). Effects of oestradiol-17β on growth and insulin-like growth factor of steers on different pasture allowances. *Livestock Prod. Sci.*, **21,** 303–8.

Bennett, G. L. & Clarke, J. N. (1984). Expected selection responses in lamb carcass composition and weight. *Proc. NZ Soc. Anim. Prod.*, **44,** 243–7.

Bennett, G. L., Johnson, D. L. Kirton, A. H. & Carter, A. H. (submitted). Genetic and environmental effects on carcass characteristics of Southdown × Romney lambs. II. Genetic and phenotypic variation. *J. Anim. Sci.*

Berg, R. T. & Butterfield, R. M. (1976). *New Concepts of Cattle Growth.* Sydney University Press, Sydney, 240 pp.

Berg, R. T., Andersen, B. B. & Liboriussen, T. (1978). Growth of bovine tissues. 1. Genetic influences on growth patterns of muscle, fat and bone in young bulls. *Anim. Prod.*, **26**, 245–58.

Bichard, M. (1968). Genetic aspects of growth and development in the pig. In *Growth and Development of Mammals*, ed. G. A. Lodge & G. E. Lamming. Butterworths, London, pp. 309–25.

Black, J. L. (1974). Manipulation of body composition through nutrition. *Proc. Aust. Soc. Anim. Prod.*, **10**, 211.

Black, J. L. (1983). Growth and development of lambs. In *Sheep Production*, ed. W. Haresign. Butterworths, London and Boston, pp. 21–58.

Bond, J., Warwick, E. J., Oltjen, R. R., Putnam, P. A., Hiner, R. L., Kotula, A. W. & Weinland, B. T. (1982). Effects of feeding on growth, composition of gain, carcass quality and mature body size in steers at ages up to six years. *Growth*, **46**, 388–403.

Botkin, M. P., Field, R. A., Riley, M. L., Nolan, J. C. & Roehrkasse, G. P. (1969). Heritability of carcass traits in lambs. *J. Anim. Sci.*, **29**, 251–5.

Bouffault, J. C. & Willemart, J. P. (1983). Anabolic activity of trenbolone acetate alone or in association with oestrogens. In *Anabolics in Animal Production*, ed. E. Meissonier. Office International des Epizooties, Seminar, Feb. 1983, pp. 155–79.

Breier, B. H., Bass, J. J., Butler, J. H. & Gluckman, P. D. (1986). The somatotrophic axis in young steers: influence of nutritional status on pubatile release of growth hormone and circulatory concentrations of insulin-like growth factor I. *J. Endocrin.*, **111**, 209–15.

Butler-Hogg, B. W. (1984). Growth patterns in sheep: changes in the chemical composition of the empty body and its constituent parts during weight loss and compensatory growth. *J. Agric. Sci., Camb.*, **103**, 17–24.

Butler-Hogg, B. W. & Johnsson, I. D. (1986). Fat partitioning and tissue distribution in crossbred ewes following different growth paths. *Anim. Prod.*, **42**, 65–72.

Butler-Hogg, B. W., Francombe, M. A. & Dransfield, E. (1984). Carcass and meat quality of ram and ewe lambs. *Anim. Prod.*, **39**, 107–13.

Butterfield, R. M., Zamora, J., James, A. M., Thompson, J. M. & Williams. J. (1983). Changes in body composition relative to weight and maturity in large and small strains of Australian Merino rams. 2. Individual muscles and muscle groups. *Anim. Prod.*, **36**, 165–74.

Butterfield, R. M., Zamora, J., Thompson, J. M., Reddacliff, K. J. & Griffiths, D. A. (1984). Changes in body composition relative to weight and maturity of Australian Dorset Horn rams and wethers. 1. Carcass muscle, fat and bone and body organs. *Anim. Prod.*, **39**, 251–8.

Cameron, N. D. & Drury, D. J. (1985). Comparison of terminal sire breeds for growth and carcass traits in crossbred lambs. *Anim. Prod.*, **40**, 315–22.

Campbell, R. G. & Dunkin, A. C. (1983). The influence of nutrition in early life on growth and development of pigs. *Anim. Prod.*, **36**, 415–23.

Campbell, R. G. & King, R. H. (1982). The influence of dietary protein and level of feeding on the growth performance and carcass characteristics of entire and castrated male pigs. *Anim. Prod.*, **35**, 177–84.

Campbell, R. G., Taverner, M. R. & Curic, D. M. (1983). The influence of feeding levels from 20 to 45 kg liveweight on the performance and body composition of female and male pigs. *Anim. Prod.*, **36,** 193–9.

Campbell, R. G., Taverner, M. R. & Curic, D. M. (1984) Effect of feeding level and dietary protein content on the growth, body composition and rate of protein deposition in pigs growing from 45 to 90 kg. *Amin. Prod.*, **38,** 233–40.

Campbell, R. G., Taverner, M. R. & Curic, D. M. (1985). The influence of feeding levels on the protein requirements of pigs between 20 and 45 kg liveweight. *Anim. Prod.*, **40,** 489–96.

Campbell, R. G., Steele, N. C., Caperna, T. J., McMurtry, J. P., Solomon, M. B. & Mitchell, A. D. (1988). Interrelationships between energy intake and endogenous growth hormone administration on the performance, body composition and protein and energy metabolism of growing pigs weighing 25 to 55 kg live weight. *J. Anim. Sci.*, **66,** 1643–55.

Charles, D. D. & Johnson, E. R. (1972). Carcass composition of the water buffalo *(Bubalus bubalis)*. *Aust. J. Agric. Res.*, **23,** 905–11.

Cockerill, W. R. (1980). The ascendant water buffalo—key domestic animal. *Wld Anim. Rev.*, No. 33, 2–13.

Cole, W. J., Ramsey, C. B., Hobbs, C. S. & Temple, R. S. (1964). Effects of type and breed of British, Zebu, and dairy cattle on production, carcass composition, and palatability. *J. Dairy Sci.*, **47,** 1138–44.

Copeland, K. C., Johnson, P. M., Kuehl, T. J. & Castracune, V. D. (1984). Estrogen stimulates growth hormone and somatomedin-c in castrate and intact female baboons. *J. Clin. Endocrin. Metab.*, **58,** 698–703.

Cotterill, P. P. & Roberts, E. M. (1976). Preliminary heritability estimates of some lamb carcass traits. *Proc. Aust. Soc. Anim. Prod.*, **11,** 53–6.

Dalrymple, R. H., Ricks, C. A., Baker, P. K., Pensack, J. M., Gingher, P. E. & Ingle, D. L. (1983). Use of the β-agonist clenbuterol to alter carcass composition in poultry. *Fed. Proc.*, **42,** 668 (Abstract).

Davies, A. S. (1983). Growth and development of pigs: a reanalysis of the effects of nutrition on body composition. *J. Agric. Sci., Camb.*, **100,** 681–7.

Davies, G. V. & Erhart, A. B. (1976). Effects of monensin and urea in finishing steer rations. *J. Anim. Sci.*, **43,** 1–8.

Davis, S. L., Ohlson, D. L., Klindt, J. & Antinson, M. S. (1977). Episodic growth hormone secretory patterns in sheep: relationship to gonadal steriod hormones. *Amer. J. Phys.*, E519–E523.

Donaldson, W. E., Combs, G. F. & Romoser, G. L. (1956). Studies on energy levels in poultry rations. 1. The effect of calorie-protein ratio of the ration on growth, nutrient utilisation of body composition of chicks. *Poul. Sci.*, **35,** 1100–5.

Dransfield, E., Nute, G. R. & Francombe, M. A. (1984). Comparison of eating quality of bull and steer beef. *Anim. Prod.*, **39,** 37–50.

Evans, D. G. & Kempster, A. J. (1979). The effects of genotype, sex and feeding regimen on pig carcass development. 1. Primary components, tissues and joints. *J. Agric. Sci.*, **93,** 339–47.

Everitt, G. C. & Evans, S. T. (1970). Classification and grading of beef and veal carcasses. *Proc. NZ Soc. Anim. Prod.*, **30,** 144–60.

Everitt, G. C. & Jury, K. E. (1966). Effects of sex and gonadectomy on the growth and development of Southdown × Romney cross lambs. *J. Agric. Sci.*, **66,** 15–27.

Farid, A., Izadifard, J., Edris, M. A. & Makarechian, M. (1983). Meat from culled old ewes of two fat-tailed Iranian breeds. II-Meat, subcutaneous fat, and bone in the wholesale cuts. *Iran Agric. Res.*, **2**(2), 93–114.
Field, R. A. (1971). Effect of castration on meat quality and quantity. *J. Anim. Sci.*, **32,** 849–8.
Fisher, C. (1984). Fat deposition in broilers. In *Fats in Animal Nutrition,* ed. J. Wiseman. Butterworths, London, pp. 437–70.
Fonk, T. J., Peel, C. J. & Bauman, D. E. (1983). Comparison of different patterns of growth hormone administration on milk production in Holstein cows. *J. Anim. Sci.*, **57,** 699–705.
Foot, J. Z. & Tulloh, N. M. (1977). Effects of two paths of liveweight change on the efficiency of feed use and on body composition of Angus steers. *J. Agric. Sci., Camb.*, **88,** 135–43.
Fortin, A., Reid, J. T., Maiga, A. M., Sim, D. W. & Wellington, G. H. (1980). Effect of energy intake level and influence of breed and sex on the physical composition of the carcass of cattle. *J. Anim. Sci.*, **51,** 331–9.
Fourie, P. D., Kirton, A. H. & Jury, K. E. (1970). Growth and development of sheep. II. Effect of breed and sex on the growth and carcass composition of the Southdown and Romney and their cross. *NZ J. Agric. Res.*, **13,** 753–70.
Fredeen, H. T. (1958). Selection and swine improvement. *Anim. Breed. Abstr.*, **26,** 229–41.
Fredeen, H. T., Martin, A. H. & Weiss, G. M. (1971). Characteristics of youthful beef carcasses in relation to weight, age and sex. II. Carcass measurements and yield of retail product. *Can. J. Anim. Sci.*, **51,** 291–304.
Gimeno, E. J. (1983). The use of hormones in Latin America present situation. In *Anabolics in Animal Production,* ed. E. Meissonnier. Office International des Epizooties, Seminar, Feb. 1983, pp. 531–7.
Gluckman, P. D. & Bass, J. J. (1987). Manipulation of growth in livestock. *The 4th Asian–Australasian Assoc. of Anim. Prod. Soc.*, pp. 119–21.
Gopinath, R. & Kilts, W. D. (1984). Growth hormone secretion and clearance rates in growing beef steers implanted with oestrogenic anabolic compounds. *Growth,* **48,** 499–514.
Grandadam, J. A., Scheid, J. P., Jobard, A., Dreux, H. & Boisson, J. M. (1975). Results obtained with Trenbolone Acetate in conjuction with estradiol 17β in veal calves, feedlot bulls, lambs and pigs. *J. Anim. Sci.*, **41**(3), 969–77.
Griffiths, L., Leeson, S. & Summers, J. D. (1977). Fat deposition in broilers: effect of dietary energy to protein balance, and early life caloric restriction on productive performance and abdominal fat pad size. *Poult. Sci.*, **56,** 538–646.
Heitman, H. & Clegg, M. T. (1957). Subcutaneous implantations in growing-fattening swine. *J. Anim. Sci.*, **16,** 901–10.
Heitzman, R. J. (1979). The efficiency and mechanisation of action of anabolic agents as growth promoters in farm animals. *J. Ster. Biochem.*, **11,** 927–30.
Henderson, W. M. (1983). Summing up In *Anabolics in Animal Production,* ed. E. Meissonnier. Office International des Epizooties, Seminar, Paris, Feb. 1983, pp. 555–8.
Hennessy, D. W., Williamson, P. J., Nolan, J. V., Kempton, T. J. & Leng, R. A. (1983). The roles of energy—or protein-rich supplements in the subtropics for young cattle consuming basal diets that are low in digestible energy and protein. *J. Agric. Sci., Camb.*, **100,** 657–66.

Hetzer, H. O. & Harvey, W. R. (1967). Selection for high and low fatness in swine. *J. Anim. Sci.*, **26,** 1244–51.

Hodges, J. (1987). Evaluation and planned use of temperate and tropical animal genetic resources in developing countries to maximise production. *The 4th Asian–Australasian Assoc of Anim. Prod. Soc.*, pp. 50–3.

Hoffmann, B. (1981). Levels of anabolic sex hormones in farm animals. In *Anabolic agents in Beef and Veal Production,* EEC—Workshop.

Hunter, R. A. & Siebert, B. D. (1980). The utilisation of spear grass *(Heteropogon contortus).* The nature and flow of digesta in cattle fed on spear grass alone and with protein or nitrogen or sulphur. *Aust. J. Agric. Res.*, **31,** 1037–47.

Jackson, S., Summers, J. D. & Leeson, S. (1982). Effect of dietary protein and energy on broiler carcass composition and efficiency of nutrient utilization. *Poultry Science,* **61,** 2224–31.

Jagusch, K. T. & Rattray, P. V. (1979). Carcass production for the consumer. Nutritional manipulation of carcass composition of lambs grown in New Zealand. *Proc. of Lincoln College Farmers Conf.*, pp. 20–30.

Johnsson, D. L., Baker, R. L., Morris, C. A., Carter, A. H. & Hunter, J. C. (1986). Reciprocal crossbreeding of Angus and Hereford cattle. 2. Steer growth and carcass traits. *NZ J. Agric. Res.*, **29,** 433–41.

Johnsson, I. D., Hathorn, D. J., Wilde, R. M., Treacher, T. T. & Butler-Hogg, B. W. (1987). The effects of dose and method of administration of biosynthetic bovine somatotrophin on liveweight gain, carcass composition and wool growth in young lambs. *Anim. Prod.*, **44,** 405–14.

Jones, S. D. M., Burgess, T. D. & Dupchak, K. (1983). Effects of dietary energy intake and sex on carcass tissue and offal growth in sheep. *Can. J. Anim. Sci.*, **63,** 303–14.

Jones, S. D. M., Rompala, R. E. & Jeremiah, L. E. (1985*a*). Growth and composition of the empty body in steers of different maturity types fed concentrate or forage diets. *J. Anim. Sci.*, **60,** 427.

Jones, R. W., Easter, R. A., McKeith, F. K., Dalrymple, R. H., Maddock, H. M. & Bechtel, P. J. (1985*b*). Effect of the β-adrenergic agonist cimaterol (CL 263, 780) on the growth and carcass characteristics of finishing swine. *J. Anim. Sci.*, **61,** 905–13.

Juny, I. & Baulieu, E. E. (1972). Testosterone cytosol 'receptor' in the rat lenator uni muscle. *Nature (New Biol.),* **237,** 24.

Kellaway, R. C. (1973). The effects of plane of nutrition, genotype and sex on growth, body composition and wool production in grazing sheep. *J. Agric. Sci., Camb.*, **80,** 17–27.

Kempster, A. J., Cuthbertson, A. & Harrington, G. (1982*a*). *Carcase Evaluation in Livestock Breeding, Production and Marketing.* Granada, London and New York, 306 pp.

Kempster, A. J., Cook, G. L. & Southgate, J. R. (1982*b*). A comparison of the progeny of British Friesian dams and different sire breeds in 16- and 24-month beef and production systems. 2. Carcass characteristics, and rate and efficiency of meat gain. *Anim. Prod.*, **34,** 167–78.

Kempster, A. J., Cook,G. L. & Southgate, J. R. (1982*c*). A comparison of different breeds and crosses from the suckler herd. 2. Carcass characteristics. *Anim. Prod.*, **35,** 99–111.

Kennedy, B. W., Johansson, K. & Hudson, G. F. S. (1985). Heritabilities and genetic correlations for backfat and age at 90 kg in performance-tested pigs. *J. Anim. Sci.*, **61**, 78–82.

Kiley, M. (1976). A review of the advantages and disadvantages of castrating farm livestock with particular reference to behavioural effects. *Br. Vet. J.*, **132**, 323–31.

Kirton, A. H. (1976). Effect of preweaning plane of nutrition on subsequent growth and carcass quality of lambs. *Proc. NZ Soc. Anim. Prod.*, **30**, 106–15.

Kirton, A. H. (1982). Carcase and meat qualities. In *Sheep and Goat Production*, ed. I. E. Coop. Elsevier Scientific Publishing Company, Amsterdam, pp. 259–74.

Kirton, A. H., Carter, A. H., Clarke, J. N., Sinclair, D. P. & Jury, K. E. (1974). Sires for export lamb production. 2. Lamb carcass results. *Proc. Ruakura Farmers' Conf.*, pp. 29–41.

Kirton, A. H., Sinclair, D. P., Chrystall, B. B., Devine, C. E. & Woods, E. G. (1981). Effect of plane of nutrition on carcass composition and the palatability of pasture-fed lamb. *J. Anim. Sci.*, **52**, 285–91.

Kirton, A. H., Clarke, J. N. & Hickey, S. M. (1982). A comparison of the composition and carcass quality of Kelly and Russian castrate, ram, wether and ewe lambs. *Proc. NZ Soc. Anim. Prod.*, **42**, 117–18.

Kirton, A. H., Winger, R. J., Dobbie, J. L. & Duganzich, D. M. (1983). Palatability of meat from electrically stimulated carcasses of yearling and older entire-male and female sheep. *J. Fd Technol.*, **18**, 639–49.

Koch, R. M., Dikeman, M. E., Lipsey, R. J., Allen, D. M. & Crouse, J. D. (1979). Characterization of biological types of cattle—Cycle 11, III. Carcass composition, quality and palatability. *J. Anim. Sci.*, **49**, 448–60.

Koch, R. M., Cundiff, L. V. & Gregory, K. E. (1982*a*). Heritabilities and genetic, environmental and phenotypic correlations of carcass traits in a population of diverse biological types and their implications in selection programs. *J. Anim. Sci.*, **55**, 1319–29.

Koch, R. M., Cundiff, L. V., Gregory, K. E. & Dikeman, M. E. (1982*b*). Characterization of breeds representing diverse biological types: carcass and meat traits of steers. Beef Research Program Progress Report No. 1, Agricultural Reviews and Manuals, US Department of Agriculture, ARM-NC-21, pp. 13–15.

Kubena, L. F., Chen, T. C., Deaton, J. W. & Reece, F. N. (1974). Factors influencing quantity of abdominal fat in broilers. 3. Dietary energy levels. *Poul. Sci.*, **53**, 974–8.

Lawrence, T. J. L. (1977). The effect of dietary nutrient density on growth in the pig. *Anim. Prod.*, **25**, 261–9.

Ledin, J. (1983). Effect of restricted feeding and realimentation on compensatory growth, carcass composition and organ growth in lambs. *Swedish J. Agric. Res.*, **33**, 175–87.

Lemenager, R. P., Marlin, T. G., Stewart, T. S. & Perry, T. W. (1981). Daily gain, feed efficiency and carcass traits of bulls as affected by early and late dietary protein levels. *J. Anim. Sci.*, **53**, 26–32.

Lewis, K. (1987). The hormonal control of lipid metabolism in sheep. PhD Thesis, University of Waikato.

McClelland, T. H. & Russel, A. J. F. (1972). The distribution of body fat in Scottish Blackface and Finnish Landrace lambs. *Anim. Prod.*, **15,** 301–6.
McClelland, T. H., Bonaiti, B. & Taylor, StC.S. (1976). Breed differences in body composition of equally mature sheep. *Anim. Prod.*, **23,** 281–93.
Machlin, L. J. (1972). Effect of porcine growth hormone on growth and carcass composition of the pig. *J. Anim. Sci.*, **5,** 794–800.
McKenzie, J. R. (1983). Growth promotion systems for sheep and cattle. A Lincoln College (NZ) Foundation study 20/10/83, pp. 5–10.
McMeekan, C. P. (1940). Growth and development in the pig, with special reference to carcass quality characters. II. The influence of the plane of nutrition on growth and development. *J. Agric. Sci., Camb.*, **30,** 387–436.
MacRae, J. C. & Lobley, G. E. (1986). Interactions between energy and protein. In *Control of Digestion and Metabolism in Ruminants,* ed. L. P. Mulligan, W. L. Curovuh & A. Dobson. Prentice-Hall, New Jersey, pp. 367–85.
March, B. E., Macmillan, C. & Chu, S. (1984). Characteristics of adipose tissue growth in broiler-type chickens to 22 weeks of age and the effects of dietary protein and lipid. *Poult. Sci.*, **63,** 2207–16.
Margan, D. E., Faichney, G. J., Graham, N. McC. & Donnelly, J. B. (1982). Digestion of a ground and pelleted diet in the stomach and intestines of young sheep from two breeds. *Aust. J. Agric. Res.*, **33,** 617–27.
Meyer, H. H. D. & Rapp, M. (1985). Estrogen receptor in bovine skeletal muscle. *J Anim. Sci.*, **60,** 294–300.
Michel, G. & Baulieu, E. E. (1983). Mode of action of anabolics. In *Anabolics Animal in Production,* ed. E. Meissonnier. Office International des Epizooties, Seminar, Feb. 1983, pp. 54–64.
Morgan, E. T. (1980). Early protein restriction of the broiler chicken and carcass quality upon later marketing. *Poult. Sci.*, **59,** 378–82.
Moseley, W. M., Krabill, L. F. & Olsen, R. F. (1982). Effect of bovine growth hormone administered in various patterns on nitrogen metabolism in the Holstein steer. *J. Anim. Sci.*, **55,** 1062–70.
Muir, L. A., Wien, S., Duquette, P. F., Rickes, E. L. & Cordes, E. H. (1983). Effects of exogenous growth hormone and diethylstilbestrol on growth and carcass composition of growing lambs. *J. Anim. Sci.*, **56,** 1315–23.
Murray, D. M. & Slezacek, O. (1976) Growth rate and its effect on empty body weight, carcass weight and dissected carcass composition of sheep. *J. Agric. Sci. Camb.*, **87,** 171–9.
Murray, D. M., Tulloh, N. M. & Winter, W. H. (1974). Effect of three different growth rates on empty body weight, carcass weight and dissected carcass composition of cattle. *J. Agric. Sci., Camb.*, **82,** 535–47.
Neumann, F. (1975). *Anabolic Agents in Animal Production,* FAO/WHO, Symposium, Rome, pp. 235–64.
Nik-Khah, N. (1984). The growth and carcass quality of Afshari, Turkey and Mehraban lambs on different diets. *Proc. Aust. Soc. Anim. Prod.*, **15,** 498–9.
O'Donovan, P. B. (1984). Compensatory gain in cattle and sheep. Commonwealth Bureau of Nutr. *Nutr. Ab. Series B,* **54**(8), 384–410.
Ollivier, L. & Sellier, P. (1982). Pig genetics: A review. *Ann. Genet. Sel. Anim.*, **14,** 481–544.
Olsen, R. F., Wangsness, P. T., Martin, R. J. & Gahagan, J. H. (1977). Effects

of zeranol on blood metabolites and hormones in wether lambs. *J. Anim. Sci.*, **45**, 1392–1396.

Orskov, E. R., McDonald, I., Fraser, C. & Corse, E. L. (1971). The nutrition of the early weaned lamb. III. The effect of *ad libitum* intake of diets varying in protein concentration on performance and on body composition at different liveweights. *J. Agric. Sci., Camb.*, **77**, 351–61.

Orskov, E. R., MacLeod, N. A., Fahmy, S. T. M., Istasse, L. & Hovell, F. D. DeB. (1983). Investigation of nitrogen balance in dairy cows and steers nourished by intragastric infusion. Effects of submaintenance energy input with or without protein. *Br. J. Nutr.*, **50**, 99–107.

Parratt, A. C., Burt, C. M., Bennett, G. L., Clarke, J. N., Kirton, A. H. & Rae, A. L. (1987). Heritabilities, genetic and phenotypic correlations for carcass traits and ultrasonic fat depth of sheep. *Proceedings of 5th Conf. Aust. Assn. Anim. Breed. and Genet.*, Perth, p. 41.

Patterson, D. L., Price, M. A. & Berg, R. T. (1985). Patterns of muscle, bone and fat accretion in three biological types of feedlot bulls fed three dietary energy levels. *Can. J. Anim. Sci.*, **65**, 351–61.

Plimpton, R. F. & Teague, H. S. (1972). Influence of sex and hormone treatment on performance and carcass composition of swine. *J. Anim. Sci.*, **35**, 1166–75.

Plouzek, C. A. & Trenkle, A. (1986). Effect of pulsatile intravenous injections of growth hormone-releasing factor on nitrogen retention in calves. *J. Anim. Sci.*, **63** (Suppl. 1), 128.

Potter, E. L., Raun, A. P., Cooley, C. O., Rathmacher, R. P. & Richardson, L. F. (1976). Effect of monensin on carcass characteristics, carcass composition and efficiency of converting feed to carcass. *J. Anim. Sci.*, **43**, 678–83.

Powers, M. L. & Florini, J. R. (1975). A direct effect of testosterone on muscle cells in tissue culture. *Endocrin*, **97**, 1043.

Price, M. A., Jones, D. M., Mathison, G. W. & Berg, R. T. (1980). The effect of increasing dietary roughage level and slaughter weight on the feedlot performance and carcass characteristics of bulls and steers. *Can. J. Anim. Sci.*, **60**, 345–58.

Purchas, R. W. & Lloyd Davies, H. (1974). Meat production of Friesian steers: The effect of intramuscular fat on palatability and the effect of growth rates on composition changes. *Aust. J. Agric. Res.*, **25**, 667–77.

Purchas, R. W., Pearson, A. M., Pritchard, D. E., Hats, H. D. & Tucker, H. A. (1971). Some carcass quality and endocrine criteria of Holstein Heifers fed melengestrolacetate. *J. Anim. Sci.*, **32**, 628–635.

Rao, M. K. & Nagarcenkar, R. (1977). Potentialities of the buffalo. *Wld Rev. Anim. Prod.*, **13**(3), 53–62.

Ravindran, V., Rajamahendran, R., Nadarajah, K. & Goonewardene, L. A. (1983). Production characteristics of indigenous sheep under traditional management systems in northern Sri Lanka. *Wld Rev. Anim. Prod.*, **19**(3), 47–52.

Read, J. L. (1982). Application of crossbreeding of sheep in the United Kingdom. In *Proceedings of the World Congress on Sheep and Beef Cattle Breeding*, Vol. II, ed. R. A. Barton & W. C. Smith. The Dunmore Press Ltd., Chicago, IL, USA, pp. 175–8.

Reid, J. F. S. (1983). Estradiol benzoate implants. *Anabolics in Animal Production,* ed. E. Meissonnier. Office International des Epizooties, Seminar, Feb. 1983, pp. 143–54.
Ricks, C. A., Dalrymple, R. H., Baker, P. K. & Ingle, D. L. (1984). Use of a β agonist to alter fat and muscle deposition in steers. *J. Anim. Sci.,* **59,** 1247–55.
Robertson, I. S. (1966). Castration in farm animals: Its advantages and disadvantages. *Vet. Rec.,* **78,** 130–5.
Rumsey, T. S. (1982). Effect of Synovex-S implants and kiln dust on tissue gain by feedlot beef steers. *J. Anim. Sci.,* **54,** 1030–4.
Ryley, J. W., Moir, K. W., Pepper, P. M. & Burton, H. W. (1970). Effect on hexoestrol implantation and body size on the chemical composition and body components of chickens. *Brit. Poult. Sci.,* **11,** 83–91.
Schoenle, E., Zapf, J. & Humbel, R. E. (1982). Insulin-like growth factor 1 stimulates growth in hypophysectomized rats. *Nature,* **296,** 252–3.
Searle, T. W. & Griffiths, D. A. (1976). The body composition of growing sheep during milk feeding, and the effect on composition of weaning at various body weights. *J. Agric. Sci., Camb.,* **86,** 483–93.
Searle, T. W., Graham, N. McC. & Donnelly, J. B. (1982). The effect of plane of nutrition on the body composition of two breeds of weaner sheep fed a high protein diet. *J. Agric. Sci., Camb.,* **98,** 241–5.
Seideman, S. C., Cross, H. R., Oltjen, R. R. & Schanbacher, B. D. (1982). Utilization of the intact male for red meat production: A review. *J. Anim. Sci.,* **55,** 826–40.
Smith, W. C. & Pearson, G. (1986). Comparative voluntary feed intakes, growth performance, carcass composition, and meat quality of Large White, Landrace and Duroc pigs. *NZ J. Exp. Agric.,* **14,** 43–50.
Smith, W. C., Ellis, M., Clark, J. B. K. & Innes, N. A. (1983). A comparison of boars, gilts and castrates for bacon manufacture. 3. Consumer reaction to bacon from boars and castrates and an economic evaluation of a non-castration policy for bacon manufacture. *Anim. Prod.,* **37,** 17–23.
Soeparno, A. B. Lloyd Davies, H. (1982). The effect of dietary energy concentration of growth and carcass in Daldale wether sheep. *Proc. Aust. Soc. Anim. Prod.,* **14,** 503.
Spencer, G. S. G. (1986). Immuno-neutralization of somatostatin and its effects on animal production. *Domestic Anim. Endocrin.,* **3**(2), 55–68.
Staigmiller, R. B., Bellows, R. A., Short, R. E. & Carr, J. B. (1978). Zeralenone implants in replacement heifers. *J. Anim. Sci.,* **47** (Suppl. 1), 392.
Sully, R. J. & Morgan, J. H. L. (1978). The influence of growth rate and diet on beef carcasses. *Proc. Aust. Soc. Anim. Prod.,* **12,** 234.
Sully, R. J. & Morgan, J. H. L. (1982). The influence of feeding level and type of feed on the carcasses of steers. *Aust. J. Agric. Res.,* **33,** 721–9.
Tatum, J. D. (1981). Is tenderness nutritionally controlled? *Proc. 34th Recip. Meat Conf.,* pp. 65–7.
Thorsteinsson, S. S. & Bjornsson, H. (1982). Genetic studies on carcass traits in Iceland twin ram lambs. I. Estimates of genetic parameters on carcass traits, liveweight at weaning and carcass weight. *Livest. Prod. Sci.,* **8,** 489–505.
Trenkle, A. (1983). Mechanisms of action for the use of anabolics in animals. In

Anabolics in Animal Production, ed. E. Meissonnier. Office International des Epizooties, Seminar, Feb. 1983, pp. 65–71.

Turton, J. D. (1969). The effect of castration on meat production from cattle, sheep and pigs. In *Meat Production from Entire Male Animals,* ed. D. N. Rhodes. J. & A. Churchill Ltd, London, pp. 1–49.

Van Der Wal, P. & Berende, P. L. M. (1983). Effects of anabolic agents on food producing animals. In *Anabolics in Animal Production,* ed, E. Meissonnier. Office International des Epizooties, Seminar, Feb. 1983, pp. 73–115.

Varner, M. A., Davis, S. L. & Reeves, J. L. (1980). Temporal serum concentrations of growth hormone, thyrotropin, insulin and glucagon in sheep immunized against somatostatin. *Endocrin.,* **106,** 1027–32.

Wagner, T. E. (1987). Prospects for the use of mammalian gene transfer in animal agriculture, *The 4th Asian-Australasian Assoc. of Anim. Prod. Soc.* pp. 115–18.

Wainman, P. & Shipounoff, G. C. (1941). The effects of castration and testosterone propionate on the striated perineal musculature in rat. *Endocrin.,* **29,** 975–78.

Walker, D. E. (1950). The influence of sex upon carcass quality of New Zealand fat lamb. *NZ J. Sci. Tech.,* **32A** 31–8.

Wilson, L. L., Varela-Alvarex, H., Rugh, M. C. & Bonger, M. L. (1972). Growth and carcass characters of rams, cryptorchids, wethers and ewes subcutaneously implanted with zeranol. *J. Anim. Sci.,* **34,** 336–8.

Winter, W. H. (1971). A study of weight-loss and compensatory gain in sheep. PhD Thesis, University of Melbourne.

Wolf, B. T. & Smith, C. (1983). Selection for carcass quality. In *Sheep Production,* ed. W. Haresign. Butterworths, London, pp. 493–514.

Wolff, J. E., Dalrymple, R. H. & Ingle, D. L. (1987). Effects of the β-adrenergic agonist cimaterol on carcass composition of lambs grazing pasture. *Proceedings of 4th AAAP Animal Science Congress,* Hamilton, NZ, pp. 490–6.

Wood, J. D. & Riley, J. E. (1982). Comparison of boars and castrates for bacon production. 1. Growth data, and carcass and joint composition. *Anim. Prod.,* **35,** 55–63.

Wood, J. D., Lodge, G. A. & Lister, D. (1979). Response to different rates of energy intake by Gloucester Old Spot and Large White boars and gilts given the same total feed allowance. *Anim. Prod.,* **28,** 371–80.

Wood, J. D., MacFie, H. J. H., Pomeroy, R. W. & Twinn, D. J. (1980). Carcass composition in four sheep breeds: The importance of type of breed and stage of maturity. *Anim. Prod.,* **30,** 135–52.

Zhang, T., Partridge, I. G. & Mitchell, K. G. (1986). The effects of dietary energy levels and protein; energy ratio on nitrogen and energy balance, performance and carcass composition of pigs weaned at 3 weeks of age. *Anim. Prod.,* **42,** 389–95.

Chapter 6

New Approaches to Measuring Body Composition in Live Meat Animals

PAUL ALLEN
National Food Centre, Dunsinea, Dublin, Ireland

INTRODUCTION

There is a general recognition of the need to reduce the average fat content of the carcasses of meat-producing species. This is particularly so in those countries, such as the UK and Ireland, where average fat levels have traditionally been higher than, for instance, the continental countries of Europe. The impetus for this change stems from an increasing awareness among consumers of the relationship between diet and health and in particular the implication of excess consumption of animal fats in the aetiology of coronary heart disease. Consumers may be offered lean meat with the amount of fat they require by trimming prior to the point of sale. But fat trimmings have a considerably lower value than lean meat so over-fat carcasses are less valuable than those with the ideal fat covering.

Body composition of livestock may be assessed *in vivo* either as chemical composition or as the relative proportions of the major tissues. In the first case proximate analysis of protein (nitrogen), water, fat and ash content is often the goal but in some studies the concentrations of certain elements such as nitrogen, calcium, phosphorus, potassium, etc. may be of interest. For studies directly related to meat production, such as breed comparisons and nutrition trials, the relative proportions and often the distribution of adipose tissue, muscle and bone are the basis of assessment. Both approaches are important in the search for methods of modifying the growth patterns of the meat-producing species to reduce the amount of unwanted fat in carcasses and improve the efficiency of lean meat production; for it is only from a knowledge of the underlying mechanisms controlling growth that meaningful progress will be made.

Most of the techniques in use for in-vivo assessment of body composition of livestock are derived from medicine, particularly the rapidly expanding diagnostic field. But the needs of medicine and of animal science are quite different, so that optimal systems for improving the efficiency of meat production are unlikely to emerge from a total reliance on developments in medicine. Taking imaging techniques as an example, a great deal of expensive development effort is devoted to improving the quality of the image. Scientists are more likely to be interested in the raw data used to construct the image and the best way of relating this to differences in composition rather than manipulation of the data to improve image quality, yet the data may not be available to us in the raw form. On a more practical level farm animals cannot be told to remain still and hold their breath during the measuring procedure so that systems which produce the best result under the conditions achievable with a co-operative patient will not necessarily be optimal for use on animals. Of course, one advantage of using medical systems is that rigorous standards of health and safety are applied so animal scientists are unlikely to run into trouble on animal welfare grounds. There is a need then to develop technology specifically for the various needs of animal production and in this respect the development work on the use of ultrasound speed at the Institute of Food Research in Langford, UK, is encouraging.

In the meat production chain there are several stages at which the need for methods of measuring composition in the live animal arises. In order to improve the genetic potential of strains or breeds for lean rather than fat deposition, individuals which are less fat under standard conditions are selected for breeding. The alternative to accurate direct measurement on the individual is the slaughter and dissection of relatives, or the slaughter of the individual itself after extracting and freezing an adequate supply of semen or ova. Both these alternatives are rather costly, and the former also lacks precision unless a large number of near relatives are used.

In most countries, payment on a deadweight basis involves classification or grading, with fat content as a major component of the quality assessment. Producers can therefore maximise their returns by assessing when their animals have reached the fat level which would command the highest payment. Furthermore, by marketing animals at the optimum level of fatness, the producer will also make considerable savings on feed costs.

In animal production research there is also a need for in-vivo methods of estimating body composition. In breed comparison studies, nutrition trials and experiments to assess the effects of treatments such as growth promoters, more information is gained from an assessment of carcass composition at various stages during the fattening period than at a single end point. Serial slaughter and dissection of contemporary groups is the traditional approach in such studies but this requires a large number of animals per trial and is costly in terms of carcass depreciation and labour. An accurate in-vivo method, allowing repeated measurement of composition in the same animals throughout the trial, would reduce the number of animals required and the dissection costs.

Although workers involved with the main meat-producing species share the common aims of reducing carcass fat and improving the efficiency of lean meat production, and all have a need for accurate in-vivo methods of estimating body composition, the possibilities, limitations and cost–benefit relationships vary among the species to the extent that a single method is unlikely to have universal application. For instance, in pigs a relatively high proportion of the total carcass fat is deposited as subcutaneous tissue, which is in a relatively even layer resulting in higher correlations between measurements of this depot and total carcass composition than in sheep or cattle. Pigs also produce larger litters than sheep or cattle which has implications for the cost–benefit analysis of in-vivo methods applied to breeding schemes. Furthermore, within each species a single method is unlikely to be suitable for all applications. The main factors affecting suitability are cost, accuracy and practical considerations such as portability. For instance sophisticated technologies such as computed tomography or nuclear magnetic resonance may be invaluable as research tools but are unlikely to find application at abattoirs or livestock markets due to their complexity.

Live evaluation methods in livestock have been the subjects of two CEC symposia (Andersen, 1982; Lister, 1984) and have also been reviewed by Alliston (1983). In this chapter recent developments in in-vivo methods will be reviewed. Factors affecting suitability of available methods to the various applications within the main meat-producing species will be discussed. Although the main emphasis will be on new techniques, reference will be made to established methods for comparison.

REQUIREMENTS OF METHODS OF MEASURING COMPOSITION IN THE LIVE ANIMAL

General

Different applications within the various meat-producing species will impose their own set of conditions on the suitability of methods of measuring composition in the live animal. Nevertheless, the main criteria involved will be common to all applications. Kempster (1984) has listed these as cost, practicability, precision and accuracy. Few studies aimed at assessing a certain technique or techniques give consideration to all these criteria, yet it is not realistic to examine the criteria in isolation when choosing the best method for a particular application. For instance, the most precise method will not be suitable for on-farm decisions about the optimum time of slaughter if its cost is greater than the potential improvement in returns to the farmer, or if it requires highly skilled operators. The real cost of a technique includes not only the initial cost of the equipment and installation but also the maintenance and running costs.

Even if the more sophisticated and costly techniques such as X-ray computed tomography and nuclear magnetic resonance can be shown to be cost-effective for wide applications such as on-farm selection of breeding stock, they will be unsuitable on grounds of practicability unless mobile systems are made available. The main factors involved in determining the practicability of a technique are portability, speed, physical requirements such as space and services needed and the level of skill required for successful operation. An example of a technique which has enjoyed wide application due to its practicability is simple A-mode ultrasonics which has become a common selection tool in pig breeding.

In most studies aimed at assessing in-vivo techniques precision has been the main concern. When quoted as the coefficient of determination (R^2) this can be misleading as it is dependent upon the variation in the measured trait in the sample. A more useful statistic, particularly for comparisons between different studies, is the residual standard deviation (RSD) for the predicted trait. This is a measure of the unexplained variation in the predicted trait among individuals with the same value for the measured trait or traits. Even so, the RSD quoted in any particular study is applicable to the animals, conditions, equipment, operators, etc. involved in that study. A lower RSD achieved in another study using a different ultrasonic probe, for instance, may not necessarily be due to the higher inherent precision of that probe but may be due to one or more other factors. The robustness of a technique to different operators and a range of conditions is therefore probably more important than precision *per se*, except in very specific applications.

Some authors have approached this problem by using equations derived from one set of animals to predict composition in another group, or in some cases the original sample is divided in two and the equation derived from one half is used as a predictor for the other half. By comparing predicted values with those measured on the same animals directly by dissection or chemical analysis, a measure of the accuracy is obtained as a function of the deviations of predicted values from actual values. When the deviations are large the accuracy is low, and when they tend to be all in the same direction the method is biased. This problem has been discussed at length by Kempster *et al.* (1982)

On-Farm Use

Farmers need techniques for assessing the fatness of their animals in order to market them at the optimum level of fatness, to select breeding animals with the greatest potential for lean production and to ensure females are in the correct body condition at mating and during lactation. All three purposes demand relatively low-cost techniques that are simple and quick to use with a high degree of practicability. Traditional methods of handling animals to assess body condition, particularly for sheep and cattle, are unlikely to be replaced by instrumental techniques. For monitoring the condition of breeding stock and as an aid to marketing animals, simple ultrasonic machines have been used by pig producers and are unlikely to be replaced by more sophisticated technology due to cost and practicability. Their application to sheep and cattle has been slower to develop due mainly to the lower precision found in these species and the need for clipping or shaving with most machines.

Centralised Breeding Schemes

The greater precision that may be achieved with more sophisticated techniques may be cost-effective if benefits are widely spread as in centralised or even national breeding schemes, preferably coupled with artificial insemination. In such circumstances precision may be all-important, especially if the number of animals that can be tested is small.

With a centralised facility portability may not be important and there is the possibility of having skilled operators, paid for out of the charges for the service.

Research

The higher cost of a more precise technique may be justified for research purposes if it replaces existing costly and laborious techniques such as serial slaughter and dissection, or if no other techniques are available for furthering basic knowledge of the physiology of growth, for instance. For basic research, cost effectiveness is much harder to establish than in more applied fields such as animal breeding and nutrition. Precision and accuracy will be more important than cost and practicability in basic research and applied research based at a research institute.

TECHNIQUES FOR MEASURING COMPOSITION IN THE LIVE ANIMAL

Ultrasound Pulse-Echo Techniques

Ultrasound waves are sound waves with frequencies above the range of the human ear (i.e. above 16-18 000 cycles/second). They can be propagated through solids, liquids and gases and behave in a similar fashion to light waves in that they display both refraction and reflection at boundaries between substances of different acoustic density. The major tissues and organs of the body have characteristically different acoustic densities so that echoes are produced from interfaces between skin, fat, muscle and the major organs. From a knowledge of the velocity of ultrasound in animal tissues it is therefore possible to map and range major tissues relative to the skin by propagating a beam of ultrasound and receiving the echoes.

Using these principles a whole range of medical diagnostic machines has been developed since the 1940s, some of which have been used to measure backfat and eye muscle dimensions in livestock while a few have been purpose built for these applications.

The principles of ultrasonic techniques have been described by Miles (1978) and their applications to predicting composition in livestock have been reviewed by Stouffer and Westervelt (1977), Simm (1983) and

Thwaites (1984). Pulse-echo techniques developed for medical diagnostic purposes have been applied to the evaluation of carcass composition in livestock since the 1950s. Initially, the simple A-mode machines (where echo-amplitude is displayed against time on an oscilloscope screen) were used to take linear fat depth mesurements and are still in use today. The more complex and costly B-mode scanners which display echoes as bright spots on a cathode-ray tube have also been evaluated. These have been developed to produce two-dimensional scans either by movement of the probe along a curved track (as in the Scanogram) or by firing an array of transducers in sequence (as in the Danscanner). From the two-dimensional scans, eye muscle and fat areas in addition to linear fat depths may be calculated, with the potential to improve the prediction of compositional traits.

Results from Pulse-Echo Techniques in Pigs

The most recent review of the application of pulse-echo techniques to prediction of pig carcass traits has been given by Thwaites (1984), who came to the conclusion that the differences in predictive ability between the several instruments that had been evaluated were small despite considerable variation in complexity and cost. Table 1 summarises the results of trials where the relationship between measurements taken ultrasonically on the live pig and by caliper or probe on the carcass has been established. The two trials presented here which were not available when Thwaites reviewed the literature do not contradict his conclusion that increased cost and complexity are not necessarily reflected in improved accuracy. Hudson and Payne-Crostin (1984) found that out of five machines tested, a simple scanner (Ilis Observer) was less accurate at predicting carcass backfat than A-scope probes. While Greer *et al.*, (1987) found the Meritrorious a simple A-mode machine, to have a lower RSD for the prediction of P_2 backfat than four digital-display units.

For backfat depths, RSD values in the range 2–2·5 mm appear to be achievable with all machines. For muscle depths and areas the errors are larger for all machines with RSD values in the range 3·3–4·3 mm for muscle depth and 3·3–5·1 mm^2 for muscle area.

For most applications, the utility of ultrasonic machines lies in their ability to predict carcass lean content from a combination of liveweight and ultrasonic measurements rather than to predict carcass measurements *per se*. Table 2 summarises the results of a number of studies where the accuracy of such prediction equations has been reported. The lower

Table 1
Prediction of carcass measurements from live measurements with different ultrasonic machines (pigs)

Reference	*Instrument*	*Measurement*[a]	*RSD*	*% reduction*[b] *in RSD*	R^2
1. Kempster *et al.* (1979)	Sonatest	FD	2·48–2·59	52–60	—
		LDD	4·34	24	—
	Scanogram	FD	1·70–2·66	52–71	—
		LDA	4·39–4·63	27–31	—
		LDD	3·45–4·08	33–39	—
	Danscanner	FD	2·05–2·87	51–61	—
		LDA	5·06	25	—
		LDD	3·34	41	—
	Ilis	FD	1·83–2·87	57–68	—
		LDA	4·80	33	—
		LDD	3·71	35	—
2. Alliston *et al.* (1982)	Sonatest	FD	1·75–2·23	32–42	—
		LDD	3·82	1	—
	Scanogram	FD	1·87–2·06	33–38	—
		LDA	3·30	8	—
		LDD	3·51	9	—
	Danscanner	FD	1·81–2·02	34–40	—
		LDA	3·50	2	—

3. Hudson & Payne-Crostin[c] (1984)	Sonatest	FD	2·31	35	0·58
	Sonalyser	FD	2·34	34	0·58
	Renco	FD	2·39	33	0·56
	Medata	FD	2·52	29	0·51
	Ilis	FD	2·68	25	0·44
4. Giles *et al.* (1981)	Sonatest	FD	—	—	0·82
	Scanoprobe	FD	—	—	0·56–0·65
	Scanogram	FD	—	—	0·88–0·89
		LDA	—	—	0·57–0·59
5. Mersmann (1982)	Scanogram	FD	—	—	0·04–0·92
		LDA	—	—	0·14–0·58
6. Greer *et al.*[c] (1987)	Meritronics	FD	1·32	47–61	0·70–0·85
	Sonalyser	FD	1·34	55	0·81
	Renco	FD	1·55	50	0·75
	New Medata	FD	1·66	28	0·47
	Old Medata	FD	2·40	20	0·37

[a] F = fat; LD = *m. longissimus dorsi*; D = depth (cm); A = area (cm^2) in this and subsequent tables.

[b] $$\frac{\text{(Sample standard deviation—RSD from prediction)}}{\text{Sample standard deviation}} \times 100$$

[c] Data refer to prediction of backfat measured with intrascope.

Table 2
Prediction of percentage lean with different ultrasonic machines (pigs)

Reference	*Instrument*	*Measurement*[a]	*RSD*	*% reduction*[b] *in RSD*	R^2
1. Kempster *et al.* (1979)	Sonatest	FD	2·72	24	—
	Scanogram	FD	2·18–2·56	31–38	—
		LDA	3·26–3·38	1–11	—
	Danscanner	FD	2·03	38	—
		LDA	3·35	0	—
	Ilis	FD	2·61	29	—
		LDA	3·40	7	—
2. Alliston *et al.* (1982)	Sonatest	FD	1·29	20	—
	Scanogram	FD	1·35	17	—
		LDA	1·62	0	—
	Danscanner	FD	1·33	18	—
		LDA	1·62	0	—
3. Sather *et al.* (1982)	Krautkramer	FD	1·72–1·73	26–27	0·39–0·40
	Scanoprobe	FD	1·71–1·73	25–27	0·39
4. Molenaar (1984)	Kontron	FD	0·84–1·86	27–56	0·47–0·83
		FD+LDA	0·87–1·50	41–55	0·74–0·87
		FD+LDA+LW	0·87–1·45	43–55	0·78–0·91

5. Molenaar (1985)	Pie-Data	FD+LDA +LW+DG+FI	1·33–1·42	41–46	0·67–0·72
	Renco	FD	—	—	0·30–0·55
6. Kanis *et al.* (1986)	Renco and Krautkramer[c]	FD	1·77–2·74	17–43	0·33–0·68
7. Morel & Gerwig (1985)	Krautkramer	FD	—	—	0·27–0·84
8. Metz *et al.* (1984)	Not known	FD	—	—	0·32–0·58
9. Stern *et al.* (1984)	Not known	FD	—	—	0·65
10. Sather *et al.* (1987)	Krautkramer	FD	1·66–2·24	7–39	0·17–0·64
	Scanoprobe	FD	1·63–2·23	7–41	0·18–0·67
	Renco	FD	1·62–2·21	8–41	0·19–0·67

[a] LW = live weight; DG = daily gain; FI = food intake.

[b] $\frac{\text{(Sample standard deviation} - \text{RSD from prediction)}}{\text{Sample standard deviation}} \times 100$

[c] Results not presented separately for the two machines.

accuracy in determining muscle areas is reflected in their poor contribution to reducing the standard deviation of lean percent.

RSD values for percentage lean predicted from one or more backfat depths have been reported in the range 0·8–2·7 with the best results achieved with the most recently developed real-time scanners (Molenaar, 1984, 1985). In the second of these studies a simple A-mode machine (Renco LM) was included for comparison. Higher correlations between lean percent and backfat depth were recorded at some locations with the real-time scanner than at a single location measured with the simpler machine. However, the improvement was not large despite the considerable increase in complexity and cost.

Results from Pulse-Echo Techniques in Cattle

The use of ultrasound to measure carcass traits in live cattle has developed more slowly than in pigs. Simm (1983) has given a comprehensive review of the literature. He concluded that marked improvements in accuracy are unlikely since correlations between ultrasonic measurements and carcass composition are often as high as those between the same measurements taken on the carcass and its composition. Tables 3 and 4 summarise some of the more recent trials and confirm that there has been no great increase in accuracy. RSD values for fat depths have been reported in the range 2·0–4·2 mm (Table 3) (Upton *et al.*, 1984; Eveleigh *et al.*, 1985) while correlations between measurements taken on the live animal and on the carcass are extremely variable in the range 0–0·86 for subcutaneous fat depth, 0·28–0·71 for subcutaneous fat area and 0·1–0·71 for *longissimus thoracis et lumborum* area. In all cases these are not higher than the best correlations already reported (Simm, 1983).

Much of this variation is undoubtedly due to the use of different machines and measuring sites and the degree of experience of operators and interpreters in addition to variation between the groups of cattle used. RSD values for percentage lean in carcass are in the range 1·78–3·9% (Table 4) with R^2 values for models including liveweight and ultrasonic measurements in the range 0–0·75. Where different machines have been compared in the same trial, differences between them in the accuracy of prediction of either carcass measurements or percentage lean have, with a few exceptions (notably the Bruel and Kjaer machine in the study by Andersen *et al.*, 1983), been small.

Table 3
Prediction of carcass measurements with different ultrasonic machines (cattle)

Reference	*Instrument*	*Measurement*	*RSD*	*% reduction*[a] *in RSD*	R^2
1. Tong *et al.* (1981)	Krautkramer	FD	—	—	0·18–0·35
	Scanogram	FD	—	—	0·14–0·37
2. Andersen *et al.* (1983)	Scanogram	FD	—	—	0·11–0·52
		FA	—	—	0·16–0·49
		LDA	—	—	0·02–0·29
	Danscanner	FD	—	—	0 –0·42
		FA	—	—	0·08–0·50
		LDA	—	—	0·17–0·46
	Philips	FD	—	—	0·05–0·25
		FA	—	—	0·14–0·46
		LDA	—	—	0·23–0·45
	Ohio	FD	—	—	0·06–0·31
		FA	—	—	0·10–0·46
		LDA	—	—	0·01–0·50
	Bruel & Kjaer	FD	—	—	0·01–0·36
3. Upton *et al.* (1984)	Scanoprobe	FD	2·7–4·2	40–48	0·53–0·74
4. Eveleigh *et al.* (1985)	Scanoprobe	FD	2·0	41	0·66
	Meritronics	FD	2·2	36	0·59
	Krautkramer	FD	2·8	19	0·35
5. Henningsson *et al.* (1986)	Danscanner	FD	—	—	0·20
		LDA	—	—	0·44

[a] $\frac{\text{(Sample standard deviation} - \text{RSD from prediction)}}{\text{Sample standard deviation}} \times 100$

Table 4
Prediction of percentage lean with different ultrasonic machines (cattle)

Reference	*Instrument*	*Measurement*	*RSD*	*% reduction*[a] *in RSD*	R^2
1. Kempster *et al.* (1981)	Sonatest	FD	2·5 –3·9	12–25	—
	Scanogram	FD	2·3 –2·5	—	—
		FA	2·0 –2·3	—	—
		LDA	2·9	—	—
2. Tong *et al.* (1981)	Krautkramer	FD	2·1 –2·7	4–7	0·56–0·75
	Scanogram	FD	2·60–2·66	8–12	0·75
3. Andersen *et al.* (1983)	Scanogram	FD	—	—	0·01–0·35
		FA	—	—	0·04–0·37
		LDA	—	—	0·03–0·07
	Danscanner	FD	—	—	0·10–0·34
		FA	—	—	0·08–0·37
		LDA	—	—	0 –0·16
	Philips	FD	—	—	0·13–0·27
		FA	—	—	0·18–0·36
		LDA	—	—	0·02–0·08
	Ohio	FD	—	—	0 –0·42
		FA	—	—	0 –0·48
		LDA	—	—	0 –0·14
	Bruel & Kjaer	FD	—	—	0·02–0·14
4. Jansen *et al.* (1985)	Danscanner	FA+LDA	1·78	6	0·16
5. Henningsson *et al.* (1986)	Danscanner	FD	—	—	0·02
		LDA	—	—	0
6. Bailey *et al.* (1986)	Danscanner	FA	—	—	0·03–0·13
		LDA	—	—	0 –0·07
		FA+LDA	—	—	0·07–0·14

[a] $\frac{\text{(Sample standard deviation—RSD from prediction)}}{\text{Sample standard deviation}} \times 100$

Results from Pulse-Echo Techniques in Sheep.

The application of pulse-echo techniques to carcass evaluation in live sheep has been reviewed by Alliston (1983). Since good acoustic contact is essential, a strip of wool has to be clipped at the scanning site for most machines, which devalues the fleece. However, a machine with a small (6 mm) diameter transducer has been reported to give sufficient acoustic contact by simply parting the wool (Gooden *et al.*, 1980). Furthermore, they report correlations between backfat depth measured ultrasonically and on the cut carcass in the range 0·72–0·91, considerably higher than those reported by Fortin and Shrestha (1986) for the Krautkramer and Scanogram equipment (Table 5). The generally lower precision of the fat depth measurements in lambs has resulted in a greater concentration on fat and muscle areas and muscle depth as predictors of composition. Correlations with actual carcass measurements for these traits are in the range 0·52–0·75 with RSD values of percentage lean in the range 1·8–3·5% (Table 6). Where more than one machine has been included in a trial, differences in precision between them have been small (Tables 5 and 6). However, Simm (1986) found that two real-time scanners (Vetscan and Technicare) were more precise than the Scanogram which was purpose-built for taking fat and muscle measurements.

Velocity of Ultrasound

Miles and Fursey (1974) first reported an ultrasound technique which overcomes some of the problems in applying pulse-echo techniques to predicting carcass compositional traits in live sheep and cattle. Compared with pigs, sheep and cattle deposit relatively less of their total fat subcutaneously and relatively more in the intermuscular fat depot. This reduced part–whole relationship is probably the main reason for the lower correlations found in sheep and cattle between subcutaneous fat depth and area measurements and carcass lean or fat content. This part–whole relationship can be improved by interrogating the inter- and intramuscular depots in addition to the subcutaneous fat by propagating a wave of ultrasound completely through the body and measuring its velocity. The principle has been described in detail by Miles *et al.*, (1984), but briefly, at body temperature ultrasound waves travel slower through adipose tissue than through lean. The time taken to travel a known distance through a mixture of lean and fat is therefore related to the proportions of the two

Table 5
Prediction of carcass measurements from live measurements with different ultrasonic machines (lambs)

Reference	*Machine*	*Measurement*	*RSD*	*% reduction[a] in RSD*	R^2
1. Gooden *et al.* (1980)		FD	0·38–1·16	31–59	0·72–0·91
2. Alliston *et al.* (1981)	Scanogram	FA	1·06	24	0·68
		LDA	1·41	13	0·62
	Danscanner	FA	1·26	10	0·75
		LDA	1·58	2	0·71
3. Fortin & Shrestha (1986)	Krautkramer	FD	—	—	0·05–0·21
		LDD	—	—	0·55–0·69
	Scanogram	FD	—	—	0·11–0·27
		LDD	—	—	0·52–0·58
		LDA	—	—	0·53–0·61

[a] $\frac{\text{(Sample standard deviation} - \text{RSD from prediction)}}{\text{Sample standard deviation}} \times 100$

Table 6
Prediction of lean percentage with different ultrasonic machines (lambs)

Reference	*Machine*	*Measurement*	*RSD*	*% reduction*[a] *in RSD*	R^2
1. Alliston *et al.* (1981)	Scanogram	FA	3·20	17	—
	Danscanner	FA	3·50	9	—
2. Cameron & Smith (1985)	Danscanner	FD	2·2–2·5	0–12	—
		FA	2·1–2·5	0–15	—
	Vetscan	FD	1·8–2·2	13–29	—
3. Leymaster *et al.* (1985)	Scanogram	FD	0·25[b]		0·23–0·26[c]
4. Fortin & Shrestha (1986)	Krautkramer	FD	2·16–2·19	—	0
		LDD	2·01–2·16	—	0
	Scanogram	FD	2·14–2·17	—	0
		LDD	2·09–2·17	—	0
		LDA	2·12–2·16	—	0
5. Simm (1986)	Scanogram	FA	2·63	14	0·30
	Technicare	LDA+FA	2·38	22	0·44
	Vetscan	LDD+FD	2·44	20	0·41

[a] $\frac{\text{(Sample standard deviation—RSD from prediction)}}{\text{Sample standard deviation}} \times 100$

[b] kg protein.
[c] Variation accounted for after removal of variation due to liveweight.

tissues in the line of flight. The equipment consists of a transmitter and receiver held facing each other by a steel frame. The distance between the two is adjustable to accommodate various animal sizes. Having selected a site where the passage of the wave will be unimpeded by bone, for instance across the hindlimb, an ultrasound pulse is propagated and the time taken to travel the known distance between the two transducers is recorded electronically and the reciprocal of the velocity is computed and displayed.

Results from Velocity of Ultrasound

The method has been shown to be highly repeatable in cattle. Within-animal standard deviations of the reciprocal of the speed of ultrasound have been reported in the range 0·0042–0·0069, representing about 0·06–0·11% of the measured values or, in terms of the predicted lipid content averaged over two sites, 4·3–7·8 ml lipid/litre tissue (Miles *et al.*, 1984).

The precision of this technique was compared with a visual assessment of fatness in the carcass as a predictor of composition by Miles *et al.*, (1983*a*). Measurement of the speed of ultrasound at two locations in the hind limb was almost as good a predictor of carcass composition as the mean fat score given by two experienced judges (Table 7), and was more highly correlated with total fat content of the carcass than was the mean of three fat depths taken on the carcass over *M. longissimus dorsi* (r = 0·60). Unfortunately, the authors do not report whether the ultrasonic measurement improved on the precision of the visual assessment when the two were combined in a single equation. In another study, however, measurements of the speed of ultrasound at two sites gave an RSD for adipose tissue proportion in the side comparable to that achieved with the

Table 7
Comparison of velocity of ultrasound and visual assessment as predictors of beef carcass composition (Miles *et al.*, 1983*a*)

Carcass component (g/kg)	*Velocity of ultrasound*		*Visual score*	
	Correlation	*RSD*	*Correlation*	*RSD*
Lean	−0·78	16·1	−0·84	14·1
Total fat	0·80	19·6	0·89	14·8
Saleable meat	−0·52	10·8	−0·65	9·6

Scanogram pulse-echo technique (0·0227 compared to 0·0223) and a further significant reduction in the RSD (to 0·020) was achieved when the two were used together to predict fatness (Miles *et al.*, 1983*b*)

In a study to compare the precision of three ultrasound scanners, X-ray computed tomography (CT) and velocity of ultrasound for predicting carcass composition, Simm (1986) found that measuring the speed of ultrasound through the hind leg gave as precise a prediction of carcass lean weight and almost as precise a prediction of carcass lean percentage as two B-mode real-time scanners (Technicare and Vetscan) (see Table 8). The purpose built animal scanner included in the study (Scanogram) was less precise than the velocity of ultrasound technique.

Table 8
Precision of predicting carcass lean weight or percentage from live animal measurements (n = 50) (Simm, 1986)

	Carcass lean weight		*Carcass lean percentage*	
Original SD	0·91 kg		3·05	
	RSD	R^2	*RSD*	R^2
LW+age	0·62	0·55	2·90	0·13
+linear and area measurements taken[a] by:				
Velocity of sound	0·57	0·64	2·50	0·37
Scanogram	0·61	0·58	2·63	0·30
Technicare	0·58	0·63	2·38	0·44
Vetscan	0·58	0·63	2·44	0·41
X-ray CT	0·50	0·73	2·00	0·60

[a] Not necessarily the same measurements included in equations to predict carcass lean weight and percentage.

X-Ray Computed Tomography (CT)

A detailed description of the principles of X-ray computed tomography may be found in textbooks (e.g. Hamilton, 1982). In brief, CT makes use of the differential that exists between the rates at which the major tissues of the body attenuate X-rays in much the same way as conventional film-based radiography. In CT however, a two-dimensional image is produced by a 360° rotation of an X-ray source around the patient. An arc-array of highly sensitive detectors measures the attenuation of the radiation beam as they rotate in synchrony with the source. During the rotation pulses of radiation are fired at discrete intervals (usually every degree or half-degree of rotation) and each detector signals to the computer the amount of radiation received. The computer processes the attenuation data from the large number of crossing pathways (512×360 = 184 320 for a system with 512 detectors firing every degree of a full rotation) to resolve a matrix of attenuation values for the target body. The matrix is displayed as an image on a monitor using either a grey-scale or colours, and may be stored on tape or disc for later recall of the image for further inspection or analysis.

Attenuation values (also known as CT numbers) are measured in Hounsfield Units (HU) after one of the joint recipients of the Nobel Prize for their contribution to the development of CT (Cormack, 1980; Hounsfield, 1980). By convention, an arbitrary scale of CT numbers is used such that air and pure water have the values −1000 and 0 HU respectively. Fatty tissue and muscle typically have values in the ranges −120 to −20 and 40 to 160 HU respectively, reflecting variation in chemical composition within and between these major tissue components; nevertheless there is good discrimination between the soft tissues and organs. Bone may vary from 300 HU for soft to 1000 HU for hard bone.

Measurements which may be taken directly using standard software include the linear distance between two points (e.g. fat depth, muscle depth), areas, the CT number at any point or the mean CT number within a region of interest. These may all be of interest and may be useful predictors of compositional traits but far more potentially useful information is available in the matrix of CT numbers. For instance, from studying frequency distribution histograms of the CT number matrices of fat and lean animals there is a clear relationship between the relative proportions of CT numbers falling within the fat and lean ranges and the fat : lean ratio of the animal. Most of the work published on the use of CT for predicting compositional traits in the live animals has focused on the optimal use of this large amount of data.

All available machines have been designed for use in human diagnostics. The patient is laid on a table which is then moved through an annular aperture housing the radiation source and detectors. The size of this aperture restricts the technique to small- to medium-sized animals such as poultry, goats, sheep and pigs. Even if the financial incentive existed for a larger purpose built machine to scan cattle there are technical barriers to such a scaling-up. As it is important to reduce movement during scanning to an absolute minimum, all animals have to be restrained in a cradle, and pigs must be anaesthetised, thereby reducing the speed of throughput. A typical scanning procedure begins by transporting the animal through the aperture with the X-ray source stationary, but firing constantly. This results in a topogram, a longitudinal image of the body in which the skeleton is readily identifiable. By this method anatomical locations for tomograms (slices through the body) may be located precisely. Once a procedure has been established requiring say two tomograms, 20–30 sheep or 10–15 pigs per hour could be scanned.

Radiation is, of course, highly dangerous, but since the equipment has been designed for use on humans the dose levels are so low as to be harmless even for breeding animals. Operators are protected by having the operating console in a room separate from the lead-shielded room housing the scanner.

Results from CT Scanning of Pigs

The potential of CT for measuring composition of farm livestock was first reported by Skjervold and his co-workers (Skjervold *et al.*, 1981). Having scanned 40 pig carcasses and established high correlations betwen the distribution of CT numbers and chemical composition, they scanned 23 live pigs which were then slaughtered and analysed for fat, protein, water and energy content of the whole carcass and of a 1 cm thick slice at the last rib. A high proportion of the variation in composition of both the slice and the carcass was explained by information from a single scan (Table 9), even though this included attenuation data attributable to non-carcass parts such as intestines and internal organs. The considerable variation in composition in this relatively small sample would inflate the R^2 values however.

Vangen (1984) reported a trial to calibrate a Siemens Somatom 2 whole body scanner for the prediction of compositional traits in boars and gilts. Ultrasonic backfat depth was also recorded on the boars for comparative purposes and the results are presented in Tables 10 and 11. Inclusion of backfat depths with weight only marginally increased the R^2 and due to the larger number of parameters in the model the RSD was actually increased (Table 10). Dividing the CT number range −200 to +199 HU (equivalent to the soft tissues) into intervals of 10 HU, enabled the frequencies of CT numbers in these 40 intervals to be used in addition to

Table 9
Prediction of carcass composition and energy content from the distribution of CT numbers (from Skjervold *et al.*, 1981)

	R^2 values for prediction of			
	% fat	*% protein*	*% water*	*Energy content*
Slice	0·89	0·80	0·85	0·85
Whole carcass	0·89	0·83	0·82	0·85

Table 10
Prediction of carcass composition and energy content from the distribution of CT numbers or from ultrasonic backfat depth (from Vangen, 1984)

	R^2 (RSD) for estimation of					
Dependent variables	*d.f.*[a]	*Protein (kg)*	*d.f.*	*Fat (kg)*	*d.f.*	*Energy content (MJ/kg)*
Weight (W)	1	0·40 (0·72)	1	0·04 (1·79)	1	0·00 (69·1)
W+one scan	41	0·76 (0·65)	41	0·90 (0·79)	41	0·88 (31·6)
W+two scans (only significant contributions)	16	0·75 (0·52)	18	0·95 (0·47)	16	0·92 (21·3)
W+backfat depth	6	0·42 (0·74)	6	0·24 (1·61)	6	0·25 (60·2)

[a]d.f. = degrees of freedom.

liveweight as predictors of protein, fat and energy content of the carcass. A single scan raised the R^2 value compared with prediction from weight alone for each trait. RSD values were reduced by using information from the scans but only including predictors which made a significant contribution to the model. A greater proportion of the variation in fat and energy content than that in protein content was explained by CT information. Table 11 shows that a very high proportion of the variation in compositional traits unexplained by weight, or by ultrasonic back fat depth was explained by models including CT information.

One criticism of this approach where a large number of variables are derived from the frequency distribution data is that it is likely to lead to

Table 11
Explanation of the residual variation in carcass composition and energy content by the distribution of CT numbers (from Vangen, 1984)

	Proportion of residual variation explained for:		
	Protein (kg)	*Fat (kg)*	*Energy (MJ/kg)*
In addition to weight:			
One scan	0·47	0·78	0·75
Two scans	0·58	0·95	0·92
Best equation	0·70	0·96	0·98
In addition to ultrasonic backfat depth:			
One scan	0·45	0·87	0·84

complex models with a relatively large number of predictors. Furthermore, differences between possible predictor variables in their contribution to increasing the R^2 or reducing the RSD of a prediction equation are small so that the actual selection of predictor variables is almost arbitrary. This might be expected to lead to prediction equations highly specific to the conditions applying to the data set from which they were derived. In other words they would lack robustness. Vangen (1985) attempted to resolve this by dividing the data set in two and using models derived from one half to predict composition in the others (cross-validation technique). He compared the RSD obtained in deriving the models from the one half of the data with the RSD obtained in applying the models to the other half of the data. These RSDs, called the standard error of estimate (SEE) and the standard error of prediction (SEP) respectively, are shown in Table 12. Prediction errors compare favourably with errors of estimation for both protein and fat with the best model in each case having as many as 40 degrees of freedom. For predicting weight of fat in the carcass the errors were more than halved in models with 40 or more degrees of freedom compared with that having only 5. Although these may appear cumbersome they would pose no problem if suitable software is developed to process the data directly from the image matrix.

A better test of stability or robustness is to test prediction equations on data sets totally unrelated to those from which the equations were derived. A report of a trial where this has been done gives some encouragement (Vangen *et al.*, 1984). Equations from the calibration trial referred

Table 12
Standard errors of estimation (SEE) and standard errors of prediction (SEP) for different regression models in prediction of carcass protein and fat content (from Vangen, 1985)

		Degrees of freedom for model						
		5	*10*	*15*	*20*	*30*	*40*	*50*
Protein (kg)	SEE	0·59	0·55	0·48	0·48	0·40	0·31	0·24
	SEP	0·59	0·59	0·57	0·53	0·55	0·51	0·59
Fat (kg)	SEE	1·06	0·93	0·83	0·75	0·60	0·42	0·35
	SEP	1·28	1·07	0·93	0·86	0·81	0·58	0·65

to earlier were used to predict composition of 18 Norwegian Landrace pigs and 19 pigs from a line selected for high backfat thickness and low rate of gain (Table 13). Unfortunately, the numbers within each group were quite small, but in experiments 1 and 3, mean predicted values for both protein and fat content were generally lower than, but quite close to, values determined by chemical analysis. In experiment 2, however, where the pigs were of lower weight than those from which the equations were derived, the bias was considerable with lower predicted values for both protein and fat content. This suggests a need for specific equations for different weight ranges probably arising from changes in the chemical composition of the tissues during growth.

One advantage of CT over ultrasonic techniques is that information about the composition of individual tissues may be obtained. Allen and Vangen (1984) investigated the utility of predicting the composition of samples of subcutaneous fat, *M. longissimus dorsi* and a mixed lean and fat sample (bacon side) from the mean CT number of equivalent areas identified on a scan image. In an attempt to compare the predictive ability of CT data with measurements that are commonly taken with an ultrasonic scanner, *M. longissimus dorsi* and the area of subcutaneous fat over this muscle were measured from the scans. The results (Table 14) show that models including CT means explained a further 4–5% of the variation in the water content of the subcutaneous fat sample and in the water, fat and protein content of the bacon side sample than did models including area measurements. Although this was not a large improvement it should be noted that due to the superior quality of CT images compared to ultrasound images, the CT area measurements were probably calcu-

Table 13
Comparison between protein and fat content determined chemically and predicted from CT data in three experiments (from Vangen *et al.*, 1984)

	Liveweight (kg)	*Protein (kg) by*		*Fat (kg) by*	
		Chemical analysis	*CT*	*Chemical analysis*	*CT*
Experiment 1:					
Fat line	97·4	8·0	7·7	24·5	24·5
Norwegian Landrace	107·7	10·4	9·2	15·8	13·4
Mean	102·1	9·1	8·5	20·6	19·5
Experiment 3:					
Fat line	81·9	6·9	6·3	20·5	19·1
Norwegian Landrace	85·0	7·5	7·0	12·2	9·4
Mean	83·5	7·2	6·7	16·7	14·7
Experiment 2:					
Fat line	63·7	7·4	5·6	23·4	12·3
Norwegian Landrace	60·3	10·1	5·3	12·5	5·5
Mean	62·3	8·5	5·5	19·1	9·6

Table 14
Prediction of composition of lean and fat samples from mean CT number and/or subcutaneous fat (SCF) and *m. longissimus dorsi* (LD areas) (from Allen & Vangen, 1984)

Dependent variable:	*R^2 (RSD) Model*			
	Wt+sex(S)	*Wt+S+Areas(A)*	*Wt+S+CT mean*	*W+S+A+CT mean*
Bacon side:				
% water	0·71 (4·46)	0·83 (3·40)	0·87 (2·97)	0·88 (2·89)
% fat	0·72 (5·66)	0·85 (4·13)	0·88 (3·60)	0·89 (3·45)
% protein	0·67 (1·48)	0·80 (1·13)	0·84 (1·02)	0·85 (0·97)
SCF:				
% water	0·59 (4·12)	0·77 (3·13)	0·82 (2·75)	0·82 (2·75)
LD:				
% fat	0·24 (0·50)	0·29 (0·49)	0·25 (0·50)	0·30 (0·48)

lated with less error than would be achieved in practice using ultrasound imaging. In a comparative study on sheep, Wright and Simm (1986) found that cross-sectional areas produced by CT were considerably more precise than ultrasonic measurements taken by three B-mode scanners, two of which were real-time scanners. For intramuscular fat content CT explained no more variation than a base model (Vange & Kolstrad, 1986), whereas muscle area explained an additional 5% of the variation and marginally reduced the RSD. The authors discuss the suboptimal use of the data and their speculation that using CT number frequency data rather than simply the CT mean may give a better prediction was proven correct. After this more sophisticated analysis the proportion of the variation explained rose to 0·39. This may be improved upon further by using techniques such as zooming and dual energy scans to increase the resolution of fat within the muscle.

Results from CT Scanning of Sheep and Goats

CT may find more application in sheep than pigs since having a fleece and having proportionally less of the total carcass fat in the subcutaneous depot are not disadvantageous as is the case using ultrasonics. Results of CT scanning of sheep have been reported by Sehested (1984) who found

Table 15

Prediction of carcass composition of live lambs from CT frequencies (from Sehested, 1984)

Dependent variable	*Predicted from:*							
	Liveweight (LW)		*LW+single scan*			*LW+four scans*		
	RSD	R^2	*d.f.*[a]	*RSD*	R^2	*d.f.*	*RSD*	R^2
Protein (kg)	0·190	0·82	7	0·142	0·90	16	0·122	0·93
Fat (kg)	0·569	0·64	5	0·323	0·89	18	0·275	0·92
Fat-free lean (kg)	0·791	0·85	8	0·571	0·92	17	0·500	0·94
Water (kg)	0·616	0·85	8	0·445	0·92	19	0·398	0·94
Energy (MJ)	23·043	0·74	4	13·628	0·91	17	11·759	0·94
Carcass wt (kg)	1·088	0·90	8	0·812	0·94	16	0·710	0·96

[a] *d.f.* = degrees of freedom.

that a high proportion (0·89–0·92) of the variation in carcass composition traits could be explained by CT frequency data from a single scan in combination with liveweight (Table 15). Further small improvements in precision were achieved by combining information from four scans. In a cross-validation exercise using the same data set (Sehested, 1984) RSD values for prediction were only 7–14% higher than those for estimation.

Simm (1986) has given preliminary results of a study comparing the precision of predicting carcass lean content from measurements taken on live ram lambs with an Elscint 902 X-ray, three B-mode ultrasonic scanners, two of which were real-time scanners (Vetscan and Technicare) and velocity of ultrasound equipment. The precision of predicting carcass lean weight or percentage from linear and/or area measurements taken from subcutaneous fat and/or muscle cross-sections is shown for each machine in Table 8. Although no attempt was made in this preliminary analysis to relate CT value means or distribution data to carcass lean content, CT was a considerably more precise technique than ultrasound-based equipment, particularly for predicting carcass lean percentage at constant liveweight and age.

In a rather different type of study, Sorensen (1984) scanned goats during pregnancy and lactation and used the data to estimate changes in adipose tissue volume. He went on to compare fat balances based on CT scans, net energy balance or weight gain and concluded that although all three models involved errors and assumptions, balances based on CT scans were probably the most reliable.

Results from CT Scanning of Poultry

CT scanning has also been investigated as a tool for ranking potential poultry breeding stock according to their fatness or meatiness. Correlations in the range 0·70–0·90 between various fat depot weights, weight of breast cut obtained by dissection and values predicted from CT data and liveweight have been reported (Bentsen & Sehested, 1986). The increase in accuracy of prediction over base models was considerable. These authors also suggest that CT data may be used to predict the chemical composition of individual tissues.

Nuclear Magnetic Resonance (NMR)

A further diagnostic technique with potential for measuring body composition *in vivo* uses the phenomenon of nuclear magnetic resonance (NMR) to image living tissues in sections of any orientation. The underlying principles are given in detail in texts such as Gadian (1981) and Hamilton (1982); a brief outline will be given here.

When a body is placed in a strong magnetic field, atomic nuclei with an odd number of protons and/or neutrons or both —the hydrogen atom with its single proton being the most common of these in the body—align with the field and spin at resonant frequencies determined by the type of nuclei and the field strength. The electromagnetic signals emitted yield information on the concentration and distribution of these nuclei. In an NMR system a strong magnetic field is produced by a large annular electromagnet or superconducting electromagnet with an aperture sufficiently large for a human to pass through. A secondary changing field is superimposed on the main field by electric currents passing through coils near the patient. The strength and orientation of the magnetism is changed in a regular pattern in order to map the locations of the spinning nuclei in the desired plane—cross-sectional, longitudinal, transverse or oblique. Data on the electromagnetic signals received are used to resolve a square matrix. Discrimination between muscle and adipose tissues result from their different proton densities. However, greater discrimination may be achieved by measuring other NMR properties.

As the magnetic field is tilted by the superimposed changing field the angular momentum of the protons delays the return of the field to its equilibrium position. Instead it will 'precess' at a frequency proportional to the magnetic field strength around an axis parallel to the magnetic field. The delay is known as relaxation of which two components can be measured. The 'spin-lattice' time (T'_1) is the longitudinal component due to the interaction of the nuclear spin system with the surrounding 'lattice'. The transverse component due to the interactions of neighbouring spins is known as the 'spin-spin' relaxation time (T_2). The magnitude of these components depends upon the chemical structure of the tissue, in particular the relative amounts of water and triglycerides. The high water content of muscle results in a moderate relaxation time (T_1) whereas a high triglyceride content gives fatty tissue a shorter T_1 value. A range of images may be formed by combining T_I and T_2 measurements with proton density information with different weightings given to the relaxation times.

A few seconds are required to produce an image with NMR, so that the technique is relatively immune to movement artefacts. As with X-ray CT, however, animals would have to be anaesthetised and strapped to some form of cradle or table. Information from NMR images through adjacent scan planes can be added together and subsequent images displayed through any desired plane within the volume. Unlike X-ray CT, there are no moving parts with NMR and the technique does not use ionising radiation. There is no known health hazard from the magnetic fields of the strength employed (up to about 2 Tesla), except to patients with certain types of cardiac pacemaker.

Results from NMR

Although the use of NMR imaging has been investigated in livestock for purposes such as foetal imaging (Foster *et al.*, 1983) and measuring mammary gland development (Foster & Knight, 1983), both in goats, as yet there are few reports where its potential for measuring compositional traits has been studied. Fuller *et al.*, (1984) obtained T_1 values for muscle and adipose tissues *in vitro* using an NMR spectrometer and concluded that T_1 times for these tissues differ by a factor of 1·5, fat having the shorter relaxation time. Using the pulse-sequence technique and a powerful NMR imager a contrast ratio of 6 between fat and muscle has been reported (Foster *et al.*, 1984). Groeneveld *et al*, (1984) reported the results of scanning a single pig and a half carcass by NMR and a single pig by X-ray CT. They concluded that while X-ray CT gave greater fat–lean and fat–bone discrimination than NMR, with a minimum contrast of 6 standard deviations for NMR, both systems gave sufficient discrimination. In the same study high correlations (0·97–0·98) between T_1 or T_2 values and fat : lean ratios for in-vitro samples were reported. As yet there are no reports of studies where body composition traits have been predicted from measurements taken by NMR.

Neutron Activation Analysis (NAA)

Neutron activation analysis has been developed as an important tool in clinical and diagnostic medicine with the ability to measure the total body content of a number of elements. To date, calcium, phosphorus, sodium, cadmium, iron, iodine, chlorine, potassium, nitrogen, hydrogen, oxygen, carbon, aluminium and silver have been measured in humans (Boddy, 1984). Of these, potassium is the only element with a naturally occurring radioisotope (^{40}K) but other elements may be made radioactive by exposing them to a source of neutrons (hence neutron activation). The induced isotopes may be measured by whole body counts of the radiation induced. The data are translated into quantitative measurements of the elements of interest by comparing the spectra and counting-rates with those obtained from phantoms containing dispensed amounts of the appropriate elements. No single facility has been developed to measure all elements but there are facilities which measure seven (Preston *et al.*, 1985).

The technique is very costly since it requires both a neutron source and a whole-body counting facility. Conventional whole-body counters employ a shielded room constructed of lead or steel to protect the detectors from background radiation. Total weight is in the region 40–100 tonnes. This may be reduced to between 2–8 tonnes in a shadow-shield system where the shield system is confined largely to the detectors. Animals would have to be anaesthetised and the limited penetration of neutrons imposes a size limit equivalent to a body weight of about 100 kg. The wide ranges in the physical and operating characteristics of systems used in human medicine have been reviewed by Chettle and Fremlin (1984).

Results from NAA

While there may be some applications in animal science requiring determinations of elemental composition, by far the greatest interest is in the measurement of the relative amounts of muscle and adipose tissue. Since these differ considerably in elemental composition, in-vivo NAA measurement of appropriate elements may provide a useful prediction of the lean : fat ratio in the body. Encouraging results have already been achieved for live rats and humans and for pig and sheep carcasses (Preston *et al.*, 1984). The same authors suggest that determination of total body protein and mineral content may be achieved with satisfactory precision from in-vivo NAA measurement of nitrogen and calcium, but assessment of body fat by difference using NAA measurement of oxygen to give total body water may not be sufficiently accurate. This has been suggested as being the cause of anomalous results in a study on the body composition of salmon (Talbot *et al.*, 1986). In human medicine, clinical applications of the determination of a range of elements *in vivo* have been reviewed by Cohn and Parr (1985), and new approaches to the derivation of total body fat, the separation of body protein into its constituents and the separation of total body water into its intra- and extra-cellular components have been discussed by Beddoe and Hill (1985). Cohn *et al.*, (1981) advocate the use of nitrogen and potassium measurement to determine lean body mass with fat being derived as the difference between this and body weight.

Photon Activation Analysis (PAA)

Photon activation analysis (PAA) is similar to NAA, except that the radioactivity is induced by photons rather than neutrons. Accordingly, rather than a neutron source a high-energy electron accelerator is required in addition to a whole-body counting facility. The technique has been used to measure total body nitrogen (TBN), total body carbon (TBC) and total body oxygen (TBO) in rats (Ulin *et al.*, 1986). Accuracy is considerably lower for TBN at only ±7% (RSD) than for TBO or TBC (±1·5%), with the result that prediction of body protein from TBN is far less accurate than prediction of body water (TBW) from TBO or body fat (TBF) from TBC. Using normal rats varying widely in body composition and comparing PAA results with chemical analysis these workers obtained correlations of 0·999, 0·992 and 0·219, respectively between TBC and percentage fat, TBO and percentage water and total body protein by both methods. Corresponding standard errors of estimates were ±0·61% fat, ±1·19% water and ±2·8% protein. In order to apply the technique to larger animals high energy X-ray sources will be required (25 MV or greater for humans) in addition to larger detection systems. Nevertheless the authors see potential for using PAA to monitor body composition in humans and possibly in animal studies.

Potassium-40

The natural occurrence of potassium-40 (^{40}K) in the body has already been referred to, and since it is more abundant in muscle than in adipose tissue its measurement gives an indication of body composition. Although the technique has been used extensively in the USA (Frahm *et al.*, 1971) for the evaluation of pig and beef breeding stock, techniques such as ultrasonics are likely to give equal precision at considerably lower cost.

Positron-Emission Tomography (PET)

This technique is analogous to X-ray CT except that rather than using an external radiation source in PET an image is constructed from radiation emanating from within the body. A chemical compound with the biological activity of interest is labelled with a radioactive isotope that decays by emitting a positron (positive electron). After administration to the patient each emitted positron combines with an electron and the two are mutually annihilated with the emission of two gamma rays which fly off in opposite directions. After penetrating the surrounding tissue they are recorded by an annular array of detectors and the data are processed by a computer to form an image of the distribution of the radioactivity within the subject. The technique is most suited to the study of specific biochemical reactions in the body, though by labelling various esters with carbon-11 it may be possible to determine tissue lipid content (Ter-Pogossian *et al.*, 1980).

Other Nuclear-Based Techniques

Other imaging techniques in use in human medicine for in-vivo measurement of body elements include X-ray fluorescence, nuclear resonance scattering and photon absorptiometry. It is unlikely that these will be of use in the measurement of the composition of meat animals though they may have limited application in some basic animal science studies.

Electrical Conductivity

Lean tissues, being more ionic, conduct electricity better than the less hydrated adipose tissues. By measuring whole-body conductivity therefore, an indication of the lean : fat ratio is obtained. In contrast to most other techniques of in-vivo assessment of body composition, the technique was originally developed for use in pigs (Domermuth *et al.*, 1973) and has more recently been examined for use in human studies (Harrison and Van Itallie, 1982). The original equipment, known as electronic meat measuring equipment (EMME) consisted of a long solenoid coil driven by a 5 MHz source generating an induced electrical field within a tunnel. The presence of an animal within the tunnel perturbates the induced field. The EMME measurement is an indication of the current needed to re-establish the field and is related to the whole-body conductivity of the animal. The method is quick to operate and totally harmless.

Results from Electrical Conductivity

As with many other techniques there have been conflicting reports of the usefulness of EMME readings in predicting body composition in the live animal. Working with pigs, Domermuth *et al.* (1976) reported that three EMME readings were of about equal value to shrunk body weight for the prediction of either carcass protein or the weight of lean cuts, with only a minimal improvement when the two were used as co-predictors. However, there was considerable variation in liveweight in this experiment and EMME readings are highly correlated with liveweight. In another experiment with less variation in body weight, the same authors found that EMME readings were more important than body weight for predicting lean cuts or protein weight (R^2 = 0·41 vs 0·18 and 0·36 vs 0·13 respectively). Stiffler *et al.*, (1976) indicated marginal utility for prediction of separable lean or fat in swine from EMME readings but Fredeen *et al.*, (1979) found that EMME readings were useful for predicting percentage total fat and protein (R^2 = 0·44 and 0·54 respectively). A more recent report by Mersmann *et al.* (1984) confirms the finding of Stiffler *et al.* (1976) that repeatability of readings on pigs walking through the tunnel is low for some individuals. Attempts to improve this by using a crate and by anaesthetisation achieved only limited success. High correlations between electrical conductivity readings and indirect estimates of body composition in humans have been reported, but using a sample of individuals varying widely in weight (45–155 kg) and fatness (9·5–53.0% body fat) (Presta *et al.*, 1983). A second generation instrument TOBEC II developed for use on humans has been evaluated, but once again using a sample varying widely in composition (6–36% body fat) (Van Loan & Mayclin, 1987). However, the improvement in the prediction of body composition from the use of first and second order Fourier coefficients may be of interest to those using this technique in pigs.

In another recently reported trial (Joyal et al., 1987) EMME number was found to be a better predictor of weight of lean in carcass than ultrasonic fat depth when weight and treatment (sex, feeding level and housing system) were included in the model, while ultrasonics was far superior to EMME in the prediction of weight of subcutaneous fat. When both were used in combination, the R^2 values for the prediction of lean weight and subcutaneous fat weight were marginally increased compared with the better single technique models. In most practical situations, however, diet and housing system are unknown, so it is more realistic to use models which exclude these. When this was done, ultrasonic backfat was a better predictor of both lean and subcutaneous fat weights and there was no improvement from using both techniques in combination. EMME may therefore be a technique with potential for predicting lean in experimental situations rather than having widespread practical applications.

Tracer Dilution Techniques

The basic theory of dilution techniques is that if a known amount of a biological tracer is injected into an animal it will become uniformly distributed throughout a compartment of the body (body water, fat, etc.). By measuring the concentration of tracer in the body pool after equilibrium is reached the size of the compartment can be calculated. Dilution techniques have most commonly been employed to measure total body water and from this to make predictions of body composition, though the theory has also been applied to the direct estimation of body fat using labelled gases such as krypton (Hytten *et al.*, 1966), albeit with only limited success. In the estimation of body water, tritiated water, antipyrene and its derivatives 4-amino antipyrene and *N*-acetyl-4-amino antipyrene have been used. Recent advances in the measurement techniques for deuterium oxide (Byers, 1979) have renewed interest in this method. Evans Blue has been used to estimate blood volume.

The main source of error in estimating carcass composition from tracer dilution in the body water arises from variation in the proportion of total body water contained in the gut contents and other non-carcass parts. This is particularly so for ruminants and is more serious in studies where treatment effects result in differences in gut-fill. A second source of error is the assumption of a constant ratio of water to protein in the lean body mass. Although this varies fairly predictably with maturity it may give rise to bias in breed comparison studies or when dietary or other treatments may affect this ratio.

Results from Dilution Techniques

Estimation of body composition by dilution of tracers in the waterspace has lead to conflicting conclusions. The principles have been outlined by Robelin (1982*a*) and the subject has been reviewed by Robelin (1973), Sheng and Huggins (1979) and more recently by Robelin (1984). In beef cattle Lunt *et al.* (1985) found that deuterium oxide (D_2O) dilution was of no use for predicting percentage separable lean (R^2 = 0·02) or fat (R^2 = 0·04) in 32 steers of a single breed fed on two diets and slaughtered at four weights. On the other hand, Robelin (1982*b*) concluded that D_2O dilution was a good method of estimating body composition in beef cattle, especially for total body lipid content. Furthermore, in this study the relationships between the measurement of D_2O space and body lipids did not vary between the two breeds (Charolais and Friesian), due to variation in gut fill being taken account of in the equations. Robelin (1982*b*) has investigated the accuracy of D_2O determinations and found an RSD for body water of 4·6 kg in 42 bulls with a mean weight of body water of 220 kg, an error of approximately 2%. He concludes that this error is divided approximately equally between the D_2O determination and the body water measurement. Since the accuracy of the lipid determination is dependent upon the accuracy of measuring body weight, several recordings of liveweight on days before and after infusion with the tracer are recommended (Robelin, 1982*b*).

While variation in the experimental accuracy and in the range and types of animals used may explain some of the conflicting results, the type of model used may also be important. Arnold *et al.* (1985) have examined the use of single-, two- and three-compartment models for determination of body composition in beef steers by the D_2O dilution technique. The two-compartment model takes account of water in the gut while the most complex model yields separate estimates for extracellular and intracellular water in addition to water in the gut-fill. Perhaps surprisingly they found that the simplest model gave the best prediction of body composition. While both single- and two-compartment models overestimated fat content of the body, in the case of the simpler model the estimate was not significantly different from the direct (ether extraction) method. Protein also was overestimated by the single compartment model (by 3·6%) and in this case the difference was significant.

Recent reports of the use of tritiated water dilution to estimate body composition in beef have been favourable. Aziz and Sani (1985) reported a correlation of 0·95 between body fat estimated by tritiated water dilution and by dissection in eight Zebu bulls with a liveweight of 300–400 kg. The estimated lean body mass was not significantly different from that measured by dissection. Bird (1984) also used tritiated water to estimate steer carcass composition and found an improvement in the prediction of fat, water and nitrogen content over prediction from liveweight alone. R^2 values were high (0·93–0·99) for all prediction equations due mainly to the large range in liveweight (90–517 kg) and total fat (4·2–85·7kg). Nevertheless, the reductions in the RSD values in water and fat content over models containing only liveweight (15 and 55%, respectively) were useful, particularly for fat content. Inclusion of a fat depth taken on the carcass further reduced both RSD values, suggesting the possibility of combining this technique with ultrasonic fat depth measurement for extra precision in determining body composition.

Conflicting results have also been reported in work on pigs. Ferrell and

Cornelius (1984) used D_2O dilution to predict body composition in obese and normal pigs. They concluded that D_2O space was little better than liveweight for predicting body composition and was influenced by pig type. In one of the few studies where several dilution techniques have been compared, Houseman and McDonald (1976) found that D_2O dilution in body water, ^{42}K dilution in lean body mass and Evans Blue dilution in blood all gave better estimates of body composition than liveweight (Table 16). Lipid weight was estimated with standard errors of 1·6 and 1·5 kg when D_2O space or ^{42}K respectively was used in combination with liveweight. One reason for the high correlations achieved in this study, however, is the wide range of fatness. Shields *et al.* (1983) also reported strong relationships between D_2O space and pig carcass composition, but the high R^2 values (0·97–0·99) reflect the very wide weight range (6·4–109 kg), and no comparison with a simpler technique is given.

Physiological Predictors

The metabolism of fat and lean individuals differs in that the former deposit a greater proportion of utilisable energy as fat. It follows therefore that the activity of enzymes and hormones regulating growth, tissue accretion rates and tissue turnover rates will also differ among individuals. Several authors have reported differences in the levels and activities of such hormones and enzymes between breeds differing in body composition. Wangsness *et al.* (1977) found differences in glucose clearance rate, plasma immuno-reactive insulin and plasma growth hormone of lean and obese pigs. Rogdakis *et al.* (1979) found breed differences in the activity of NADPH-generating enzymes in subcutaneous fat tissue. The same workers (Strutz and Rogdakis, 1979) and Muller (1986) both obtained responses to selection for this activity together with correlated responses in backfat thickness. In the second of these studies, selection for enzyme activity was almost as effective as selection for ultrasonic backfat in changing backfat depth and carcass composition. In Norwegian Landrace pigs selected for rate of gain and thickness of backfat, line differences in lipid mobilisation (Standal *et al.*, 1973) and in the serum levels of free-fatty acids (EFA) and glucose (Bakke, 1975) have been reported. However, Gregory *et al.* (1980) found differences in the concentration of plasma free fatty acid (FFA) only after 16 hours without food and concluded that changes in FFA are not important in the reduced fat deposition of the line selected against backfat thickness since the pigs had not been reared with periods longer than 16 hours without food. In a study of

Table 16
Correlation coefficients between individual predictors and body composition measurements in 24 pigs (weight range 72·3–92·3 kg) (Houseman & McDonald, 1976)

Method	*Fat-free wt*			*Lipid wt*		
	Total corr.	*Partial corr.*[a]	*CV est. (%)*[b]	*Total corr.*	*Partial corr.*[a]	*CV est. (%)*[b]
Live wt	0·62	—	8·0	−0·01	—	—
D_2O dilution	0·96	0·95	3·3	−0·66	−0·96	2·9
^{42}K dilution	0·98	0·96	2·9	−0·77	−0·95	3·3
Evans Blue	0·83	0·72	7·2	−0·56	−0·71	7·4

[a] Liveweight constant.

[b] $\frac{RSD}{Mean} \times 100$

genetically obese and lean lines of pigs, Mersmann & MacNeil (1985) found significant correlations between backfat thickness and plasma lipid concentration but concluded that the low magnitude and inconsistency of these correlations did not indicate that plasma lipid concentration was a reliable indicator of adiposity in pigs. This was confirmation of an in-vitro study of lipolytic rate in adipose tissue from the same lines which indicated that obese pigs do not have low adipose tissue lipolytic rates compared with lean pigs (Mersmann, 1985).

Recent work from the USSR indicates that age (or maturity) has an important effect on variation in enzymatic activity (Bazhov & Bakhireva, 1987). These workers report that the heritabilities (h^2) of the activities of four lymphocyte dehydrogenases declined from about three to six months of age. Sire- and dam-component h^2s ranged from 0·26–0·88 and from 0·02–0·42 respectively. Furthermore, activities of succinate, lactate and glycerophosphate dehydrogenases at three months of age were significantly correlated with age at slaughter (0·34, 0·38 and 0·39 respectively) and backfat thickness (0·51, 0·55 and 0·59 respectively).

Plasma very low density lipoprotein concentration has been shown to be a useful indirect method of estimating fatness in turkeys (Griffin & Whitehead, 1985). Birds with the lowest concentrations of plasma very low density lipoproteins had significantly less abdominal fat and/or total fat than those with the highest concentrations in each of three groups.

Adipose Tissue Characteristics

Adipose Cell Size

Accretion of adipose tissues occurs mainly through an increase in cell size. The possibility of using adipose cell diameter as an indicator of beef carcass composition has been investigated by Robelin (1982*c*) and Robelin and Agabriel (1986). Although the fat samples were taken after slaughter in these studies, the method could be applied *in vivo* using a simple biopsy technique. In the first of these studies adipose cell diameter was as accurate a predictor of total body fat as was body water estimated by the D_2O dilution technique (RSD = 10·3 and 11·5 g/kg respectively). While the author accepts that the results need verifying in different groups of animals, he points out that it is considerably cheaper than the dilution technique and is not influenced by gut-fill. When the method was used on finishing bulls and mature cows of four breeds (Robelin & Agabriel,

1986) the regression equations relating adipocyte diameter to body fat, carcass fat and carcass muscle proportion varied considerably among the sex × breed sub-groups. The accuracy of the estimates was about 3% of body weight (RSD as a percentage of the mean) for each of the three traits which led the authors to conclude that the technique gave a useful indication of fatness. From the variation in the parameters of the regression equations it would appear that robust equations applicable to different groups of animals are unlikely. However it may be a useful technique for ranking animals of similar background, for instance at the beginning of an experiment.

Adipose Tissue Composition

The chemical composition of adipose tissue also changes in a predictable way as animals become fatter. The lipid droplet in each cell increases in size causing a dilution in relative terms of the intracellular components, mainly water and protein. So adipose tissue from fatter animals contains more lipid and less water and protein than that from lean animals. The relationship between subcutaneous adipose tissue composition and composition of the ham joint has been investigated in pigs by Aberle *et al.*, (1977). Percent lipid in the ham was negatively correlated with percent adipose tissue moisture ($r = -0{\cdot}83$) and positively correlated with percent adipose tissue lipid ($r = 0{\cdot}80$). Correlations with percent moisture in the ham were of similar magnitude, though of opposite sign, while correlations with percent protein in the ham were lower. Inclusion of fat thickness and muscle area measurements as covariates with percent adipose moisture improved the prediction of percent lipid in the ham ($R^2 = 0{\cdot}84$–$0{\cdot}92$). This technique could be applied *in vivo* by biopsy, but obtaining a representative sample of sufficient size may be a problem.

Infrared Interactance (Apparent Reflectance)

Near-infrared (NIR) reflectance spectroscopy was originally developed to determine the composition of grain samples but has since been demonstrated to give readings that are highly correlated with moisture, fat, protein and caloric content of raw pork and beef (Lanza, 1983). Of more interest is its use in predicting fatness in humans (Conway *et al.*, 1984) and its possible application to in-vivo measurement in meat animals. NIR spectra are highly influenced by the amounts of water and fat in the reflecting body. Conway *et al.* (1984) used a machine with a single beam rapid scanning monochromator and a fibre optic probe to obtain spectra at five sites where ultrasonic fat depth measurements were also taken. Body fat was also predicted from skinfold measurements and D_2O water space. Correlations between body composition predicted by infrared interactance and the other methods were sufficiently high (0·89–0·94) to warrant further investigation of this rapid, simple non-invasive objective method. The technique merits investigation for use in meat-producing species, though wool clipping or shaving may be necessary.

COMPARISON OF TECHNIQUES

There are several problems in making definitive statements about the relative merits of the techniques available for in-vivo measurement of body composition. Not least of these is the small number of trials where a range of techniques has been compared on the same group of animals. Furthermore, in many trials relatively complex methods have not even been compared with simple, readily available indicators of composition such as liveweight, sex and growth rate. Even when results are presented in terms of the marginal improvement over prediction from liveweight alone, no attempt is usually made to determine liveweight accurately by making several weighings to take account of fluctuations due to gut fill, as is recommended for dilution techniques (Robelin, 1982*a*).

Many trials, particularly those involving new techniques, have been carried out using animals with much wider variation in age, weight and composition than would be the case in practice. While this may be justified in order to check the linearity of relationships over a wide range, the high correlation coefficients reported can be misleading. Too often such studies have not been followed by validation exercises on more homogeneous groups of animals.

Making comparisons between techniques across different studies is also complicated by factors such as differences in the experience of operators, the choice of dependent variables, the size and variability of the sample and the presentation of the results. In the last respect, the RSD of the dependent variable and the percentage reduction achieved over prediction from liveweight and other readily available predictors should be reported as a minimum.

Table 17
Summary of trials where more than one technique has been used to predict composition *in vivo*

Reference	*Species*	*Technique*	*Dependent variable*	*Precision*	*Ranking*
Simm (1986)	Sheep	Velocity of ultrasound	% lean	RSD = 2·50	2 =
		Ultrasound scan	% lean	RSD = 2·38–2·63	2 =
		X-ray CT	% lean	RSD = 2·00	1
Vangen & Kolstad (1986)	Pigs	Ultrasound scan	Protein wt	RSD = 0·74	2
		X-ray CT	Protein wt	RSD = 0·52	1
Cameron & Smith (1985)	Sheep	Ultrasound scan	% lean	RSD = 2·20–2·50[a]	1
		FFA	% lean	RSD = 2·50[a]	0
		Very low density lipoprotein	% lean	RSD = 2·50[a]	0
		Food conversion efficiency	% lean	RSD = 2·50[a]	0
Domermuth *et al.* (1976)	Pigs	Electrical conductivity	Lean cuts wt	R^2 = 0·80	1 =
		^{40}K	Lean cuts wt	R^2 = 0·82	1 =
		Electrical conductivity	Protein wt	R^2 = 0·78	1 =
		^{40}K	Protein wt	R^2 = 0·78	1 =
		Electrical conductivity	Lipid wt	R^2 = 0·67	1 =
		^{40}K	Lipid wt	R^2 = 0·65	1 =
Robelin (1982*c*)	Cattle	D_2O dilution	% Fat	RSD = 10·3	1 =
		Adipose cell size	% Fat	RSD = 11·5	1 =
Houseman & McDonald (1976)	Pigs	Corrected feed conversion[b]	Lipid wt	R^2 = 0·96	1 =
		Ultrasonic scan	Lipid wt	R^2 = 0·82	2
		Evans Blue	Lipid wt	R^2 = 0·71	3

		D_2O	Lipid wt	$R^2 = 0.96$	1 =
		^{42}K	Lipid wt	$R^2 = 0.95$	1 =
		Corrected feed conversion[b]	Fat wt	$R^2 = 0.94$	1 =
		Ultrasonic scan	Fat wt	$R^2 = 0.81$	2
		Evans Blue	Fat wt	$R^2 = 0.73$	3
		D_2O	Fat wt	$R^2 = 0.94$	1 =
		^{42}K	Fat wt	$R^2 = 0.90$	1 =
		Corrected feed conversion[b]	Lean wt	$R^2 = 0.87$	1 =
		Ultrasonic scan	Lean wt	$R^2 = 0.72$	2
		Evans Blue	Lean wt	$R^2 = 0.83$	1 =
		D_2O	Lean wt	$R^2 = 0.84$	1 =
		^{42}K	Lean wt	$R^2 = 0.84$	1 =
		Corrected feed conversion[b]	Fat-free wt	$R^2 = 0.96$	1 =
		Ultrasonic scan	Fat-free wt	$R^2 = 0.88$	2 =
		Evans Blue	Fat-free wt	$R^2 = 0.84$	2 =
		D_2O	Fat-free wt	$R^2 = 0.97$	1 =
		^{42}K	Fat-free wt	$R^2 = 0.98$	1 =
Fredeen *et al.* (1979)	Pigs	Electrical conductivity	% diss. fat	RSD = 2·2–2·4	2
		Backfat probe	% diss. fat	RSD = 1·6–1·9	1
		Electrical conductivity	% lean	RSD = 2·2–2·3	1 =
		Backfat probe	% lean	RSD = 2·1	1 =
Metz *et al.* (1984)	Pigs	Ultrasonic scan	% lean	$r^c = -0.76$	1
		Adipose tissue composition	% lean	$r^c = -0.37\text{–}(-0.05)$	2 =
		Fat cell size	% lean	$r^c = -0.51$	2 =
		Ultrasonic scan	% fat	$r^c = 0.82$	1
		Adipose tissue composition	% fat	$r^c = 0.47\text{–}0.65$	2 =
		Fat cell size	% fat	$r^c = 0.55$	2 =

[a] No improvement over weight and age.
[b] Corrected for maintenance.
[c] Partial correlation at fixed carcass weight.

Some trials where more than one technique have been included are summarised in Table 17. The 'rankings' are intended as a guide only as they are somewhat arbitrary, being not always based on statistical tests but sometimes on the general conclusions in the paper or on this author's judgement. The first conclusion from the table is that many more comparative trials have been done on pigs than on the other main meat-producing species. All available evidence suggests that X-ray CT is more accurate at predicting composition in sheep and pigs than ultrasonic methods (Simm, 1986, & Vangen & Kolstad, 1986). Moreover the extra accuracy is quite considerable, particularly for predicting weight of carcass fat in pigs. Whether this increase in accuracy is justified by the higher cost will depend upon the particular use of the in-vivo measurements (see below). The only other instrumental method included in these comparisons is the electrical conductivity measurement (EMME). In the study of Fredeen *et al.* (1979) this was no better at predicting percent lean in the carcass when used in combination with liveweight than was the simple backfat probe (Hazel & Kline, 1959) and was less accurate than the probe in predicting percent dissected fat at constant weight. It seems unlikely, therefore, that EMME would be superior to ultrasonic scanners.

The remainder of the trials summarised in Table 17 give an indication of the relative accuracies of some of the 'laboratory-based' methods compared with ultrasonic scanning and electrical conductivity. EMME readings were no better than ^{40}K counts in predicting weights of carcass protein, chemical fat or lean cuts (Domermuth *et al.*, 1976). Measurements taken by ultrasound were better than adipose tissue composition or adipose cell size in predicting either percent muscle or lipid in pigs (Metz *et al.*, 1984) but were worse than feed conversion corrected for maintenance or the D_2O or ^{42}K dilution techniques (Houseman & McDonald, 1976). In sheep, however, food conversion efficiency, non-esterified fatty acids, and very low density lipoproteins were of no value in predicting lean content at constant liveweight whereas the best ultrasonic measurements reduced the RSD by nearly 30% (Cameron & Smith, 1985). It is pertinent to note here that the worst performing ultrasonic machine made no marginal reduction in the RSD either, so the conclusion would have been different had two other machines not been included. Among the dilution techniques the Evans Blue method appears to be inferior to D_2O and ^{42}K particularly for predicting parameters of fatness (Houseman & McDonald, 1976). In cattle, Robelin (1982*c*) found adipose cell size to be almost as good a predictor of percent separable fat as D_2O dilution.

Selection of a Technique

King (1982) has given an account of the important criteria in selecting an in-vivo technique for use in breeding schemes and Kempster (1984) has given a wider discussion relating to other uses. The main criteria are cost, practicability, precision and accuracy. Total cost includes running, as well as capital, costs. Practicability includes factors such as mobility, physical requirements such as size, power supply, shielding, simplicity of operation and speed of throughput. Public acceptability is also included in the list by King (1982) and this may well rule out large scale use of biopsy techniques to measure adipose tissue characteristics. Precision is the RSD of the predicted characteristic and accuracy is the lack of bias in predicting carcass characteristics when a prediction equation is used on different groups of animals to those from which the equation was derived. A more complete discussion of this with actual examples may be found in Kempster *et al.*, (1982).

The various techniques are listed in Table 18 and 'scored' for the important criteria. This table is of necessity somewhat arbitrary, based on a judgement of the available information. Accuracy is omitted from the table as this has been established in only a few cases. The inclusion of infrared interactance is speculative as it has not, to this author's knowledge, been used on animals.

The main situations where in-vivo techniques are required to improve the efficiency of lean meat production are on the farm, in the marketing chain, i.e. livestock sales and slaughterhouses, in large scale breeding programmes, either national or large company schemes, and finally at research institutions. Each of these the criteria listed above will have different weightings in the selection of suitable techniques. On the farm, low cost and practicability will be most important both for monitoring breeding stock and for selecting stock for slaughter. At livestock sales and slaughterhouses it is likely that large numbers of animals will be measured in a day so the cost will be spread over a larger number of animals. However, large outlays for capital and running costs are unlikely here as the financial returns will not be so easily identifiable. Factories with specialised markets may find use for a relatively simple technique for rejection of unsuitable animals before slaughter. Here also, relatively low cost, simple-to-operate methods will be needed.

In large scale breeding schemes where the benefits of any genetic gain will be disseminated widely, the cost/benefit ratio of the more expensive techniques will be favourable if they offer an increase in precision over

Table 18
Relative performance of in-vivo techniques and suitability for various uses

Method	*Cost*	*Portability*	*Simplicity*	*Potential precision*	*Species[a] availability*	*Potential applications[b]*
Ultrasonics	****	*****	****	****	C, P, S	F, M, B, E
Velocity of ultrasound	*****	*****	*****	***	C, P, S	F, M, B, E
X-ray CT	*	*	*	*****	P, S	B, E
NMR	*	*	*	*****	P, S	B, E
NAA	*	*	*	*****	P, S	B, E
Dilution techniques	***	***	**	***	C, P, S	B, E
Electrical conductivity	**	**	**	***	P, S	B, E
Physiological predictors	***	***	**	*	C, P, S	B, E
Adipose cell characteristics	****	***	**	**	C, P, S	B, E
Infrared interactance	***	*****	****	***?	C, P, S?	F, M, B, E?

Scores: * = least favourable; ***** = most favourable.
[a] C = cattle; P = pigs; S = sheep.
[b] F = on farm; M = markets and slaughterhouses; B = large breeding programmes; E = experimental stations.

other techniques. Kempster (1984) has outlined a cost/benefit approach to this topic. A cost/benefit analysis should form an important part of the decision-making process when choosing an in-vivo technique for particular application. The problem is that some of the essential information, for example the comparative accuracy of various techniques or genetic parameters, is either lacking or not well established. In national centralised breeding schemes precision is probably more important than accuracy since the test environment is common and it is the ranking of individuals that is important.

For basic research studies comprehensive cost/benefit analysis is even more difficult. Notwithstanding budgetary considerations, precision and accuracy are most important since in-vivo techniques allow investigations that would otherwise either not be possible or would use the relatively imprecise serial slaughter technique with consequent increase in numbers of animals. This forms the basis for a type of cost/benefit analysis since savings on the cost of animals and labour for dissections can be quantified. This is not to say that expensive, precise techniques are always required for research purposes. Often treatment differences are sufficiently large for a significant difference to be detected with a simpler technique.

REFERENCES

Aberle, E. D., Etherton, T. D. & Allen, C. B. (1977). Prediction of pork carcass composition using subcutaneous adipose tissue moisture or lipid concentration. *J. Anim. Sci.*, **46**, 449–56.

Allen, P. & Vangen, O. (1984). X-ray tomography of pigs—some preliminary results. In *In Vivo Measurement of Body Composition in Meat Animals*, ed. D. Lister. Elsevier Applied Science Publishers, London, pp. 52–66.

Alliston, J. C. (1983). Evaluation of carcass quality in the live animal. In *Sheep Production*, ed. W. Haresign. Butterworths, London, pp. 75–94.

Alliston, J. C., Barker, J. D., Kempster, A. J. & Arnall, D. (1981). The use of two ultrasonic machines (Danscanner and Scanogram) for the prediction of body composition in crossbred lambs. *Anim. Prod.*, **32**, 375 (Abstract).

Alliston, J. C., Kempster, A. J. & Owen, M. G. (1982). An evaluation of three ultrasonic machines for predicting the body composition of live pigs of the same breed, sex and live weight. *Anim. Prod.*, **35**, 165–9.

Andersen, B. B. (1982). In *In Vivo Estimation of Body Composition in Beef*, ed. B. B. Andersen. Beretning fra statens Husdyrbrugs Forsog No. 524, Copenhagen.

Andersen, B. B., Busk, H., Chadwick, J. P., Cuthbertson, A., Fursey, G. A. J., Jones, D. W., Lewin, P., Miles, C. A. & Owen, M. G. (1983). Comparison of ultrasonic equipment for describing beef carcass characteristics in live cattle (report of a joint ultrasonic trial carried out in the U.K. and Denmark). *Livest. Prod. Sci.*, **10,** 133–47.

Arnold, R. N., Hentges, E. J. & Trenkle, A. (1985). Evaluation of the use of deuterium oxide dilution technique for determination of body composition in beef steers. *J. Anim. Sci.*, **60,** 1188–200.

Aziz, D. M. & Sani, R. A. (1985). Total body water measurement for estimation of body fat and lean body mass in cattle. *MARDI Research Bulletin,* **13,** 98–102.

Bailey, C. M., Jensen, J. & Andersen, B. B. (1986). Ultrasonic scanning and body measurements for predicting composition and muscle distribution in young Holstein × Friesian bulls. *J. Anim. Sci.*, **63,** 1337–46.

Bakke, H. (1975). Serum levels of non-esterified fatty acids and glucose in lines of pigs selected for rate of gain and thickness of backfat. *Acta Agric. Scand.*, **25,** 113–16.

Bazhov, G. M. & Bakhireva, L. A. (1987). Activity of cellular dehydrogenases in pigs, its heritability and relationship with performance. *Animal Breeding Abstracts,* **55,** 625 (Abstract).

Beddoe, A. H. & Hill, G. L. (1985). Clinical measurement of body composition using in vivo neutron activation analysis. *J. Parenteral and Enteral Nutrition,* **9,** 504–20.

Bentsen, H. B. & Sehested, E. (1986). Computerised tomography of chickens, 28th British Poultry Breeders' Roundtable, Cambridge, UK.

Bird, P. R. (1984). Prediction of components of steer carcasses using titrated body water space, fat depth and fasted liveweight or carcass weight. *Aust. J. Agric. Res.*, **35,** 435–42.

Boddy, K. (1984). Measurement of body elements and their metabolism. In *In Vivo Measurement of Body Composition in Meat Animals,* ed. D. Lister. Elsevier Applied Science Publishers, London, pp. 36–8.

Byers, F. M. (1979). Extraction and measurement of deuterium oxide at tracer levels in biological fluids. *Anal. Biochem.*, **98,** 208–13.

Cameron, N. D. & Smith, C. (1985). Estimation of carcass leanness in young rams. *Animal. Prod.*, **40,** 303–8.

Chettle, D. R. & Fremlin, J. H. (1984). Techniques of in vivo neutron activation analysis. *Phys. Med. Biol.*, **29,** 1101–43.

Cohn, S. H. & Parr, R. M. (1985). Nuclear-based techniques for the in vivo study of human body composition. *Clin. Phys. Physiol. Meas.*, **6,** 275–301.

Cohn, S. H., Ellis, K. J., Vartsky, P., Sawitsky, A., Gartenhaus, W., Yasmura, S. & Vaswani, A. N. (1981). Comparison of methods of estimating body fat in normal subjects and cancer patients. *Am. J. Clin. Nutr.*, **34,** 2839–47.

Conway, J. M., Norris, K. H. & Bodwell, C. E. (1984). A new approach to the estimation of body composition: infrared interactance. *Am. J. Clin. Nutr.*, **40,** 1123–30.

Cormack, A. M. (1980). Early two-dimensional reconstruction (CT scanning) and recent topics stemming from it. *J. Computer Assisted Tomography,* **4,** 658–64.

Domermuth, W. F., Veum, T. L., Alexander, M. A., Hedrick, H. B. & Clark, J. L. (1973). Evaluation of EMME for swine. *J. Anim. Sci.*, **37,** 259 (Abstract).

Domermuth, W. F., Veum, T. L., Alexander, M. A., Hedrick, H. B., Clark, J. & Eklund, D. (1976). Prediction of lean body composition of live market weight swine by indirect methods. *J. Anim. Sci.*, **43,** 966–76.

Eveleigh, C. F., Thwaites, C. J., Hassab, P. B., Paton, P. G., Smith, R. J. & Upton, W. H. (1985). A note on the ability of three portable ultrasonic probes to predict backfat thickness in cattle. *Anim. Prod.*, **41,** 247–8.

Ferrell, C. L. & Cornelius, S. G. (1984). Estimation of body composition of pigs. *J. Anim. Sci.*, **58,** 903–12.

Fortin, A. & Shrestha, J. N. B. (1986). In vivo estimation of carcass meat by ultrasound in ram lambs slaughtered at an average live weight of 37 kg. *Anim. Prod.*, **43,** 469–75.

Foster, M. A. & Knight, C. H. (1983). Nuclear magnetic resonance (NMR) imaging, a non-invasive technique for measuring mammary gland development. *J. Physiol.*, **341,** 82P.

Foster, M. A., Knight, C. H., Rimmington, J. E. & Mallard, J. R. (1983). Fetal imaging by nuclear magnetic resonance: a study in goats. *Radiology,* **149,** 193–5.

Foster, M. A., Hutchison, J. M. S., Mallard, J. R. & Fuller, M. (1984). Nuclear magnetic resonance pulse sequence and discrimination of high- and low-fat tissues. *Magnetic Resonance Imaging,* **2,** 187–92.

Frahm, R. R., Walters, L. E. & McLellan, C. R. (1971). Evaluation of ^{40}K count as a predictor of muscle in yearling beef bulls, *J. Anim. Sci.*, **32,** 463–9.

Fredeen, H. T., Martin, A. H. & Sather, A. P. (1979). Evaluation of an electronic technique for measuring lean content of the live pig, *J. Anim. Sci.*, **48,** 536–40.

Fuller, M. F., Foster, M. A. & Hutchison, J. M. S. (1984). Nuclear magnetic resonance imaging of pigs. In *In Vivo Measurement of Body Composition in Meat Animals,* ed. D. Lister. Elsevier Applied Science Publishers, London, pp. 123–33.

Gadian, D. G. (1981). *Nuclear Magnetic Resonance and its Application to Living Systems.* Oxford University Press, Oxford.

Giles, L. R., Murison, R. D. & Wilson, B. R. (1981). Backfat studies in growing pigs 2. A comparison of ultrasound and ruler probe predictors of backfat and eye-muscle measurements in the live pig. *Anim. Prod.* **32,** 47–50.

Gooden, J. M., Beach, A. D. & Purchas, R. W. (1980). Measurement of subcutaneous backfat depth in live lambs with an ultrasonic probe. *NZ J. Agric. Res.*, **23,** 161–5.

Greer, E. B., Most, P. C., Lowe, T. W. & Giles, L. R. (1987). Accuracy of ultrasonic backfat testers in predicting carcass P_2 fat depth from live pigs measurement and the effect on accuracy of mislocating the P_2 site on the live pig. *Aust. J. Exp. Agric.*, **27,** 27–34.

Gregory, N. G., Wood, J. D., Enser, M., Smith, W. C. & Ellis, M. (1980). Fat mobilisation in Large White pigs selected for low backfat thickness. *J. Sci. Food Agric.*, **31,** 567–72.

Griffin, H. D. & Whitehead, C. C. (1985). Identification of lean or fat turkeys by measurement of plasma very low density lipoprotein concentrations. *British Poultry Science,* **26,** 51–6.

Groeneveld, E., Kallweit, E., Henning, M. & Pfau, A. (1984). Evaluation of body composition of live animals by X-ray and nuclear magnetic resonance computed tomography. In *In Vivo Measurement of Body Composition in Meat Animals,* ed. D. Lister. Elsevier Applied Science Publishers, London, pp. 84–8.

Hamilton, B. (Ed.) (1982). *Medical Diagnostic Imaging Systems—Technology and Applications.* F and S Press, New York.

Harrison, G. G. & Van Itallie, T. B. (1982). Estimation of body composition: a new approach based on electromagnetic principles. *Am. J. Clin. Nutr.,* **35,** 1176–9.

Hazel, L. N. & Kline, E. A. (1959). Ultrasonic measurement of fatness in swine. *J. Anim. Sci.,* **18,** 815–19.

Henningsson, T., Ral, G., Anderson, O., Karlsson, U. & Martinsson, K. (1986). A study on the value of ultrasonic scanning as a method to estimate carcass traits on live cattle. *Acta Agric. Scand.,* **36,** 81–94.

Hounsfield, G. N. (1980). Computed medical imaging. *J. Computer Assisted Tomography,* **4,** 665–74.

Houseman, R. A. & McDonald, I. (1976). The comparative precision of estimates of body composition in living pigs obtained from numerous different predictors applied severally or jointly. *J. Agric. Sci.,* **87,** 499–510.

Hudson, J. E. & Payne-Crostin, A. (1984). A comparison of ultrasonic machines for the prediction of backfat thickness in the live pig. *Aust. J. Exp. Agric. Anim. Husb.,* **24,** 512–15.

Hytten, F. E., Taylor, K. & Taggart, N. (1966). Measurement of total body fat in man by absorption of 85 Kr. *Clin. Sci.,* **31,** 111–19.

Jansen, J., Andersen, B. B., Bergstrom, D. L., Busk, H., Langerweij, G. W. & Oldenbroek, J. K. (1985). In vivo estimation of body composition in young bulls for slaughter. 2. The prediction of carcass traits from scores, ultrasonic scanning and body measurements. *Livest. Prod. Sci.,* **12,** 231–40.

Joyal, S. M., Jones, S. D. M. & Kennedy, B. W. (1987). Evaluation of electronic meat measuring equipment in predicting carcass composition in the live pig. *Anim. Prod.* **45,** 97–102.

Kanis, E., Steen, H. A. M. van der., Roo, K. de & Groot, P. N. de (1986). Prediction of lean parts and carcass price from ultrasonic backfat measurements in live pigs. *Livest. Prod. Sci.,* **14,** 55–64.

Kempster, A. J. (1984). Cost benefit analysis of *in vivo* estimates of body composition in meat animals. In *In Vivo Measurement of Body Composition in Meat Animals,* ed. D. Lister. Elsevier Applied Science Publishers, London, pp. 191–203.

Kempster, A. J., Cuthbertson, A., Owen, M. G. & Alliston, J. C. (1979). A comparison of four ultrasonic machines (Sonatest, Scanogram, Ilis Observer and Danscanner) for predicting the body composition of live pigs. *Anim. Prod.,* **29,** 175–81.

Kempster, A. J., Cuthbertson, A., Jones, D. W. & Owen, M. G. (1981). Prediction of body composition of live cattle using two ultrasonic machines of different complexity—a report of four separate trials. *J. Agric. Sci.,* **96,** 301–7.

Kempster, A. J., Cuthbertson, A. & Harrington, G. (1982). *Carcass Evaluation in Livestock Breeding, Production and Marketing*. Granada Publishing, St Albans.

King, J. W. B. (1982). Potential use of in vivo techniques for breeding purposes. In *In Vivo Estimation of Body Composition in Beef*, ed. B. B. Andersen. Beretning fra Statens Husdyrbrugs Forsog, Copenhagen, pp. 86–93.

Lanza, E. (1983). Determination of moisture, protein, fat and calories in raw pork and beef by Near Infrared Spectroscopy. *J. Food Sci.*, **48**, 471–4.

Leymaster, K. A., Mersmann, H. J. & Jenkins, T. G. (1985). Prediction of the chemical composition of sheep by use of ultrasound. *J. Anim. Sci.*, **61**, 165–72.

Lister, D. (Ed.) (1984). *In Vivo Measurement of Body Composition in Meat Animals*. Elsevier Applied Science Publishers, London.

Lunt, D. K., Smith, G. C., McKeith, F. K. Savell, J. W., Riewe, M. E., Horn, F. P. & Coleman, S. W. (1985). Techniques for predicting beef carcass composition. *J. Anim. Sci.*, **60**, 1201–7.

Mersmann, H. J. (1982). The utility of ultrasonic measurements in growing swine. *J. Anim. Sci.*, **54**, 276–84.

Mersmann, H. J. (1985). Adipose tissue lipolytic rate in genetically obese and lean swine. *J. Anim. Sci.*, **60**, 131–5.

Mersmann, H. J. & MacNeil, M. D. (1985). Relationships of plasma lipid concentrations to fat deposition in pigs. *J. Anim. Sci.*, **61**, 122–8.

Mersmann, H. J., Brown, L. J., Chai, E. Y. & Fogg, T. S. (1984). Use of electronic meat measuring equipment to estimate body composition in swine. *J. Anim. Sci.*, **58**, 85–93.

Metz, S. H. M., Verstegen, M. W. A., de Wilde, R. O., Brandsma, H. A., van der Hel, W., Brascamp, E. W., Lenis, W. P. & Kanis, E. (1984). Estimation of carcass and growth composition in the growing pig. *Netherlands J. Agric. Sci.*, **32**, 301–18.

Miles, C. A. (1978). Note on recent advances in ultrasonic scanning of animals, European Meat Research Workers Conf., Kulmbach, W13:3.

Miles, C. A. & Fursey, G. A. J. (1974). A note on the velocity of ultrasound in living tissue. *Anim. Prod.*, **18**, 93–6.

Miles, C. A., Fursey, G. A. J., Fisher, A. V. & Brown, A. J. (1983*a*). Predicting carcass composition from the speed of ultrasound in live Hereford bulls. *Anim. Prod.*, **36**, 526 (Abstract).

Miles, C. A., Fursey, G. A. J. & Pomeroy, R. W. (1983*b*). Ultrasonic evaluation of cattle. *Anim. Prod.*, **36**, 363–70.

Miles, C. A., Fursey, G. A. J. & York, R. W. R. (1984). New equipment for measuring the speed of ultrasound and its application in the estimation of body composition of farm livestock. In *In Vivo Measurement of Body Composition in Meat Animals*, ed. D. Lister. Elsevier Applied Science Publishers, London, pp. 93–105.

Molenaar, B. A. J. (1984). Results of real time ultrasonic scanning in estimating lean tissue ratios in live pigs, 35th Annual Meeting of EAAP, Netherlands.

Molenaar, B. A. J. (1985). The use of real-time linear array ultrasound scanners for evaluation of live body composition, 36th Annual Meeting of EAAP, Greece.

Morel, P. & Gerwig, C. (1985). Prediction of meatiness at end of fattening on the strength of ultrasonic measurements at 25 kg live weight, 36th Annual Meeting of EAAP, Greece.

Muller, E. (1986). Physiological and biochemical indicators of growth and composition. In *Exploiting New Techniques in Animal Breeding,* ed., C. Smith, J. W. B. King & J. C. McKay. Oxford University Press, Oxford, pp. 132–9.

Presta, E., Segal, K. R., Gutin, B., Harrison, G. G. & Van Itallie, T. B. (1983). Comparison in man of total body electrical conductivity and lean body mass derived from body density: validation of a new body composition method. *Metabolism,* **32,** 524–7.

Preston, T., East, B. W. & Robertson, I. (1984). Body composition measurements of rats, sheep, pigs and humans by neutron activation analysis. In *In Vivo Measurement of Body Composition in Meat Animals*, ed. D. Lister. Elsevier Applied Science Publishers, London, pp. 181–4.

Preston, T., Fuller, M. F., East B. W. & Bruce, I. (1985). Preliminary experiments to assess the suitability of whole-body neutron activation for body composition analysis in 70 kg pigs. *Proc. Nutr. Soc.,* **44,** 109A (Abstract).

Robelin, J. (1973). Estimation de la composition corporelle des animaux a partir des espaces de diffusion de lean marquee. *Ann. Biol. Anim. Bioch. Biophys.,* **13,** 285–305.

Robelin, J. (1982*a*). Measurement of body water in living cattle by dilution technique. In *In Vivo Estimation of Body Composition in Beef,* ed. B. B. Andersen. Beretning fra Statens Husdyrbugs Forsog, Copenhagen, pp. 156–64.

Robelin, J. (1982*b*). Estimation of body composition by dilution techniques in nutrition experiments. In *In Vivo Estimation of Body Composition in Beef,* ed. B. B. Andersen. Beretning fra Statens Husdrybrugs Forsog, Copenhagen, pp. 107–17.

Robelin, J. (1982*c*). A note on the estimation in vivo of body fat in cows using deuterium oxide or adipose cell size. *Anim. Prod.,* **34,** 347–50.

Robelin, J. (1984). Prediction of body composition in vivo by dilution technique. In *In Vivo Measurement of Body Composition in Meat Animals,* ed. D. Lister. Elsevier Applied Science Publishers, London, pp. 106–12.

Robelin, J. & Agabriel, J. (1986). Estimation de l'etat d'engraissement des bovins vivants a partir de la taille des celules adipeuses. *Bulletin Technique, Centre de Recherches Zootechniques et Veterinaires de Theix,* **66,** 37–41.

Rogdakis, E., Ensinger, U. & Faber, H. V. (1979). Hormonspiegel, Plasma und Ensymaktintaten in Festgewebe von Pietrain und Edelsehureinen. *Z. Tiersuchtg. Zuchstbiol.,* **96,** 108–19.

Sather, A. P., Fredeen, H. T. & Martin, A. H. (1982). Live animal evaluation of two ultrasonic probes as estimators of subcutaneous backfat and carcass composition in pigs. *Can. J. Anim. Sci.,* **62,** 943–9.

Sather, A. P., Tong, A. K. N. & Harbison, D. S. (1987). A study of ultrasonic probing techniques for swine. II Prediction of carcass yield from the live pig. *Can. J. Anim. Sci.,* **67,** 381–9.

Sehested, E. (1984). Evaluation of carcass composition of live lambs based on computed tomography, 35th Annual Meeting of EAAP, Netherlands.

Sheng, H. P. & Huggins, R. A. (1979). A review of body composition studies with emphasis on total body water and fat. *Am. J. Clin. Nutr.*, **32,** 630–47.

Shields, R. G. Jr, Mahan, D. C. & Cahill, V. R. (1983). A comparison of methods for estimating carcass and empty body composition in swine from birth to 145 kg. *J. Anim. Sci.*, **57,** 55–65.

Simm, G. (1983). The use of ultrasound to predict the carcass composition of live cattle–a review. *Animal Breeding Abstracts,* **51,** 853–75.

Simm, G. (1986). In vivo estimation of carcass composition in breeding programmes for large animals, 28th Poultry Breeders' Roundtable, Cambridge, UK.

Skjervold, H., Gronseth, K., Vangen, O. & Evensen, A. (1981). In vivo estimation of body composition by computerised tomography. *Z. Tierzuchtg. Zuchtsbiol.*, **98,** 77–9.

Sorensen, M. T. (1984). Computerised tomography of goats during pregnancy and lactation. In *In Vivo Measurement of Body Composition in Meat Animals,* ed. D. Lister. Elsevier Applied Science Publishers, London, pp. 75–83.

Standal, N., Vold, E., Trygstad, O. & Foss, I. (1973). Lipid mobilisation in pigs selected for high and low fatness. *Anim. Prod.*, **16,** 37–42.

Stern, S., Anderson, K., Peterson, H. & Sundgren, P. E. (1984). Performance testing of boars using ultrasonic measurements of fat and muscle, 35th Annual Meeting of EAAP, Netherlands.

Stiffler, D. M., Walters, L. E. & Johnson, R. K. (1976). Use of the EMME as a measure of leanness in swine. In *Meat and Carcass Evaluation.* Oklahoma State University, MP-96, p. 137.

Stouffer, J. R. & Westervelt, R. G. (1977). A review of ultrasonic applications in animal science—reverberations and echocardiogram. *J. Clinical Ultrasound,* **5,** 124–8.

Strutz, C. & Rogdakis F. (1979). Phenotypic and genetic parameters of NADPH-generating enzymes in porcine adipose tissue. *Z. Tierzuchtg. Zuchtsbiol.*, **96,** 170–85.

Talbot, C., Preston, T. & East, B. W. (1986). Body composition of atlantic salmon (*Salmo salar* L.) studied by neutron activation analysis. *Comp. Biochem. Physiol.*, **85A,** 445–50.

Ter-Pogossian, M. M., Raichle, N. E. & Sobel, B. E. (1980). Positron-Emission Tomography. *Scientific American,* **243,** 140–4.

Thwaites, C. J. (1984). Ultrasonic estimation of carcass composition. *Australian Meat Research Committee Review,* **47,** 1–32.

Tong, A. K. W., Newman, J. A., Martin, A. H. & Fredeen, H. T. (1981). Live animal ultrasonic measurements of subcutaneous fat thickness as predictors of beef carcass composition. *Can. J. Anim. Sci.*, **61,** 483–91.

Ulin, K., Meydani, M., Zamenhof, R. G. & Blumberg, J. B. (1986). Photon activation analysis as a new technique for body composition studies. *Am. J. Clin. Nutr.*, **44,** 963–72.

Upton, W. H., Ryan, D. M., Mansfield, B. W. & Sundstrom, B. (1984). An evaluation of the Scanoprobe for measuring fat depth of beef cattle. *Animal Production in Australia,* **15,** 764 (Abstract).

Vangen, O. (1984). Evaluation of carcass composition of live pigs based on computed tomography, 35th Annual Meeting of EAAP, Netherlands.

Vangen, O. (1985). Computerised tomography in pig improvement, British Pig Breeders' Roundtable, Wye College, UK.

Vangen, O. & Kolstad, N. (1986). Genetic control of growth, composition, appetite and feed utilisation in pigs and poultry, *3rd World Congress on Genetics Applied to Livestock Production,* Vol. XI, Saur, London.

Vangen, O., Standal, N. & Walach-Janiak, M. (1984). Tissue deposition rate in genetically lean and fat pigs estimated by computed tomography, 35th Annual Meeting of EAAP, Netherlands.

Van Loan, M. & Mayclin, P. (1987). A new TOBEC instrument and procedure for the assessment of body composition: use of Fourier coefficients to predict lean body mass and total body water. *Am. J. Clin. Nutr.,* **45,** 131–7.

Wangsness, P. J., Martin, P. J. & Gahagan, J. H. (1977). Insulin and growth hormone in lean and obese pigs. *Am. J. Physiol.,* **233,** E104–8.

Wright, I. A. & Simm, G. (1986). Composition in the live animal and its carcass. *J. Sci. Food Agric.,* **37,** 431–2.

Chapter 7

New Approaches to Measuring Fat in the Carcasses of Meat Animals

A. V. FISHER
*Department of Meat Animal Science, University of Bristol, Langford, Bristol, UK**

INTRODUCTION

In 1926, Lush wrote 'An indicator of fatness, which could be expressed quantitatively and which would be free from the personal opinion of the men who judge the live cattle and the dressed meat, would be very helpful in interpreting the results of feeding trials. . . . A definite knowledge of the degree of fatness of the dressed meat would add much to the knowledge of its value for human food, although it may be true . . . that the desirability of fatness has been overemphasised in much of the literature on the subject'. More than sixty years later, the goal remains the same, namely the objective measurement of fatness to replace, or at least supplement, visual assessment, and there have been developments to achieve this aim. But the pressures which mould these technological advances have themselves changed markedly: the previously expedient task of producing a fat carcass is now condemned, and the slight doubts concerning the desirability of fatness have been replaced by categorical statements that carcass fatness should be reduced (Department of Health and Social Security, 1984), primarily because of the acceptance of evidence connecting dietary

*Formerly Carcass and Abattoir Department of the AFRC Institute of Food Research—Bristol Laboratory, Langford, Bristol, UK.

fat intake and the incidence of cardiovascular disease. This change of policy in animal production has intensified the need to identify levels of fatness in livestock and carcass meat. The future of the livestock industry, during times when meat is likely to be in increasing competition with other foods, will increasingly depend on the ability of the industry to meet the new requirements of the modern consumer. There are already derivatives of soya bean protein and fungal protein which have been formulated into meat replacers and are being marketed as individual ingredients or incorporated into ready meals. Such products are either low in total fat and/or low in saturated fatty acids; their appeal to the consumer, even if based solely on these considerations, should not be underestimated.

In this chapter, the nature of carcass fatness is examined and methods of estimation are reviewed. In assessing the merits of different techniques, due accord must be given to the different applications, which broadly divide into research and academic needs to further scientific understanding of the factors influencing carcass composition, and the commercial need to more fully describe a marketable product. It is possible only to speculate which of these applications will be the more important in changing the composition of meat in future years. There is already evidence of a trend for the beef industry to trim more external fat prior to merchandising, and it has been claimed that eventually the retail food industry will be selling beef with no subcutaneous or intermuscular fat (Coleman *et al.*, 1988). Clearly, the need to estimate the fat content of carcasses is lessened if much of the fat is removed 'at source' and is no longer part of the product being marketed; and if practically all fat is removed prior to sale there will be virtually no incentives to specify fatness levels from the retailing end of the chain. However, this is probably an over-simplification, for it is practically impossible to produce beef which classifies as good quality from other aspects without fat deposition occurring in the carcass, using many breed types and production systems currently in operation, and the trimming of fat is an expensive and wasteful process affecting overall profitability of the trade. The need for improved carcass description will therefore continue to be a priority, necessary to improve marketing operations between producers and abattoirs during times when the link between quality and price may well be strengthened. Scientific developments, on the other hand, are actively pursuing the control of fatness, and hence its reduction, through various immunological and endocrine pathways, and in many studies indirect estimates of their effects on carcass composition are employed. The control of fatness in the growing animal and the effects that market pressures can bring to bear on changing carcass composition are the subjects of other chapters in this book, and will not be considered further here.

KEY OBJECTIVES IN ESTIMATING CARCASS COMPOSITION

Choice of End-Point

Although they are to a large extent complementary, a distinction is to be made between 'fatness' and 'leanness' of a carcass. The proportion of lean meat in a carcass has usually been the focal point in scientific studies, primarily because the yield of lean meat is the single most important factor determining carcass value (Harries *et al.*, 1975). At the same level of carcass fatness, there is commercially important variation in leanness owing to variation in the third major carcass component, bone. It is unusual to see this relation expressed in reciprocal form, namely that at the same level of carcass leanness, there is important variation in fatness, but human nutritionists might argue that this is a more relevant expression of carcass quality in today's market.

The option of which end-point to choose in prediction methodology has relevance in the selection of carcass characteristics which are to be measured. There are some techniques that provide indices of both fat and lean mass, but there are others that quantify an index (usually thickness) of only one tissue (usually fat). The relation between a measurement of fatness and carcass leanness is an indirect one and, *a priori*, such a relation is less stable than a direct one linking an index of lean development to the total mass of that same tissue. The main sources of error in this relation include not only the error ascribed to differences in fat partition between body depots and variation in fat distribution (which could also be error sources in the prediction of total fat content), but also the variation in bone content. These error components can be attributable to breed type, sex, or environmental factors operating during the growth of the animal.

These indirect relations can be regarded as accepted limitations in the application of some prediction techniques used to estimate lean content. Although there is no parallel in the prediction of fat content (the author is not aware of any study in which the key objective was estimation of fat content through measurement of an index of muscularity), which is the subject of this review, this matter is of consequence since most studies are aimed at the prediction of lean content, and the data are reported in terms of the precision of *its* estimation. They are thus included in this chapter.

Accuracy and Precision

The need for an accurate prediction is self-evident, the only exception to this being investigations which aim to rank animals or carcasses according to their body composition. For the majority of applications, it is obviously an advantage if a prediction method is less subject to the effects of animal or environment. The testing of the accuracy of a method has not been undertaken in carcass evaluation studies as often as it should have been, but more recent investigations have explored the extent of bias and its determining factors. For example, Diestre & Kempster (1985) found that derived overall prediction equations to estimate lean percent in pigs, based on carcass weight, and a number of fat and muscle thicknesses, produced bias when applied to separate source populations. Moreover, increasing the number of independent variables did not reduce the size of the bias.

The degree of precision required in the prediction of carcass composition is not so self-evident as the need for high accuracy, and the reasoning underpinning the choice is more complex. In scientific studies where factors influencing carcass composition are being evaluated, the level of precision required is generally high. Here, the compositional data form part of complex interactions of cause and effect, and the need to minimise uncertainty in these equations demands precise measurement. For marketing purposes, the arguments are different. For example, if carcass value and payment are based on carcass composition, it can be argued that because animals are usually sold in batches, precision need not be high because the producer, or wholesaler, will receive a price which reflects the mean value, or composition. But for many market outlets, the demand is for a carcass or joint having a lean-to-fat ratio optimal for a specific purpose, and uniformity of product is essential. This goal of meeting a specification is becoming an increasingly common stipulation in marketing agreements and it is advantageous to have a measuring system with a greater, rather than lesser, degree of precision.

An example of how small increases in precision have been identified and incorporated into a carcass classification system is provided by the development in automatic light reflectance probes used for pig carcass classification. These probes, in addition to measuring backfat depths and thus providing the information required to classify according to P2 (backfat depth at the last rib), also measure the depth of the muscle *longissimus thoracis*, lying directly underneath the backfat in the rib region. The inclusion of this muscle depth provides a statistically significant improvement in the precision of predicting carcass lean proportion in commercial British pigs, but the improvement is small (Kempster *et al.*, 1985). Nevertheless, muscle depth is incorporated into the prediction equations used to estimate lean percent in the EC pig carcass grading scheme which became operative on January 1st 1989 (Meat and Livestock Commission, 1986).

The statistic used as a criterion of precision is the residual standard deviation, abbreviated to RSD in the text. However, this statistic is partly dependent on the variance of the dependent variable in the particular study used to derive each prediction equation, and comparison between studies is always a matter of some judgement. Within a study, a comparison of prediction equations is aided by calculation of the C_p statistic, which identifies the relative contribution to the residual variance made by squared true error (the difference between the mean of observed values of Y, at a given value of X, and the individual Y values) and squared lack of fit (the difference between the mean of the observed values of Y at a given value of X and the value calculated from the regression equation). MacNeil (1983) has recommended that C_p is the best single criterion within a single study, and the use of C_p is illustrated in the prediction of pork belly composition by Johnson *et al.* (1984).

When the precision of prediction is examined for individual carcasses, the probable error depends on the deviation of the value of the X (predictor) variable from the mean value of the X variable used in establishing the regression (prediction) equation. Timon and Bichard (1965) gave examples of the precision of estimation of percentage composition of individual lamb carcasses lying at the extremes of an observed range in X, and compared this probable error with that obtained for a group of 12 lambs. In this example, the X variables were the percentages of fat, muscle and bone in two sample joints (loin and best end of neck) which were used to predict these same total percentages in the carcasses of 83 castrated male lambs. The standard error of prediction (maximum probable error, Sy) for individual estimates was approximately 3·5 times greater than that based on 12 observations, and the 95% confidence limits differed by a factor of similar magnitude. These data show that, despite the high correlations observed between the X and Y variables (Table 1), differences between individual animals cannot reliably be quantified even by a relatively precise technique.

Table 1
Examples of the expected error in prediction of tissue percentages of individual lambs at the limits of the observed range, and of the mean of groups of 12 lambs, based on sample joint (loin, best-neck) dissections (from Timon & Bichard, 1965)

		Fat %		*Muscle %*		*Bone %*	
		$\bar{x}$	*SD*	$\bar{x}$	*SD*	$\bar{x}$	*SD*
Means ($\bar{x}$) and standard deviations (SD) of % tissues in carcass and sample joints	Carcass	27·9	3·9	55·2	3·1	16·3	1·6
	Loin	35·0	5·5	51·4	2·5	13·1	1·9
	Best-neck	34·3	5·7	47·5	4·9	17·4	2·6
		r	*RSD*	*r*	*RSD*	*r*	*RSD*
Correlation coefficients (*r*) and residual standard deviations (RSD) from regression of tissue percentages in carcass on same tissue in sample joint	Loin	0·96	1·1	0·93	1·2	0·84	0·8
	Best-neck	0·94	1·4	0·92	1·2	0·75	1·0
		Sy individual	*95% limits*	*Sy individual*	*95% limits*	*Sy individual*	*95% limits*
Estimates of maximum probable error (*Sy*) and 95% confidence limits for individual and group ($n = 12$) estimates of percent fat, muscle and bone	Loin	1·12	±2·22	1·23	±2·44	0·82	±1·62
	Best-neck	1·39	±2·76	1·27	±2·52	1·14	±2·26
		Sy group	*95% limits*	*Sy group*	*95% limits*	*Sy group*	*95% limits*
	Loin	0·32	±0·63	0·36	±0·71	0·24	±0·47
	Best-neck	0·40	±0·79	0·37	±0·73	0·33	±0·65

In the future, precision in estimating composition may receive a new emphasis as the result of a more detailed description of individual animals. It is conceivable that animals may be identified using new technologies (e.g. transponders) and that essential details of breeding, management and health are logged on the system. The linking of a precise estimate of composition in a system that supplies these other details would provide a valuable database to examine factors affecting composition.

Kempster (1984) has assessed the relative importance of different criteria in different applications of techniques for estimating body composition *in vivo*. He concludes that for valuation in the marketing chain, precision and accuracy are equally important, and it is a fair assumption also that they should have about equal weighting in carcass valuation procedures.

The Need for Objectivity

The third key objective in estimating carcass composition is to make the systems entirely objective. It can be argued that the use of visual assessment schemes, in a formalised classification system involving training of assessors and the use of photographic standards as reference points, results in an evaluation system which is not subjective, but rather sensory in nature. However, unlike instrumental measurement which can almost invariably be calibrated and made to operate at a known resolution for a given set of environmental parameters, visual assessment is in every way less stable, so that differences between operators, or between periods of assessment when made by the same operator, are major causes for concern even when photographic standards theoretically are calibration tools. This lack of consistency undermines the confidence that those using (and paying for) commercial classification of beef and lamb carcasses have in its application.

It should be noted that some techniques are partly objective, but require some human element of interpretation in their operation. Thus the measurement of backfat thickness using the Intrascope or Optical probe is objectively made by sliding a graduated cylinder over a barrel carrying a pointer, but the positioning of the barrel is decided by the operator who identifies a fat/lean boundary. Even sophisticated technologies such as pulse-echo ultrasound tissue mapping require some interpretative input.

DEFINITIONS OF CARCASS FATNESS

Chemical Versus Physical Entities

Two definitions of 'fat' in the carcasses of meat animals are frequently encountered, and are necessary to describe meat composition in nutritional or biochemical terms on the one hand, and anatomical or butchers' terms on the other. Strictly speaking, 'fat' is the term reserved for the group of solidified lipids known as glycerides (esters of glycerol and fatty acids), the most abundant in animal tissues being the triglycerides in which each of the three alcohol groups of glycerol is esterified with fatty acids; but 'chemical fat' is the label attached to that mass of extractable lipid using a specified solvent. This commonly comprises predominantly the triglycerides but other lipids will be present in amounts which vary depending on the solvent used, and lipids themselves are a very varied group of compounds that differ markedly in their chemical nature (Masoro, 1968).

Triglycerides are present in many tissues of the body, but are stored mainly in adipose tissue. The main masses of adipose tissue lying outside the musculature constitute the physically separable fat, the concept of more relevance to the marketers of meat. This adipose tissue consists of fat cells which develop around a capillary bed, the whole being supported in a connective tissue matrix, and its development has been reviewed by Robelin (1986). The proportion of the total weight of adipose tissue contributed by triglyceride varies with the growth and development of the animal and differs between depots. However, this proportion increases very rapidly after birth (Wood, 1984), and in meat animals of normal slaughter weights there is thus a high correlation between the weight of chemical fat in the adipose tissue and the weight of dissectible fat.

The relation between the amount of dissectible fat and its lipid content in cattle, sheep and pigs has been calculated by Kempster *et al.* (1986*a*) for carcasses covering the United Kingdom commercial fat range. Only in the equations for beef was there due allowance made for the relative amounts of subcutaneous and intermuscular fat. The lipid content of intermuscular fat is substantially lower than that of subcutaneous fat, particularly at low overall carcass fatness levels, and the ratio of these depots is influenced by breed (particularly in cattle) and degree of total fatness. The estimates provided are, nevertheless, useful guides to the fat content not only of adipose tissue but also lean tissue and the combination of both (= total edible tissues) in carcasses of different composition.

The relations between the amount of separable fat in the carcass and total carcass (including bone) lipid for cattle, sheep and pigs are shown in

Table 2
Regressions of separable fat (subcutaneous plus intermuscular unless otherwise stated) on total carcass lipid in cattle and pigs (both variables as percentages) and in cattle and sheep (both variables in kg)

	Reference	*Equation*	*r*	*RSD*
As percentages				
Cattle[a]	Fisher *et al.* (1983)	$y = 0{\cdot}934x + 1{\cdot}271$	0·986	1·20
Pigs[b]	Smith & Pearson (1985)	$y = 1{\cdot}01x - 3{\cdot}84$	0·98	0·74
As absolute weights				
Cattle[a]	Fisher *et al.* (1983)	$y = 1{\cdot}018x + 1{\cdot}181$	0·989	N/A
Sheep[c]	Ulyatt & Barton (1963)	$y = 1{\cdot}04x - 0{\cdot}68$	0·99	0·42

[a]Excludes kidney and channel fat.
[b]Excludes head, feet and flare fat.
[c]Includes kidney and channel fat.
N/A signifies 'not available'.

Table 2. The two variables exhibit a high degree of correlation and, furthermore, these relations appear to be very similar in all three species.

Location of Fat in the Carcass

In most mammals, there is very little adipose tissue present at birth (Adolph & Heggeness, 1971), but during growth to commercial slaughter weights there are important changes in composition of meat animals with the proportion of fat increasing, markedly so in the later stages. However, this growth is not uniform throughout the carcass, so the distribution of fat is constantly changing, and this has implications for the prediction of total fatness. The detailed description of fat tissue development and its anatomical variation is problematical because fat, unlike muscle tissue, is not formed in discrete anatomical units. As stated by Pond (1984), there is a widespread assumption that adipose tissue 'has no comparative anatomy' because of these limitations coupled to the difficulty of identifying adipose tissue cells in young, or poorly nourished animals, in which insufficient amounts of triglyceride have been deposited. In meat animals, most studies have examined differences in the *partition* of fat between confluent volumes of the carcass which have anatomical identities defined with respect to adjacent tissues. Thus, the so-called *depots* lie (a) peripheral to the main muscle mass occupying the volume under the skin in the living animal (the subcutaneous depot); (b) between the muscles, and between muscles and bones, lying within the periphery formed by the superficial faces of the most superficial muscles (the intermuscular depot); and (c) the fat lying within the epimysial boundaries of skeletal muscles (the intramuscular fat). In addition to the partition between depots, the *distribution* between *sites* (within a depot) has also been the subject of study, and usually the sites have been defined with respect to butcher's joints (e.g. Kempster *et al.*, 1976; Butler-Hogg & Wood, 1982), less frequently by anatomical reference (e.g. Williams & Bergstrom, 1980).

Many of the simple predictors of overall carcass composition used in breeding programmes and classification schemes involve the measurement of subcutaneous fat thickness or areas of fat. The important relations from the viewpoint of prediction are, therefore, those involving the amount of fat at the measurement site and the total amount of carcass fat. These relations have usually been examined by workers employing the allometric equation, originally developed by J. S. Huxley (1932) to describe the growth of an organ or part relative to that of the remainder of the body. In studies on fat development, the procedure has almost invariably been to relate growth of a depot, or site (y) to the growth of the total fat mass, or depot mass (x) (not the total less the depot or site in question) using the equation $\log y = \log a + b \log x$, where a is a constant.

Although this 'corrupted' form can be physiologically misleading (e.g. in considerations of nutrient supply during growth), it has advantages in facilitating the conceptual appreciation of development of parts since the rates of relative growth (the growth coefficients, or b values in the equation) can be directly compared. Fat location has been reviewed by Kempster (1981) and Wood (1984), and their salient points and other relevant information are summarised below.

Partition of Carcass Fat

In cattle, sheep and pigs, the growth of the subcutaneous depot relative to total carcass or total body fat is higher than that of intermuscular fat, and the difference is greatest in the ruminant species over a wide range from near-birth to near-maturity. Intramuscular fat appears to have the lowest growth coefficient of all three carcass depots (Wood, 1990).

There are important breed effects on the partitioning of fat, the extremes of variation occurring in beef with the traditional British beef breeds such as the Hereford and Aberdeen Angus having relatively high amounts of subcutaneous fat and low amounts of abdominal fat compared with milk-producing breeds (e.g. Friesian and particularly Jersey) at the same level of total fatness. Of greater importance in prediction methodology is the bias introduced through a variable subcutaneous/intermuscular fat ratio, and although this is higher in the extreme British beef breeds than in the extreme dairy breeds, the data of Kempster (1981) show that it is also low in some beef-type breeds such as the Lincoln Red and the Galloway. Charolais crosses also had a low ratio, whereas the Limousin appears to have low levels of intermuscular fat.

There are also indications that the variability *within* depots is related to breed type, and may affect prediction precision. Fisher and Bayntun (1984) showed that in certain of the British beef breeds, the variation in subcutaneous fat weight was substantially greater than the variation in intermuscular fat, whereas the reverse pattern was observed in samples from the dairy or dairy-cross breeds. These effects, in a sample of Herefords and Friesians matched for fat proportion, resulted in a higher prediction error for total dissectible fat in the dairy breed (RSD = 3·0 kg) than in the Hereford (RSD = 2·4 kg) when subcutaneous fat weight was used as the predictor. It has not been established how important this greater instability is in the dairy types when indirect measures of subcutaneous fat mass, e.g. fat depths, visually assessed fat cover, are used in prediction.

Breed effects on fat partitioning in sheep and pigs appear to be smaller than those in cattle, but in sheep there is a similar trend in that the meat-type breeds have more carcass fat and less body cavity fat than the ewe-type breeds (Wood *et al.*, 1980), particularly when comparisons are made with high milk-yielding breeds such as the Friesland (Butler-Hogg & Whelehan, 1986). However, in a comparison of studies, involving mainly crossbred lambs, Kempster (1981) found inconsistencies within breeds, and suggested that environmental factors or dam breed effects were responsible. In pigs, lines selected for improved performance characteristics, which included reduced backfat, tended to show a redistribution of body fat, with more fat occurring in the intermuscular depot, and less in the subcutaneous depot (Kempster & Evans, 1979*a*; Rook *et al.*, 1987).

Kempster and Evans (1979*a*) compared, at the same level of total carcass fat, the amounts of subcutaneous fat and intermuscular fat in pig genotypes which exhibited the greatest difference in fat partitioning, by expressing the amounts as a ratio. For subcutaneous fat the value was 1·03, and for intermuscular fat, 1·06.

Sex Effects

The comparison between sexes in the amount of fat in the body at a common body weight shows that in the ruminants, entire males have less than castrates which in turn have less than non-breeding females; in pigs, castrated males are generally fatter than females (Martin *et al.*, 1972). But the partition of fat is not very different between the sexes. Berg *et al.* (1979) found that there were no significant differences between sexes (heifers, steers and bulls) in the allometric growth coefficients from regressions of depot weight on total carcass fat weight, and mean values for the depot weights adjusted to a common total side fat weight were very similar in the three sexes. In their data, heifers had slightly more subcutaneous fat than steers (9·82 and 9·36 kg respectively) and less intermuscular fat (10·39 and 10·60 kg respectively), a finding confirmed by Fisher *et al.* (1988) in animals from a much narrower weight range.

Data for pigs (Richmond & Berg, 1971) lead to a similar conclusion that differences between sexes in fat partition are very small, but again there is evidence that females have slightly more subcutaneous fat and less intermuscular fat than castrated males (Kempster & Evans, 1979*b*).

The data for sheep show more diverse relations. Thompson *et al.* (1979) found no difference between ewe lambs and wethers in fat partition at the same total carcass fat, but Jones (1982) found that, although there were no differences in the rates of fat deposition in the carcass depots relative to total carcass fat, ewes had slightly more intermuscular fat than rams at the same total carcass fat weight. Butterfield *et al.* (1985) found a marked effect of castration in the Dorset Horn breed, so that at a stage of fat development approximating to 0·6 of mature total body fat weight, rams had approximately 80 g/kg total fat less in the subcutaneous depot, and 70 g/kg more in the intermuscular depot.

Distribution of Depot Fats Between Sites

The pioneering work on the development of fat and its distribution among different anatomical regions was that of Hammond (1932), who postulated growth gradients similar to those occurring in muscle where the lowest growth rates occur in the distal limbs and the fastest post-natal growth occurs in the central region of the trunk. This general pattern is common to both subcutaneous and intermuscular fat, and applies across the species cattle, sheep and pigs. But there is a certain amount of conflicting evidence when the detail of relative growth rates in the different anatomical regions of the later maturing trunk is examined. Because of its importance in prediction accuracy, this is examined in more detail under 'Estimation of Carcass Composition—Measurement of Tissue Thicknesses'.

ESTIMATION OF CARCASS COMPOSITION

Quantifying Total Subcutaneous Fat

Visual Appraisal

When slaughtered cattle are dressed and the hide removed, the subcutaneous fat covering most of the carcass surface is exposed, the exception being the ventro-lateral areas, predominantly in the forequarter, over which the *cutaneous trunci* remains on the carcass. For sheep, the skin separates from the underlying tissues more readily and the entire cutaneous muscle remains on the carcass. Even so, it is possible to perceive the development of subcutaneous fat, particularly in the hind limb, shoulder and neck regions, so that in sheep and cattle the visible surface provides information on the degree of fattening, whereas in pigs and poultry, the skin is left on the carcass and this information is not available.

In visual appraisal systems, the degree of fat cover is assessed by an operator who scores this development on a scale of points. The most detailed description of the amount of fat cover on beef carcasses has been given by Roy and Dumont (1975) who related these descriptions to a scale of points in which there were five main classes, ranging from 1 (lean) to 5 (fat), and in which each main class had additional subclasses either side of the class mean to account for lesss (−) or more (+) fat development. There are thus fifteen points in this scale, which contrasts with the optimum of seven points suggested by Williams (1969) on the grounds that '(seven) was considered the psychologically optimum number of categories which could be conceptualised as statements'. In a study of the prediction of carcass composition of different cattle breeds and crosses, Kempster *et al.* (1986*b*) found that the overall sample standard deviation of lean percentage (3·34) was reduced to a residual standard deviation of 2·47 using a 7-point fatness scale, and to 2·28 using a system in which subcutaneous fat was estimated to the nearest percentage point (SF_e)(in practice, virtually a 15-point scale). Similarly, in a trial involving 1478 lambs sired by ten sire breeds, the overall carcass lean percentage of 3·83 was reduced to a residual standard deviation of 2·97 using carcass weight plus a 5-point fatness score as predictors and to 2·61 when the latter score was substituted by SF_e (Kempster *et al.*, 1981*a*). Essentially similar results were obtained in a study of a larger number (2808) of lambs (Kempster *et al.*, 1986*c*).

The difficulties involved in standardising assessments and which influence accuracy and consistency have been discussed by Kempster *et al.* (1982*a*). However, in commercial classification, scales are rarely divided into more than seven classes, presumably because it was felt that more categories would be difficult to operate successfully, and perhaps because the role of a subjective classification in marketing has not demanded greater discrimination. Certainly price differences among beef carcasses of different fat class (same conformation class) in the United Kingdom do not adequately reflect the effect of variation of fatness on saleable meat yield (Meat and Livestock Commission, 1983), and while such meagre price differentials exist there is little incentive to provide the deadweight market with a more discriminatory, albeit more complicated visual classification system.

Video Image Analysis

The system involving the human visual sense, comparison with a set of known standards (either stored in the memory complex of the human brain, or available to the assessor at the time of operation in the form of photographs) and decision-making, can be mimicked using a combination of camera/computer technologies.

The principles of the technique are well established. Briefly, light reflected from the object (carcass) strikes the photo-sensitive surface of the tube/semi-conductor in the video camera, and individual elements in the tube become charged to a degree that is proportional to the intensity of light at that position. An electron beam scans the elements and each of these discharges a voltage which is proportional to the light intensity. In this way, an electronic 'map', comprising 240 000 elements, or pixels, in the full screen image (600 lines each containing 400 pixels) is constructed. This map can then be interpreted by the analyser in a number of ways. Pre-set voltages (thresholds) can be used to differentiate the object from the background and high intensity areas (fat) from lower intensity areas (lean). Carcass profiles, areas, and perimeters can be determined, and the use of additional cameras strategically placed, allows information to be generated from different aspects of a three-dimensional object.

Video image analysis (VIA) systems have already been developed for use on boneless processing meats in which lean and fat proportions are quantified (Newman, 1984), for minced meats (Newman, 1987), and a schematic plan for estimation of fat cover on beef carcasses has been given by Sorensen (1984). VIA of the cross-sectional tissue map between the 12th and 13th ribs to estimate carcass yield has been attempted by Cross *et al*. (1983), and by Wassenberg *et al*. (1986). To date, no published results exist on the use of VIA to quantify carcass surface fat, but there are claims that such a system can be applied to beef and lamb (Newman & Wood, 1989). The *a priori* advantages of VIA are that the method does not involve carcass contact or penetration and there is therefore no possibility of cross-contamination, and it is extremely quick, all carcass data being captured within 0·02 s of being positioned in front of the camera and processed within 3 s. Equally, there are potential problems that can be envisaged; for example, the variable extent of the area of the carcass

Table 3
The precision (residual standard deviations) of estimating lean percentage (unless otherwise indicated) in the carcass from measurements of subcutaneous fat depths at different sites

Position	Residual standard deviation		
	PIGS		
Reference: Kempster & Evans (1979*a*)			
	Pork weight	*Bacon weight*	*Heavy weight*
(a) Over the dorsal mid-line			
Shoulder	3·08	3·32	3·27
Mid-back	3·05	3·18	3·29
Loin 2	2·66	2·89	2·87
(b) 6·5 cm from the dorsal mid-line			
4/5th cervical	3·07	3·30	3·45
7th rib	2·97	2·96	2·92
10th rib	2·63	2·68	2·86
13th rib	2·25	2·20	2·42
Last rib	2·25	2·36	2·32
2/3rd lumbar	2·41	2·45	2·53
5/6th lumbar	2·64	2·76	2·83
Reference: Fortin *et al.* (1984)			
	Distance from dorsal mid-line		
	5 cm	*7 cm*	*9 cm*
(a) Last rib[a]	2·44	2·41	2·39
3/4th from last rib[a]	2·30	2·26	2·23
4/5th from last rib[a]	2·34	2·26	2·31
5/6th from last rib[a]	2·41	2·31	2·30
Maximum loin[a]	—	—	2·54[b]

(b) Over the dorsal mid-line	
Cranial)	2·53
Middle) edges of *m.gluteus medius*	2·32
Caudal)	2·49
Maximum loin	2·60
Maximum shoulder	2·67

Reference: Diestre & Kempster (1985)

	Within sex	*Within source population*	*Overall*
P_2	2·42	2·31	2·52
Over dorsal mid-line			
Shoulder	3·00	2·87	3·18
Mid-back	3·07	2·94	3·23
Loin 2	2·72	2·61	2·85
On cut surface at level of last rib			
C	2·30	2·18	2·35
J	2·84	2·76	2·98
K	2·48	2·32	2·51
E	3·07	2·96	3·22
CP_1	2·26	2·16	2·31
CP_2	2·27	2·14	2·31
CP_3	2·21	2·07	2·24

All above RSD values from regression including carcass weight.

C = over max. depth of *m.longissimus*; J = innermost layer of fat at C; K = at dorso-lateral corner of *m.longissimus*; E = fat+skin level with dorsal edge of cut *m.transversus abdominis*; CP_1 = 45 mm from dorsal mid-line; CP_2 = 65 mm from dorsal mid-line; CP_3 = 80 mm from dorsal mid-line.

Table 3—*contd.*

Position	*Residual standard deviation*		
Reference: Sterrenburg (1989)			
Position[c]	*Distance from dorsal mid-line*		
	4 cm	*7 cm*	*10 cm*
40 (approx. 5th lumbar vertebra)	4·61	4·51	4·45
42	3·98	4·08	3·99
44	3·45	3·93	3·79
46	3·19	3·73	3·47
48	2·83	3·24	2·93
50	2·62	2.77	2·51
52	2·46	2·50	2·43
54	2·48	2·47	2·50
56 (approx. 12th rib)	2·66	2·27	2·49

CATTLE

Reference: Johnson & Vidyadaran (1981)	
	Dissectible fat %
Sacral crest	3·07
Biceps femoris	2·81
10th rib	3·12
12th rib	2·81
2nd sternebra	4·80
5th sternebra	3·78

Reference: Bass *et al*. (1982)

			Trimmed, deboned, saleable meat (kg)
(a)	mm ventral to ventral edge of *m.longissimus thoracis* over 11th rib	0	3·38
		2	3·52
		4	3·94
		6	3·21
		8	3·13
		10	2·91
		12	3·58
		14	3·82
		16	3·92
		18	4·00
		20	3·95
		22	3·97
(b)	At 13th rib		
	Over centre of *m.longissimus thoracis* 'C'		2·81
	8 cm ventral to neural spine		3·00
(c)	'GR' measurement with steel ruler	—hot	3·30
		—cold	3·38

All above RSD values from regression including side weight.

Reference: Chadwick & Kempster (1983)

	Hot	*Cold*
6th rib, 7·5 cm from mid-line	3·03	2·63
10th rib, 10 cm from mid-line	2·96	2·94
10th rib, 12·5 cm from mid-line	2·84	2·87
13th rib, 10 cm from mid-line	2·93	3·19
13th rib, 12·5 cm from mid-line	3·03	3·36

All above RSD values are the weighted means from four trials. Carcass weight included in all regressions.

Table 3—*contd.*

Position	*Residual standard deviation*
Reference: Jones *et al.* (1986)	
(a) 11/12th rib	
0·25 width of *m.longissimus thoracis*	2·36
0·50 width of *m.longissimus thoracis*	2·29
0·75 width of *m.longissimus thoracis*	2·18
Minimum fat thickness	2·16
(b) 12/13th rib	
0·25 width of *m.longissimus thoracis*	2·34
0·50 width of *m.longissimus thoracis*	2·26
0·75 width of *m.longissimus thoracis*	2.21
Minimum fat thickness	2·19
All above RSD values from regression including carcasss weight and sex/year effects.	
Reference: Kempster *et al.* (1986*a*)	
(a) At the 6th rib	
5 cm from mid-line	2·80
7·5 cm from mid-line	2·82
10 cm from mid-line	2·87
(b) At the 10th rib	
5 cm from mid-line	2·73
7·5 cm from mid-line	2·75
10 cm from mid-line	2·84
Over maximum width of *m.longissimus thoracis*	2·89
0·25 width of *m.longissimus thoracis*	2·89
0·50 width of *m.longissimus thoracis*	2·98
0·75 width of *m.longissimus thoracis*	2·84

All above RSD values are overall values combining data from dairy-bred and suckler bred cattle. Side weight included in all

SHEEP	
Reference: Chadwick *et al*. (1986)	
4·5 cm from mid-line	
8th rib	3·9
12th rib	2·9
3/4th lumbar	2·7
6/7th lumbar	2·8
All above RSD values from regression including carcass weight.	
Reference: Kempster *et al*. (1986*b*)	
6th rib, 4·5 cm from mid-line	3·68
12th rib, 4·5 cm from mid-line	3.47
3/4th lumbar, 4·5 cm from mid-line	3·42
J (ventral to *m.longissimus lumborum* at 3/4th lumbar vertebrae)	2·91
Scapular	3·63
5th thoracic, ventral edge of scapula	3·68
Ventral mid-line, 2nd sternebra	3·77
Ventral mid-line, 6th sternebra	3·72
Inflection (9 cm distal to caudal end of sacrum)	3·45
Flank	3·79
All above RSD values from regression including carcass weight.	

[a]Mean of two values obtained from two light reflectance probes.
[b]8 cm from mid-line.
[c]Percentage of carcass length measured from distal hind foot to tip of snout.

covered by *m.cutaneous trunci*, the shadow effects produced by irregularities on the subcutaneous fat surface, and carcass blemishes produced by bruising and marks resulting from contamination by blood, etc. However, software development in artificial intelligence will probably allow most of these to be overcome (expert system).

Measurement of Tissue Thicknesses

For pig carcasses, in which as much as 80% of the carcass fat lies in the subcutaneous depot, thickness measurements of this accessible depot provide useful information on its mass and, through a significant part-whole relation, on the mass of total carcass fat. In sheep, and particularly cattle, subcutaneous fat forms a much smaller proportion of the total, being of the order of 35–65% and 20–50% for lean and fat examples of the species, respectively. In addition, the skin is left on pig carcasses, contrasting with the situation in sheep and cattle, and this smooth outer boundary reduces the error in recording fat thickness. In beef particularly, the surface contours of fat can be very variable, as some fat is removed with the hide. Because of this, in the yield grade prediction equations operated by the United States Department of Agriculture, the subcutaneous fat thickness measured at the 12th rib is adjusted by the grader to correct for unevenness in distribution. In the study by Abraham *et al.* (1980), the R^2 value from the relation between retail cuts yield and unadjusted fat thickness was 0·79, whilst that using adjusted fat thickness was 0·83. There is also a suggestion that the adjustment should be greater in a positive direction (i.e. increased estimate of fat thickness) for heifers compared with steers (Murphey *et al.*, 1985). Despite these limitations, the appeal of a simple objective measurement to classify beef and sheep carcasses has been such that the variation in thickness of subcutaneous fat has been the subject of detailed study, as it has in pigs. In particular, researchers have examined the precision of the relation between fat thickness at various sites on the carcass and the composition of the carcass obtained by dissection. Some estimates of this precision using different fat thicknesses are summarised in Table 3 for pigs, cattle and sheep.

The most precise esimates of pig carcass composition are provided by fat depths measured lateral to, and not directly over, the dorsal mid-line. Generally, precision increases as the measurement site moves cranially from the caudal loin region to the last rib region, and remains high up to the 12th rib or so. In the detailed study by Sterrenburg (1989), in the region from about the 3rd lumbar cranially to the last rib region, measurements taken at 4 cm from the dorsal mid-line were consistently more precise than measurements at 7 cm for the mid-line, and in most cases more precise than at 10 cm from the mid-line (Fig. 1).

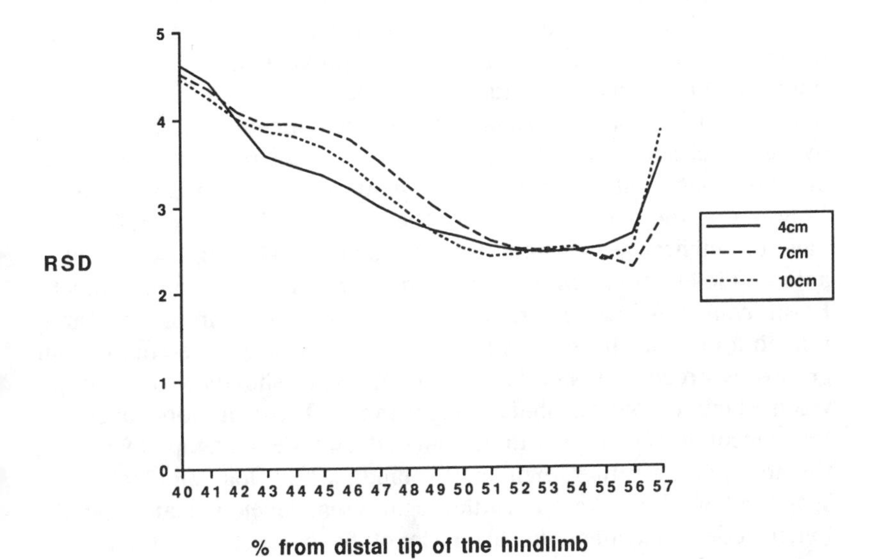

Fig. 1. The precision (RSD) of the relation between backfat depth in pigs at 4, 7 and 10 cm from the dorsal mid-line and percentage lean. Position along the carcass expressed as a percentage of the length from distal hindlimb to tip of snout. (Data of Sterrenburg, 1989.)

Kempster and Evans (1979*b*) found similar results in pork weight (61 kg live weight), bacon weight (91 kg live weight) and heavy weight pigs (118 kg live weight). Error of estimation of lean proportion decreased from over 3% (RSD) at the 4th/5th rib to a minimum of approximately 2·3% at the 13th-last rib region, and then increased as progression was made caudally to the 5th/6th lumbar region where the RSD was approximately 2·7% (all measurements at 6·5 cm from the dorsal mid-line). The last rib site is known as P_2, and when measured on the hot carcass it explained 50% of the variation in lean percentage overall. P_2 measured on the cold carcass was 0·3, 1·2 and 1·7 mm larger than P_2 hot in pork, bacon and heavyweight pigs respectively, and was more highly correlated with lean percentage, explaining 62% of the variation. The authors state that their results support the hypothesis, or at least the prediction from it, of Hammond (1933) and McMeekan (1941), that the junction of the loin and thorax is the best region for prediction purposes since it is the latest developing part of the carcass. There is, however, no fundamental mathematical reason why the relation between the amount of fat in a late developing region and total fat is stronger than that between the same total fat and the fat in an early-maturing region, provided fat accretion is continuing in all regions. The evidence for junction of loin and thorax being the latest to develop its subcutaneous fat is not unequivocal. In the study by Kempster and Evans (1979*a*), assessment of the relative growth rates of subcutaneous fat within pig carcass joints using the allometric relation showed that the dorsal region over the lumbar and sacral vertebrae did have the highest growth rate over the range 61 –118 kg live weight in animals fed either *ad libitum* or a restricted regimen. But in the study by Fortin *et al.* (1987), the corresponding region of the carcass (hindloin) ranked fourth or fifth behind the belly, foreloin, leg (three out of four groups by breed and sex) flank (two groups) or shoulder (one group), when examined over a similar weight range. Those authors concluded that subcutaneous fat growth 'terminated' (*sic*, i.e. developed latest) in the forequarter rather than in the hindquarter (loin and pelvis) as suggested by McMeekan. In cattle, regions that developed later than the loin included the cranial, dorsal quarter of the thorax, the ventral thorax, and the abdominal region (Kempster *et al.*, 1976), or, in the study by Truscott *et al.* (1983), in the dorsal and ventral caudal half of the thorax. In sheep, Butler-Hogg (1984) obtained regional growth coefficients (b) from the allometric relation with total subcutaneous fat which ranked pelvic limb (lowest), followed by lumbar and abdominal ($b = 1{\cdot}01$), *brisket* ($b = 1{\cdot}04$) and crop and neck ($b = 1{\cdot}09$).

Using McMeekans's data (1940), allometric growth coefficients can be calculated for subcutaneous fat in the loin region and in the hind limb (both have similar masses over the growth range studied, and so part-whole effects on correlations will be similar) over the range from birth to seven months of age. The values obtained are $b = 1{\cdot}038$ and $0{\cdot}982$ for the loin and hind limb respectively, i.e. the loin is later developing. But the correlations between subcutaneous fat weight in the joint and total subcutaneous fat are $r = 0{\cdot}979$ and $0{\cdot}998$ for the loin and hind limb respectively, i.e. there is a more precise relation between total subcutaneous fat and hind limb subcutaneous fat.

The second important relation between fat thickness and fat mass can be examined using the data from McMeekan's fourth paper in his series on growth and development in the pig (McMeekan, 1941). In pigs of the same live weight (91 kg), the correlation coefficient from the relations between fat thickness on a joint and the weight of subcutaneous fat in that joint are $r = 0{\cdot}806$ and 0·874 for the shoulder and loin, respectively: the measurement of fat thickness in the loin is more highly correlated with the mass of fat deposited locally than the corresponding fat thickness and mass in the shoulder. The implications from this are that the pattern of development of subcutaneous fat in the loin is in some way more 'regular' than in the shoulder, and probably in other regions, and/or practical aspects of measuring fat thickness differ between regions. It is a truism that the best predictor variables are ones that are changing rapidly with respect to the predicted component so that the level of resolution of the measurement technique is less critical. Thus, measurement of fat thickness to the nearest millimetre, as achieved in McMeekan's data (McMeekan, 1941), will result in a more precise relation with subcutaneous fat when measured in the loin than the corresponding measurement in the shoulder (standard deviation of loin fat thickness = 7·33, shoulder = 6·44). It is interesting to note that fat thickness '*C*' (over *m.longissimus* 1·5 inches from the median line at the last rib—does not include skin) was even more variable than mid-line loin fat thickness (standard deviation = 8·22) and was more highly correlated with total carcass fat than the latter.

For cattle and sheep, the relative merits of particular site fat thicknesses are less clear than they are for pigs (Table 3). A review of American work in the 1960s indicates that in most studies, a fat depth measurement over *m.longissimus thoracis* in the 12th rib region was most highly correlated with indices of composition, but sites over the rump and thoracic limb were equally, or more, valuable in some studies (e.g. Lewis *et al.*, 1964; Wallace *et al.*, 1977). Evidence from Jones *et al.* (1986) showed that a minimum fat depth over *m.longissimus thoracis* is preferable to depths at fixed locations defined by the dimensions of the muscle—an advantage in commercial situations where minimum depth position could be estimated very quickly by eye.

The increasing recent use of mechanised hide-pullers sometimes results in increased variation in subcutaneous fat thickness in particular carcass regions, including the caudal rib area. Australian work (Johnson & Vidyadaran, 1981) examined the precision of prediction of total dissected carcass fat from fat thickness measurements at seven sites, and also examined 'reliability' at each site in providing an accurate thickness. Their data, on a limited number of carcasses (36), suggested that a fat depth 30 mm lateral to the dorsal mid-line horizontal to the 2/3rd sacral vertebrae (the 'sacral crest' site) provided information which was about as precise in predicting total fat as a depth over *m.longissimus thoracis* at the 12th rib. Although it is not clear how the authors estimated 'reliability', their conclusion was that this sacral site was more reliable than the 12th rib site in six abattoirs where hide-pullers were used, the failure rate in obtaining a measurement being approximately 2% and 17% respectively.

Probes to Measure Tissue Thickness—Different Types

The measurement of subcutaneous fat thicknesses, originally on pig carcasses but now actively studied on lamb and beef carcasses, has formed the basis of objective classification schemes in many countries. The technology required to measure fat thickness, and more latterly the underlying muscle thickness as an additional predictor, to enable such measurements to be made at increasingly rapid rates of carcass throughput in modern abattoirs, and to reduce operator fatigue, has been directed towards automation of these measurements.

Originally, backfat was measured at various points along the mid-line of the split carcass, and in Denmark this had been implemented as early as the 1930s. Subsequently, McMeekan (1941) found that the single fat thickness measurement most closely correlated with total fatness in 90 kg live weight pigs reared on four widely differing treatments was measurement '*C*', lying laterally to the mid-line and over the central portion of *m.longissimus*. From the late 1950s on, the use of fat thickness measurements at sites lateral to the dorsal mid-line became the norm, and were particularly advantageous in grading or classifying pork weight carcasses in the United Kingdom which are not split into sides prior to wholesale marketing.

The simplest form of probe was simply a sharpened steel rule which was pushed in through the backfat until the greater physical resistance to penetration of the collagenous epimysium surrounding the underlying muscle was detected. Of much greater significance, in that it has been used in national classification schemes, including that of the UK since 1960, is the optical probe (OP) or intrascope, manufactured by SFK Limited, Hvidovre, Denmark. There is a light source near the tip of the probe and the light is emitted through a lens carrying a clearly marked line at right angles to the probe shaft. This line can be observed through the top of the probe, adjacent to the handle, by the operator, by means of an internal mirror system. A graduated sliding barrel indicates the probe depth when the operator judges the line to coincide with the fat/lean boundary.

The first automated detection of the tissue boundaries was based on the differential electrical conductivity of fat and lean tissues, and the Danish probe, the KS meter (K = kod (meat), S = spaeck (fat)), was used in Denmark from 1970. The automated version of this probe, the KSA (sometimes called the MFA [Meat-Fat Automatic] in English texts) was introduced in 1974, and the electronic signals from the probe are interfaced to a microcomputer, which calculates the lean meat percentage from fat and 'meat' thicknesses and based on data involving carcass dissections. The 'meat' thicknesses are measured as a result of the drop of conductivity, from that in muscle tissue, to zero as the probe emerges through the body wall into the body cavity. This is not the depth of muscle alone, but its underlying tissue as well (e.g. deep to *m.longissimus*).

The other probes which function as stand-alone systems for estimating lean percentage via tissue thicknesses are based on the optical properties of lean and fat tissues. These are known as light reflectance probes, and they may utilise visible light or light in the near-infrared part of the spectrum. This generated light is emitted from an LED near the tip of the probe and reflected light from the tissues is logged via an adjacent window. Reflectance signals are logged with distance from the skin surface by means of a spring loaded base plate which can move in relation to the probe but which always makes contact with the skin/carcass surface during operation.

The first probes to utilise this principle were the Ulster Probe (UP), developed by the Wolfson Opto-Electronics Unit in the Queens University of Belfast, and used in all Ulster bacon factories, and the Fat Depth Indicator (FDI), developed by Hennessy and Chong Limited, Auckland, New Zealand.

There are three more recent versions of the automatic light probes, manufactured by different companies. The Hennessy Grading Probe (HGP) is produced by Hennessy, New Zealand and uses light in the green-yellow range (= 570 nm). The Fat-O-Meater (FOM) is produced by SFK Limited in Denmark, and uses light in the near-infrared (= 915 nm). The Destron probe (DST) is manufactured by Destron Technologies Inc, Canada, and uses light in the near-infrared. The FOM differs from the other two in that electronic signals are transmitted along a cable connecting the probe to a microprocessor, keyboard and display. In the HGP and DST, the microprocessor is an integral part of the probe, and the relevant information (tissue depths, lean percent) are displayed on the probes themselves. The DST also has a keyboard on the probe body.

Development of these probes is on-going. They differ in design detail but there have been many accusations of infringement of patents. The present thrust is to refine the hardware and software to enable the manufacturers to claim that aspects of meat quality (pale, soft, exudative/ dark, firm, dry muscle; intramuscular fat content) can also be measured using this technology.

There have been several investigations to evaluate the precision and accuracy of lean prediction using different automatic probes, and to compare different probes or probes and other techniques. Most have been carried out on the pig.

The Use of Probes on Pig Carcasses

In an evaluation of the KSA, Pedersen and Busk (1982) compared the means and standard deviations of fat depths at various positions along the back and muscle depth between the 3rd from last and 4th from last rib (3/4 last rib) recorded by the probe on hot carcasses, with the same tissue thicknesses measured from photographs of the cross-sections obtained by cutting the cold sides. There were two experiments: 367 randomly selected carcasses subsequently dissected to establish regression parameters to use in the prediction of lean percent; and in the second experiment, 472 carcasses chosen to represent in equal numbers gradations of fatness over a wide range, designed to check the regression coefficients previously obtained. Among the tissue depths, there were some which differed significantly between the two methods of measurement. For example, in both experiments the backfat 8 cm from the mid-line at the level of the 3rd lumbar vertebra was thinner in the cold carcass, whereas at 6 cm from the mid-line between the 3/4 last ribs is was increased when cold. The authors claim that these changes in fat thickness due to cooling are consistent with other observations. Moreover, the authors claim, quite correctly, that the differences between KSA and cold carcass measurements are not, in themselves, important, as the objective of the technique is to predict lean content. However, for some tissue thicknesses, notably muscle depth between the 3rd and 4th from last ribs, the bias between KSA and control measurements was not consistent between experiments—in the first experiment the control depths were significantly lower than the KSA (51·0 versus 53·3 mm respectively, $p < 0{\cdot}001$), whereas in the second experiment there was no significant difference (50·1 versus 50·4 mm, respectively). The effect of these inaccuracies in the prediction of lean percent in these two experiments was small, as the regression coefficient for muscle depth was of the order of 0·13, being only one of five independent variables in the equation, hence a 2 mm error would only alter predicted lean percent by 0·26. Additionally, the correlation coefficients between KSA and cold carcass muscle depths was only of the order 0·55–0·72 in the two experiments, compared with 0·8–0·9 for the fat depths. In the study of the 472 carcasses by Pedersen & Busk (1982), the sample standard deviation of 3·54% lean was reduced progressively by prediction using tissue thicknesses, with resulting residual standard deviations ranging from 2·29% using fat depth at the 3/4 last rib only, to 1·85% using three fat depths plus *m.longissimus thoracis* depth plus carcass weight. The results suggest that utilising the principle of conductivity to estimate tissue thicknesses provides the basis of a workable system, but there are some question marks over its ability to measure muscle depths accurately.

A comparison of the KSA, UP, and OP was carried out by Kempster *et al.* (1979). There was little to choose between the instruments in predicting percentage lean in a group of pigs including pork weight, bacon and heavy weight categories, either when examined in the group as a whole or within weight ranges. The KSA-recorded muscle depths, and the calculated muscle depths using the other two probes (total tissue depth was measured using a steel rule and the muscle depth estimated by subtracting fat depth from this total), did not materially improve precision in this study over and above that obtained from fat depth measurement alone. Neither did combinations of fat depth measurements markedly improve precision. There was, however, a quite considerable discrepancy between the overall mean muscle depths (3/4 last rib and last rib) measured with the KSA and the other instruments, and with the other instruments and the depth measured on the cut surface with callipers at the 3/4 last rib. The authors claim that, at the last rib position, the boundary between *m.longissimus* and the underlying muscles (especially *m.psoas major*) was not always identified using the KSA. But the precision of estimation of lean percent using the KSA was not better at the 3/4 last rib site where it is claimed that measurement of *m.longissimus* depth is less ambiguous. The precision of prediction using the different probes at different sites are given in Table 4, together with those from other studies.

The FDI was only marginally superior in its precision of estimation of lean percent compared with the UP and OP in the study by Kempster *et al.* (1981*b*). There was no clear advantage in measuring fat thicknesses at the 3/4 last rib compared with the last rib, as in the previous study (Kempster *et al.*, 1979). In fact, for the UP and the OP, precision was greater at the last rib.

Table 4

The precision of different probes and measurement sites[a] in the prediction of percentage lean in pig carcasses

Predictors	*Residual standard deviation*		
Reference: Kempster *et al.* (1979)			
Sample standard deviation = 4·09			
	KSA	*UP*	*OP*
W	3·54	3·54	3·54
W+3/4F	2·24	2·44	2·23
W+LRF	2·55	2·75	2·30
W+3/4F+3/4M	2·18	2·44[b]	2·15[b]
Reference: Kempster *et al.* (1981*a*)			
Sample standard deviation = 4·04			
	FDI	*UP*	*OP*
W	3·46	3·46	3·46
W+3/4F	2·16	2·56	2·37
W+LRF	2·20	2·27	2·21
Reference: Fortin *et al.* (1984)			
Sample standard deviation = 3·62			
	HGP	*FOM*	
LRF	2·39	2·42	
LRF+LRM	2·27	2·25	
3/4F	2·24	2·28	
3/4F+3/4M	2·15	2·14	
4/5F	2·26	2·34	
4/5F+4/5M	2·11	2·15	
All above RSD values are the means of three regressions obtained from measurement at 5, 7 and 9 cm from the dorsal mid-line.			
Reference: Kempster *et al.* (1985)			
Sample standard deviation = 6·85			
	HGP	*FOM*	*OP*
W	6·19	6·19	6·19
W+LRF	4·06	3.71	3·08
W+3/4F	3·66	3·14	3·08
W+LRF+LRM	3·99*	3.68	—
W+3/4F+3/4M	3·59*	2·91***	—

	Estimated % lean from partial dissection			
Reference: Matzke *et al.* (1986)				
Sample standard deviation = 3·88				
	FOM			*SKG II*
Multiple: (FOM 'German Formula'; SKG 'Integrated' and 'ham/body[c] formulae)	2·15			2·49 2·41[c]
Reference: Cook *et al.* (1989)				
Sample standard deviation = 4·93				
	FOM	*HGP*	*DST*	*OP*
P_2	2·64	2·85	3·27	2·56
3/4F	2·72	2·83	2·84	2·67
3/4F+3/4M	2·41	2·58	2·58	—
P_2+3/4F+3/4M	2·23	2·37	2·56	2·50
W+P_2	2·53	2·80	3·28	2·44
W+3/4F+3/4M	2·39	2·58	2·58	2·54[d]
W+P_2+3/4F+3/4M	2·18	2·37	2·55	2·31[d]
Reference: Diestre *et al.* (1989)				
Sample standard deviation = 4·6				
	FOM	*HGP*	*DST*	
LRF	2·97	2·90	3·06	
3/4F	2·91	2·95	2·53	
3/4F+3/4M	2·48	2·63	2·31	
LRF+3/4F+3/4M	2·23	2·45	2·25	

[a]W = carcass weight; 3/4F = fat depth at 3/4th from last rib; 3/4M = muscle depth at 3/4th from last rib; LRF = fat depth at last rib; LRM = muscle depth at last rib; 4/5F = fat depth at 4/5th from last rib; 4/5M = muscle depth at 4/5th from last rib.

[b]Muscle depth calculated as total depth (ruler probe) − fat depth.

[c]Includes $\frac{\text{ham width}}{\text{loin width}}$

[d]Muscle depth not included.

*,***significance ($p < 0{\cdot}05$, $< 0{\cdot}001$, respectively), of added precision achieved by including muscle thickness measurement with corresponding fat thickness.

In the later trial by these workers (Kempster *et al.*, 1985), the conclusion was that results from automatic probes were better at 3/4 last rib than at last rib, and overall the FOM was more precise in predicting lean content (g/kg) than the HGP although there was an interaction between abattoir and probe (Table 4). The authors also make the point that in Britain there is a need to calibrate the automatic probes against the OP, since the introduction of automatic probes is likely to be gradual, and most medium or small abattoirs will continue to use the intrascope, at least in the short term. The current trend, observed since the publication of that paper, would suggest that the replacement of the intrascope by automatic probes will be extended over a considerable period. The FOM was again a more precise predictor of the OP values that the HGP at the 3/4 last rib site, and was equal to the HGP at the last rib. However, for both probes there were significant differences between abattoirs in the regressions of fat thickness measurements taken by the OP on those taken by the automatic probes, indicating that a separate calibration will be required at each abattoir.

In a study of 224 pig carcasses chosen to cover the market range in hot carcass weight found in Canada (58·5–94·5 kg), Fortin *et al.* (1984) reported that the measurement of last rib and 3/4 last rib fat thickness at 50 mm from the dorsal mid-line was significantly lower when measured with the HGP compared with the FOM. The HGP measurement also underestimated the corresponding thickness of fat measured with a ruler after cutting the fat down to the level of the lean. In addition to these bias effects which differed between probes, the FOM was more repeatable than the HGP in all but one measurement (muscle thickness at last rib). However, these two probes differed little in the precision of percentage lean prediction (Table 4), and the authors conclude that the choice of a particular probe will depend on factors other than the degree of precision of prediction, notably cost and operational efficiency under commercial abattoir conditions. Also, in that study, the position of measurement at either 5,7 or 9 cm from the dorsal mid-line in a given anatomical transverse reference plane, did not have any influence on the precision of prediction, which contrasts with the results of Sterrenburg (1989) referred to earlier. The authors may therefore be mistaken in suggesting that automated grading would be easier to achieve since the precise definition of probing site need not be necessary.

A conclusion that seems to be universally supported by the investigations examining the relation between fat depths, measured by probes, and carcass leanness is that improvement in the precison of prediction obtained from the best fat depth measurement, when other additional fat, or muscle, depths are added, is small. Using the automated light probes, the *m.longissimus* depth at the same site as the best fat measurement can be obtained with no increased labour or time inputs, and so the marginal improvement in precision is considered to be worthwhile. Another approach is to automate the measurements so that these considerations are irrelevant, but the necessity to probe at many carcass sites in the light of the scientific evidence on precision is questionable. Kempster and Evans (1979*b*) did find that P_2 and R13 (6·5 cm from the dorsal mid-line at the head of the 13th rib) was the best pair of fat depths in the prediction of percentage lean in each of the joints. P_2 plus the fat thickness measured on the joint in question was a second-best pair of predictors. Nevertheless, the new system for classifying pigs in Denmark obtains data from seventeen light reflectance probes which are inserted. The system, known as the pig carcass classification centre (KC), is fully automated (Luthje *et al.*, 1988) and can classify 360 hogs per hour (Hök, 1988). Total carcass length, and the distances from the gambrels to the lower edge of the fore limb and to the pubic bone are measured, the last two in one of the stations which are modules containing carcass supporting frames arranged in a carousel. These dimensions determine the positioning of the probes which are operated in the probe measuring station, and the following measurements are recorded: hind limb—5 fat thicknesses; loin—3 fat thicknesses, 3 muscle thicknesses; belly—5 fat thicknesses, 5 total tissue thicknesses; shoulder—4 fat thicknesses. Not all of these measurements are actually used in the prediction of total carcass leanness, or lean percentage in the three primal joints (shoulder, middle and leg). Stepwise regression analysis showed that significant improvements to the equations were not made when more than eight parameters were used. However, there are ten possible equations which can be used, involving different measurements because in some cases an individual probe may fail to measure through hitting a bone.

The results obtained using the KC and the KSA probe on 99 carcasses in three weight categories (55·0–64·9, 65·0–74·9, 75·0–85·0 kg dressed weight) are shown in Table 5. In the regression on lean percentage, the KSA had a residual standard deviation of 1·8 percentage points, and this was reduced to 1·63 using the KC. Even bigger reductions in residual variation were observed using the KC to estimate the lean percentage of individual primal joints, particularly in the middle and leg. Additionally, Luthje *et al.* (1988) demonstrated that in a sample of 9847 carcasses classified using both MFA and KC, there was good agreement of predicted

Table 5
Residual standard deviations (RSD) of percentage lean obtained from the KC[a] and the KSA probe (data from Luthje *et al.*, 1988)

	Lean %		*KC*		*KSA*	
	$\bar{x}$	*SD*	R^2	*RSD*	R^2	*RSD*
Carcass	56.7	3·62	0·82	1·63	0·75	1·80
Shoulder	64·4	3·36	0·76	1·77	0·53	2·33
Middle	60·6	5·43	0·85	2.27	0·63	3·34
Leg	69·6	3·75	0·81	1·73	0·54	2·58

[a]KC = pig classification centre.

lean percentage using both methods in the region of the population mean value (54–58%), but relative to the MFA, the KC underpredicted lean percentage in the fatter pigs, and overpredicted in the leaner pigs, and the authors claim that the KC values are nearer to the true meat content.

Other techniques utilising fat thickness measurement

A technique that has attempted to objectively measure the traditional parameters assessed visually in many European countries, namely mid-line backfat thickness and carcass shape or conformation, has been developed in the Federal Republic of Germany and the equipment is known as the SKG (Schlachtkörper–Klassifizierungs–Gerät). The instrument was developed by Breitsameter (Aichach) and is marketed by Tecpro GmbH. Two procedures are involved in its operation. Mid-line backfat thickness at its thinnest point over *m.gluteus medius* (measurement *a*) is determined using a probe and sliding cylinder whose travel distance is measured by an integral potentiometer. The operator pushes the probe along the cut surface in the sagittal plane until a mark on the shaft coincides with the boundary of *m.gluteus medius* and the subcutaneous fat. The second procedure in the SKG II employs a series of electromechanical or electropneumatic probes to make contact with the ham and provide objective measurements of certain dimensions and shapes. These comprise (*b*) the width of the leg at its widest point in the medio-lateral direction, (*c*) the width of the carcass at its narrowest point in the caudal lumbar region/flank, again in the medio-lateral direction, and (*d*) the angle (degrees from horizontal) of the medial profile of the ham (over *m.semimembranosus*) when viewed from the dorsal aspect. These parameters are combined in a multiple regression equation to predict lean percentage (y) using the following:

$$y = 73{\cdot}193 - 0{\cdot}6683a + 0{\cdot}29555b - 0{\cdot}26037c - 18{\cdot}02302b/c - 0{\cdot}07303d.$$

The SKG II is particularly attractive to users in those areas where extreme meat-type pigs are produced in the belief that the better conformation and high yielding carcasses will be identified. It is not clear how, as claimed by Casteels *et al.* (1984), it gives information on the percentages of primal cuts (ham, cutlet (= middle) and shoulder).

In Belgian trials, the correlations between the lean meat proportion estimated by the Belgian reference method (primal cuts as a percentage of half carcass weight, the middle or cutlet joint only being trimmed of fat) and the SKG II measurements compared with HGP measurements were 0·87 and 0·81 respectively (Casteels *et al.*, 1984). Other results are presented in Table 4.

The 'Porkitron', produced by Zimmermann, Bahlingen a.K. is a simpler version of the SKG II and does not measure ham angle (Walstra, 1989). It can be interfaced to a computer.

Three other 'techniques' have been developed for pigs in the Federal Republic and Democratic Republic of Germany. These are the LSQ (Lenden–Speck Quotient), the SFQ (Speck–Fleisch–Quotient) and the ZP (Zwei–Punkt) procedure. They are all based on fat and muscle depths in the caudal loin region, measured on the sagittal plane cut surface, and differ only in detail. In the LSQ calculation, the minimum fat thickness over *m.gluteus medius* is measured (a_1), together with the fat thickness at the cranial edge of that muscle (a_2) and the muscle depth underlying a_2 down to the spinal column (b). Lean meat is estimated from the quotient

$$\frac{a_1 + a_2}{2b}.$$

The SFQ is the ratio between a_1 and the muscle depth underlying a_1. ZP, the most important of the three as it has been approved by the EC, uses the measurements a_1 and b in the following formula.

$$\text{Lean \%} = 47{\cdot}978 + \left(26{\cdot}0429 \times \frac{a_1}{b}\right) + (4{\cdot}5154 \times \sqrt{b}) - (2{\cdot}5018 \times \log a_1) - (8{\cdot}4212 \times \sqrt{a_1}).$$

There is a special apparatus to measure SFQ based on a mechanical design produced by the Research Centre Dummerstorf–Rostock and an

Table 6
Results of approval tests for methods of pig carcass classification according to EC Commission regulation 2967/85 (modified after Walstra, 1989)

Country	*Probe*	*RSD*	R^2	*Independent variables*
Belgium	SKG II	1·99	0·71	$X_7, X_9, X_{10}, X_{11}, X_{12}$
Denmark	KSA	1·79	0·65	X_1, X_2, X_3, X_6, X_7
	KC	1·34–1·47	0·77–0·81	(combinations of 8 variables—see text)
	FOM	1·66	0·70	X_1, X_2, X_3, X_6
Federal Republic of Germany	FOM	1·79	0·93	X_2, X_3, X_4, X_6
	SKG II	2·19	0·89	$X_7, X_9, X_{10}, X_{11}, X_{12}$
	ZP	2·45	0·86	$X_{14}, X_{15}, X_{16}, X_{17}$
	SSD 256	2·16	0·77	X_{19}, X_{20}
France	FOM (8 mm)[a]	2·10	0·79	X_2, X_3, X_6
	FOM (6 mm)[a]	2·13	0·79	X_2, X_3, X_6
	SPC-Sydel	2·11	0·79	X_2, X_3, X_6
	DST	2·14	0·78	X_2, X_3, X_6
Great Britain	OP	2·44	0·76	X_1, X_5
	OP	2·31	0·78	X_1, X_3, X_5
	FOM	2·23	0·90	X_3, X_5, X_6
	HGP2	2·37	0·77	X_3, X_5, X_6
Greece	DST	2·30	0·73	X_3, X_5, X_6
	FOM	2·12	0·76	X_3, X_5, X_6
	HGP2	2·18	0·75	X_3, X_5, X_6
Italy	FOM (60–120 kg)	2·37	0·70	X_2, X_3, X_4, X_6
	FOM (120–180 kg)	2·30	0·79	X_2, X_3, X_4, X_6
	DST (60–120 kg)	2·44	0·68	X_2, X_3, X_4, X_6
	DST (120–180 kg)	2·36	0·77	X_2, X_3, X_4, X_6
	OP (60–120 kg)	2·38	0·70	X_3, X_7, X_8
	OP (120–180 kg)	2·45	0·75	X_3, X_7, X_8
Luxembourg	HGP2	2·42	0·86	X_3, X_5, X_6
Netherlands	HGP2	2·19	0·75	X_3, X_6
Northern Ireland	OP	1·92	0·70	X_1, X_5
	OP	1·82	0·74	X_1, X_3, X_5
	UP	2·00	0·68	X_1, X_5
	UP	1·90	0·70	X_1, X_3, X_5
Republic of Ireland	OP	2·07	0·67	X_3, X_7, X_{17}
	FOM	2·04	0·68	X_3, X_6, X_{17}
	HGP2	2·09	0·66	X_3, X_6, X_{17}

Table 6—*contd.*

Country	*Probe*	*RSD*	R^2	*Independent variables*
Spain	FOM	2·23	0·77	X_3, X_5, X_6
	HGP2	2·45	0·72	X_3, X_5, X_6
	DST	2·25	0·76	X_3, X_5, X_6

[a] = Probe diameter.

X_1 = Cold carcass weight (warm carcass weight excluding flare fat, kidneys, and diaphragm in Danish KSA and FOM formulae).

X_2 = Fat thickness between 3rd and 4th lumbar vertebrae, 8 cm from dorsal mid-line.

X_3 = Fat thickness between 3rd and 4th from last ribs (3/4 LR), 6 cm (or 6–8 cm in FRG, 7 cm in Luxembourg, 8 cm in France and Italy) from dorsal mid-line.

$X_4 = \frac{(X_2+X_3)^2}{2}$

X_5 = Fat thickness at last rib (LR), 6 cm (or 8 cm in Greece and Luxembourg) from the dorsal mid-line.

X_6 = Muscle thickness at 3/4 LR.

X_7 = Fat thickness over *m.gluteus medius*, thinnest depth at the dorsal mid-line.

$X_8 = \frac{(X_3+X_7)^2}{2}$

X_9 = Thickest width of ham.

X_{10} = Thinnest width of 'waist'.

$X_{11} = X_9/X_{10}$

X_{12} = Angle between medial ham contour and horizontal.

X_{13} = Muscle thickness from cranial edge of *m.gluteus medius* to dorsal edge of vertebral column.

$X_{14} = X_7/X_{13}$.

$X_{15} = \sqrt{X_{13}}$

$X_{16} = \text{Log } X_7$.

$X_{17} = \sqrt{X_7}$.

$X_{18} = X_3^2$.

X_{19} = Fat thickness between 2/3rd from last ribs (2/3 LR) 7 cm from dorsal mid-line.

X_{20} = Muscle thickness at 2/3 LR.

electronic version (EQM 1) can be interfaced to computers (Walstra, 1989). Similarly Tecpro GmbH have produced instruments to measure ZP, interfaced to a terminal.

European Community pig carcass grading schemes

On January 1st 1989 the European Community introduced a revised pig carcass grading scheme (Council regulation (EEC) No. 3220/84, Commission regulations 2967/85 and 3530/86) to provide a common and more accurate basis for classifying carcasses and thus facilitate inter-community trade. There was concern over the variation in lean percentage associated with different grades in different countries. Hence, the basis of the scheme is that classification must be founded on objective criteria that enable an estimation of the lean meat content to be made (Council regulation No. 3220/84); the efficacy of the measurements must be established on a sample of at least 120 pigs representative of the national or regional production; the coefficient of determination (R^2) is greater than 0·64; the residual standard deviation is less than 2·50% (Council regulation No. 2967/85). The member states can choose which technique to use, but it must be approved by the Pigmeat Management Committee of the EC.

By September 1987, four Member States had notified the Commission of the grading methods which they intended applying (Germany, Denmark, The Netherlands, Republic of Ireland), five had notified the Commission of their plans in relation to implementation of the new scale (France, Great Britain, Italy, Luxembourg, Greece), and three other states had not supplied information on the manner of intended implementation (Belgium, Spain, Portugal). By July 1988, the approved methods had been accepted for Great Britain, Northern Ireland, and Belgium. The use of particular probes in different European countries was summarised by Walstra (1989), but the list is evolving rapidly, and the regression parameters emerging from the approval tests according to Commission regulation No. 2967/85 at the time of going to press have been summarised in Table 6.

It is not valid to compare the different probes between countries, because different types of animals and different measurement parameters are used. But within a country the best equations for different probes do indicate the probes' relative precision. Thus, the Federal Republic of Germany data suggest that the ZP and SKG II techniques are not such good predictors as the FOM. Differences between the OP, HGP, FOM and UP are smaller, but the additional information on *m.longissimus* depth using the HGP and FOM gave these probes the edge over the OP in the Republic of Ireland test, and the FOM only in the British tests.

The Use of Probes on Beef Carcasses

The FDI was used by Kutsley *et al.* (1982) to measure fat thickness between the 12th and 13th ribs, at a point approximately 0·75 of the *m.longissimus* width measured from the dorsal mid-line, on 400 beef carcasses. There was a large discrepancy between hot carcass probe measurement (mean depth±standard deviation = 15·3±5·5 mm) and cold carcass probe measurement (23·8±8·7 mm). Cold carcass measurement using a ruler (15·5±6·6 mm) was of a similar magnitude to hot probe mean value, but the correlation was higher between ruler depth and cold probe (r = 0·81) than that between ruler depth and hot probe (r = 0·67).

In a second trial involving 50 carcasses from mixed breeds of cattle, Kutsley *et al.* (1982) measured cold carcass fat thickness with the FDI at seven sites extending from the brisket, and 5th thoracic vertebrae (20 cm lateral to the dorsal mid-line) in the forequarter to the 4th sacral vertebra (10 cm from mid-line) in the hindquarter. Correlations with actual fat thickness measured by inserting a ruler through an incision at each site were moderate to high (r = 0·56–0·99), but correlations with percentage boneless trimmed cuts, or less severely trimmed bone-in cuts, were low (r = 0·24–0·52).

Jones and Haworth (1982) also found that the FDI consistently over-predicted fat thickness measured with a ruler, and the degree of over-prediction tended to increase as fatness itself increased. The reverse trend was reported by Phillips *et al.* (1987) for HGP. In that study, measurements at 0·25 *m.longissimus* width measured from the dorsal mid-line 10 mm cranial to the 11th rib, in heifers and steers was more repeatable and predicted the corresponding depth measurements with less bias than measurements at 0·5 and 0·75 *m.longissimus* width.

Chadwick and Kempster (1983) compared different probes and included visual assessment as a bench mark in the prediction of lean and subcutaneous fat percentages of 182 beef carcasses. The probes tested were the KS, the UP, the FDI and a ruler probe (RP) (a sharpened steel rule with sliding cursor). Visual assessment of subcutaneous fat (to the nearest estimated percentage point, SF_e) was the most precise predictor of both lean and fat percentages. The probe measurements were gener-

Table 7
The precision of different probes and measurement sites in the prediction of percentage lean or saleable meat in beef carcasses

Predictors	*Residual standard deviation*							
Reference: Chadwick & Kempster (1983)								
	Sample standard deviation	*Lean %*						
		Fat score	*MF*		*RP*		*FDI*	
A. Side wt (W)+single measurement[a]								
(a) Trial I, cold	5·98	2·77						
Fat depth, 13th rib, 12·5 cm from mid-line			2·77		4·92		—	
(b) Trial III, cold	3·11	2·69						
Fat depth, 13th rib, 12·5 cm from mid-line			—		2·89		2·60	
(c) Trial III, cold	3·11	2·69						
Fat depth, 6th rib, 7·5 cm from mid-line			—		2·67		2·61	
(d) Trial III, hot	3·11	2·69						
Fat depth 10th rib, 12·5 cm from mid-line			—		2·91		2·59	
B. W+best (2) or (3) measurements								
			(2)	(3)	(2)	(3)	(2)	(3)
(a) Trial I, cold	5·98	2·77	2·38	2·33	2·65	2.52	—	—
(b) Trial III, cold	3·11	2·69	—	—	2·61	NI	2·37	2·29
(c) Trial III, hot	3·11	2·69	—	—	2·88	NI	2·53	2·48

Reference: Kirton *et al*. (1987)

	Saleable meat % HGP			
	Bulls	*Cows*	*Steers*	*Heifers*
Sample standard deviation	1·71	1·96	3·09	2·91
Hot carcass wt (HCW)	1·65	1·81	2·92	2·93
HCW+fat thickness over *m.biceps femoris* 5–7 cm lateral to perianal region (BF)	1·56	1·80	1·96	2·37
HCW+fat thickness at 2/3rd sacral vertebrae, 8cm from mid-line (SC8)	1·60	1·80	2·49	2·50
HCW+fat thickness at 3rd lumbar vertebra, 6–8 cm from mid-line (3L)	1·62	1·71	2·80	2·48
HCW+fat thickness, 10–13th ribs, 6–9 cm from the mid-line (10/13R)	1·63	1·81	2·55	2·77

Reference: Jones *et al*. (1988)

	Lean % HGP	
	Hot	*Cold*
Sample standard deviation = 3·61		
Fat depth, 5/6th ribs, 12 cm from mid-line	3·37	3·39
Fat+lean depth, 11/12th rib, 9 cm from mid-line	3·26	3·27
Fat+lean depth, 12/13th rib, 9 cm from mid-line	3.18	3·15
Fat depth, 2/3rd lumbar vertebrae, 9 cm from mid-line	3·33	3·33
P8	3·30	3·47
Fat score[b]	3·03	2·93

[a]The only data presented are those where the best of any single-probe measurement(s) equals or exceeds fat score (7-point scale) in precision of estimation, together with the corresponding RSD for other probes.
[b]7-point scale.
NI = no significant improvement over corresponding value using two measurements.

ally better predictors when taken on cold, rather than warm, carcasses. Certain fat depths used individually with carcass weight were more highly correlated with carcass lean percentage than fat class, estimated visually on a 7-point scale in conjunction with carcass weight, but there was little indication that specific fat measurements were particularly precise predictors when taken by specific probes. For the subsets of animals involved, the MF on cold carcasses and the FDI on both hot and cold carcasses gave the best results, particularly where more than one fat thickness was used in prediction. The main point of significance to emerge from these trials was that fat thickness in combination with visual fat score improved the precision of prediction in most cases, particularly for estimating lean percentage, and the authors suggest that probe fat measurements have a role in classification to improve visual scores, particularly for carcasses which are borderline between two fat classes. The main findings are shown in Table 7, together with the results from other studies on beef.

The HGP, which generates data on muscle as well as fat depths, was used in the study by Kirton *et al.* (1987) (Table 7). Five sites were probed, including two in the lateral sacral region where subcutaneous fat is relatively undamaged during removal of the hide when mechanised hide-pullers are used (Johnson & Vidyadaran, 1981) and the depth of fat at four of these sites, *m. longissimus* depths at two, and total tissue depth at one site were related to saleable meat (90% visual lean), fat trim and bone. The best predictor of fat trim was the depth 5–7 cm lateral to the perianal region, overlying *m. biceps femoris* (BF site) with R^2 values of 41–74% in different sex groups. Addition of other tissue depths did not reduce the RSD obtained for BF alone in the bull or steer groups, and only in the cow group did a lean depth measurement (6–8 cm from midline in the region of the 3rd lumbar vertebra) improve precision. The highest R^2 value (61%) occurred in the steer group, but this was also the most variable group in terms of saleable meat proportion. The sample standard deviations of this characteristic were 1·71, 1·96, 3·09 and 2·91% units for bulls, cows, steers and heifers, respectively. Corresponding RSD values using BF were 1·61, 1·86, 1·95 and 2·60% units.

There continues to be uncertainty whether the relation between HGP generated fat depths and the ruler depths in the cut-and-measure techniques (CM, used in Australia) is linear or quadratic. However, as noted by Phillips *et al.* (1987), the issue of importance is whether there is bias in the prediction of carcass lean yield using the HGP compared with the CM.

The P8 measurement site (defined as the intersection of a line parallel to the dorsal mid-line and passing through the dorsal tuberosity of the tuber ischium with a line through the crest of the 3rd sacral vertebra at right angles to the vertebral column) is favoured in Northern Australia because of the widespread use of hide-pullers. Fat depth measured with the HGP at P8 was not materially different in the precision of predictions of lean percentage from fat depth at the 5th/6th ribs, or 2nd/3rd lumbar vertebrae, in the study by Jones *et al*. (1988). The best predictors were combined lean and fat depths at the 11/12th and 12/13th ribs. Data for fat depth alone at these sites were not given, even though the authors state that lean depth did not improve precision. Johnson and Ball (1988) also found little difference between P8 and fat thickness at the 12th rib, 0·75 of the width of *m.longissimus* from the dorsal mid-line, in the relations with saleable meat percentage. They concluded that the fat thickness measurements were relatively poor predictors of meat yield in groups of carcasses of similar weights.

The Use of Probes on Lamb Carcasses

The use of a combined fat and muscle depth, the so-called GR measurement, has proved valuable in the New Zealand lamb carcass classification scheme to allocate carcasses to a particular class when they are judged as borderline between adjacent classes according to visual assessment of fat cover (Kirton, 1982). This measurement is defined as the total soft tissue thickness overlying the rib 11 cm from the dorsal mid-line in the region of the 12th rib, and is measured with a sharpened metal rule with base plate. According to Kempster *et al*. (1982*a*), the GR is proving difficult to measure at line speeds of 6–8 carcasses per minute, and it penalises heavier carcasses which are less fat at a given value of GR. There is thus much interest in improving the ease of measurement and its precision, and hence in automatic probes. A total depth indicator (TDI), manufactured by Hennessey and Chong Limited, and similar in principle to the FDI was assessed by Kirton *et al*. (1984). They found that a total tissue depth between the 11th and 12th ribs, 11 cm from the dorsal mid-line, was as closely related to carcass fatness (petroleum-ether extracted lipid as a percentage of hot carcass weight) as the GR measurement or measurement C (the fat over the greatest depth of *m.longissimus* at the last rib). Furthermore, the FDI-measured total tissue thickness was as highly correlated with percent fat as the same measurement made with a ruler. The results of this investigation also showed that addition of hot carcass weight did little to improve the precision using fat depths, or total tissue depths, alone. This

Table 8
The precision of different probes and measurement sites in the prediction of lamb carcass composition

Predictors	Residual standard deviation			
Reference: Kirton *et al.* (1984)				
Sample standard deviation = 3·3				
		Chemical fat % *FDI*		
Hot carcass weight (HCW)		2·96		
Fat thickness 11/12th ribs (11/12F)		2·19		
11/12F+HCW		2·20		
Reference: Kirton *et al.* (1985)				
	Dissectible fat %	*Chemical fat %*	*Lean %*	*Bone %*
Group 1, 70 lambs				
Sample standard deviation	6·20	6·0	3·9	2·9
(HCW)	3·77	3·69	3·12	1·83
GR (ruler probe)	3·11	3·34	2·91	1·50
HCW+GR	3·10	3·10	2·92	1·51
Group 2, 837 lambs				
Sample standard deviation	4·2	4·3	3·1	2·0
HCW	3·31	3·39	2·85	1·58
GR	2·91	3·12	2·70	1·43
HCW+GR	2·86	3·06	2·70	1·40

Reference: Chadwick *et al*. (1986)

	Lean %	
	HLP	*RP*
Sample standard deviation = 4·8		
(a) 4·5 cm from mid-line		
8th rib fat	3·9	3·9
8th rib muscle	4·1	
12th rib fat (12F)	2·9	3·2
12th rib muscle	4·1	
3/4th lumbar fat (3/4F)	2·7	3·1
3/4th lumbar muscle	4·1	
6/7th lumbar fat	2·8	3·3
(b) 12·0 cm from mid-line		
12th rib total tissue depth (12TT)	3·4	3·0
3/4th lumbar total tissue depth	4·0	3·8
6/7th lumbar total tissue depth	3·6	3·8
(c) Best multiple regression including carcass weight		
12F+3/4F	2·5	
3/4F+12TT		2·5
(d) Visual fat score		
Fat class[a]	2·3	
SF_e[b]	2·3	

Reference: Kempster *et al*. (1986*b*)

	Fat class[a]	*SF_e*[b]	*Sample standard deviation*	Lean %		
				OP[c]	*RP*[c]	*HGP*[c]
Trial 1 (n = 105)	3·16	2·57	4·23	3·28(2·87)	3·49(2·85)	—
Trial 2 (n = 67)	3·18	2·68	3·45	2·85(2·86)	2·92(2·90)	3·13(3·08)
Trial 3 (n = 49)	2·68	2·53	3·03	—	2·76(2·55)	2·64(2·46)
Combined trial estimates	3·0	2·6	3·5	—	3·0	—

[a]6-point commercial scale. [b]SF_e = subcutaneous fat assessed to the nearest 1%.
[c]The best single site values shown. Not all measurements were recorded with each probe in each Trial. The best sites were: fat depth 3/4 lumbar, 4·5 cm from mid-line (OP Trial 1; RP Trial 1); fat depth 12th rib, 4·5 cm from mid-line (OP, Trial 2; RP Trials 2+3; HGP Trial 3); fat depth 12th rib, 12·0 cm from mid-line (HGP Trial 2). Values in parentheses are the RSD values from the regression of percent lean on the fat class plus the corresponding probe measurement.

was confirmed in a later study (Kirton *et al.*, 1985). Although statistically a significant improvement in the prediction of fat and bone percentage, the effects were very small, and there was no improvement in the relation with lean (Table 8).

In a study by Kempster *et al.* (1986*d*), fat (and muscle) thickness measurements were recorded using the OP, a ruler probe and the pig version of the HGP. Caliper measurements were also taken on the cut surface at three exposed cross-sections. These measurements were compared with visual assessment of fatness on a 6-point commercial scale, and also subcutaneous fatness estimated to the nearest percentage point (SF_e), in the prediction of tissue percentages. SF_e was the most precise predictor, whilst the best probe fat thickness measurements were in most cases at least as good as the commercial fat class. Fat plus muscle depth at the 12th rib recorded with the HGP was not as good a predictor as two fat depths in combination. Measurements at 4·5 cm from the dorsal mid-line were generally more precise than those 12 cm from the mid-line, and this finding contrasts with those reported earlier by Kirton *et al.* (1984). However, in the study by Kempster *et al.* (1986*d*) measurement 'J' was recorded on a subset of carcasses and was the most precise predictor in that data set. 'J' is highly correlated with GR in the New Zealand data. Certain fat thickness measurements added to prediction precision when combined with commercial fat class, but a smaller improvement when combined with SF_e.

A later study (Chadwick *et al.*, 1986) included a specific probe for lambs, the Hennessy Lamb Probe (HLP) and a simple ruler probe (RP). None of the tissue depth measurements equalled the precision of the 6-point commercial fat score or SF_e (which were equal, RSD = 2·3%), but HLP measurements of fat depth were more precise predictors than RP.

Techniques Using Ultrasound

The use of ultrasound to estimate body or carcass composition in live animals began with studies on pigs, and the application of pulse-echo techniques has become an integral part of the performance testing of pigs in several countries. This technology, which provides information on the thickness and distribution of superficial tissues, is now being applied to live cattle and sheep as a selection tool in stock breeding, and there are several machines currently in use. The use of ultrasound to assess compostion by measurement on the carcass has, by contrast, been very little exploited. One reason for this is that other techniques, which involve less costly apparatus and which are easier to use, e.g. fat thickness probes, can provide comparable information on some tissue depths. Practical problems are also involved. Miles *et al.* (1970) found that a B-mode scanner with a moving transducer assembly running on a track (the Scanogram) would not operate successfully on hot lamb carcasses because of excessive movement of the surface soft tissues over the skeleton as the transducer housing dragged over the surface. Air pockets within the surface fat layers also posed a problem as they did in beef carcasses and the quality of image in the latter was impaired.

The objectives in the study by Fortin *et al.* (1980) were mainly directed at determining the variability of backfat in pig carcasses with a view to defining an area where variation was relatively small. Identification of such an area would be an important consideration in the future use of live ultrasonic work, and would facilitate its automation since precise location of the measuring head would not be required. The need to evaluate the method as a carcass classification tool seemed to be a secondary objective. Backfat thickness was estimated from ultrasound reflected from tissue boundaries at ten locations extending from 25 cm cranial to 20 cm caudal to the last rib, and measured at three positions on the dorsal midline and at 2·5 and 5·0 cm lateral to it. It is assumed that the boundaries were between the innermost subcutaneous fat layer and underlying muscle. Three areas were identified as having relatively invariable backfat thickness, but no carcass composition data were generated to determine if the measurements were useful predictors of tissue proportions.

Similarly, in the study by Fredeen and Weiss (1981), the main thrust of the enquiry was an evaluation of the Danish KSA probe, which utilises differences in electrical conductivity of fat and lean tissues, to estimate the lean content of pig carcasses. Ultrasonic measurements (pulse-echo), using the same machine (Krautkramer USM # 2) as in the study by Fortin *et al.* (1980) were included as a comparative tool, together with ruler measurements of fat depths. The ultrasonic fat depths, although differing in magnitude quite considerably in some cases from those recorded by ruler or the KSA, were more highly correlated than either of the latter measurements with different yield end-points—either commercially trimmed joints, boneless, or boneless defatted joints. It was only in multiple regression analysis that the KSA measurements slightly exceeded the precision of ultrasonic fat depths, and the variables used were different.

Forrest *et al.* (1988) used an Aloka Corometrics 210 DX Real-Time

Ultrasound Scanner to measure fat depth and *m.longissimus* area at the 10th and last ribs on the warm carcasses of commercial weight gilts and barrows. The correlations with percentage dissectible lean adjusted to 10% lipid content ranged from 0·44 to 0·47 for muscle areas, and from −0·66 to −0·81 for fat depths. When used as independent variables together with warm carcass weight in a multiple regression equation to relate to weight of lean (adjusted to 10% fat), ultrasound measurements were nearly as precise as actual measurements of fat depths and muscle areas made on the carcass cut surface (RSD values = 2·20 and 2·14 kg, respectively). However, these values, which appear in the results section of the paper, are not consistent with those given in the summary (corresponding values 0·95 and 0·89 kg) and because different samples of carcasses were evaluated using different techniques, including electromagnetic scanning which measures body electrical conductivity, it is difficult to compare the comparative precision of different methods, although the authors have apparently attempted to adjust the data to do that.

Cross *et al.* (1989) reported the historical background to instrumental carcass grading in the United States. The United States Department of Agriculture, in cooperation with the National Aeronautics and Space Administration (NASA) and the Jet Propulsion Laboratory, began a project in 1978 to develop an instrument for the objective evaluation of beef carcass quality and yield grade traits. Video image analysis was identified by NASA as having the greatest potential for that purpose, but following preliminary studies which indicated promising potential for a VIA system to estimate yield, but less promising results for quality traits, industry representatives switched the emphasis to techniques that could operate on unchilled and intact beef sides. The argument was that grading of such material would be necessary if new technology such as hot processing were to be implemented. Ultrasound was then identified as the technology to pursue, and a Johnson and Johnson Model 210 DX linear array utilising 3·5 MHz transducers was used to obtain images of fat and muscle at the 12–13th rib site on 52 bulls, 196 steers and 33 heifers. Comparisons between estimated subcutaneous fat depth, estimated *m.longissimus* area, and the same traits measured directly on the carcass appeared promising, but the data were only presented as means (overall, or within sex, weight group, etc). Similarly, the yield of boneless, closely trimmed cuts from the prime joints estimated using ultrasound data was very close to the actual value from cut-out data when the means of ten carcasses were compared.

Terry *et al.* (1989) used the same real-time scanner as Cross *et al.* (1989) to measure nine fat thicknesses and *m.longissimus* area at the 10th rib in both live and the unsplit carcasses of twenty hogs. The most appropriate equation for predicting the percentage of four lean trimmed cuts in the carcass from carcass ultrasound measurements utilised a single fat thickness over the dorsal mid-line at the 1st rib (RSD = 1·68, R^2 = 0·82). There was little to choose between live and carcass ultrasonic measurements in terms of their precision in estimating the percent of lean cuts.

In the Federal Republic of Germany, an ultrasound scanner—the Aloka SSD 256—has been approved as a method under the EC grading of pig carcass scheme. However, its apparent intended use is not so much as a practical system for use in abattoirs as a reference method which must be used to calibrate other apparatus which may be intended for estimating the lean content of pig carcasses in Germany.

The critical measurements to be used in the new German system, based on the ultrasound prediction, are the thickness of backfat, 7 cm from the dorsal mid-line between the 2nd from last and 3rd from last ribs, and the thickness of muscle tissue underlying this defined fat thickness. The correlations between the ultrasonically obtained estimates of fat and muscle thickness, the same tissue thicknesses measured with the Fat-O-Meater (FOM) and those estimated from magnetic resonance imaging (assumed to be accurate and used as the reference method), have been determined by Branscheid *et al.* (1989). Relative to the reference method, the ultrasonic scanner estimates showed a smaller mean deviation than the FOM for fat thickness (0·06 and 0·35 units—not stated but presumed to be cm) and for muscle thickness (1·7 and 1·8 cm(?), respectively). In a second trial in which the carcasses of 393 pigs of mixed breed and sex were dissected, the ultrasonic scanner measurements were related to lean meat percentage, and compared with FOM estimates (lean and fat thickness measured at the 3/4 last rib). The residual standard deviations were 2·17 and 2·28 for the scanner and the FOM, respectively.

Although some of the results of the investigations described above are promising, there are limitations of the pulse-echo technique when applied to the prediction of composition. These include image interpretation (which is performed by an operator, although there are developments to automate this process using image analysis coupled to artificial intelligence); and reliance on relatively superficial tissue thicknesses and areas as indices of composition. A system has been devised which overcomes these disadvantages, based on the measurement of the velocity of ultra-

Table 9
Comparative precision of the velocity of sound technique[a], visual score, and subcutaneous fat depth measurements in the prediction of tissue percentages of beef carcasses

Predictors	*Residual standard deviation*		
Reference: Miles *et al.* (1987)			
	Lean %	*Subcutaneous fat %*	*Intermuscular fat %*
Sample standard deviation	3·40	2·40	1.80
Reciprocal speed, sites 3,4	1·97	1·72	1·18
Fat score[b]	2·18	1·08	1·36
Fat depth	2.20	1·20	1·38
Lipid thickness sites 3,4	1·79	1·45	1·11
Reference: Miles *et al.* (1990)			
Group A			
Sample standard deviation	4·3	2·6	1·9
Reciprocal speed, sites 3,4,6	1·68	1·21	1·16
Fat score[b]	2·26	1·10	1·44
Fat depth	3.07	1·83	1·50
Group B			
Sample standard deviation	4·5	2·5	2·2
Reciprocal speed, sites 3,4,6 (1)	1·89	1·21	1·10
Fat score[b] (2)	2·39	0·91	1·30
Fat depth (3)	3·46	1·71	1·82
(1)+conformation score[b]	1·58	—	—
(2)+conformation score	2.05	—	—
(3)+conformation score	2·92	—	—
(1)+$\frac{d_3^3}{m}$+$\frac{d_3}{d_6}$ (4)[c]	1·35	—	—
(4)+sex	1.26	—	—

[a]Site 3 = through shoulder, cranial to 1st rib; 4 = 10/11th ribs, 6 cm from mid-line; 6 = 7/8th ribs, 6 cm from mid-line; lipid thickness = estimated from ultrasound measurements and tissue thickness.
[b]EAAP scales (1–15).
[c]For explanation see text.

sound through the tissues. The equipment was first developed for use on live beef animals (Miles *et al.*, 1984), and is based on the observation that the velocity of ultrasound is approximately 10% faster through lean tissue than through adipose tissue at live body temperature. Subsequently, Miles *et al.* (1987) established that the technique could be applied to warm beef carcasses within one hour of slaughter, and measurement at two sites (mean) was better correlated with lean tissue proportion in 72 beef carcasses of mixed breed and sex than visually assessed fatness and conformation using 15-point scales. The advantage of this method is that the ratio of total fat (i.e. subcutaneous plus intermuscular plus intramuscular) to fat-free tissue at the measurement sites is given in the readings, and the ultrasound speed measurements were better correlated with lean proportion than either fat score or subcutaneous fat thickness (Table 9), even though the latter two were more highly correlated with total carcass fat percentage. In the study by Miles *et al.* (1987), a lipid thickness, calculated by multiplying the volume fraction of lipid at the measurement site (proportional to the reciprocal of the velocity of ultrasound) by the thickness of tissue at the site (logged as the distance between the transducers), was more highly correlated with tissue proportions than ultrasound speed. This adjustment, in part, removes some of the effect of variation in size and/or conformation on the relation between the tissue proportions at the site and total carcass composition; addition of side weight or conformation score did not improve the precision of prediction of lean percentage based on the calculated lipid thickness. This additional information of tissue thickness at the measurement site was explored more fully by Miles *et al.* (1990). In a group of 27 bulls and 23 steers including British Friesian and Belgian Blue crossbred animals and thus exhibiting a wider range in conformation than in the earlier study, the better correlations with lean percentage were obtained using reciprocal velocity measurements rather than calculated lipid thickness, and conformation was a significant variable in reducing the residual variation. However, the effect of variation in conformation was removed by including inter-transducer tissue thicknesses as two ratios: thickness at site 3 (immediately cranial to 1st rib), cubed, divided by side mass, and the thickness at site 3 divided by the thickness at site 6 (through *m. longissimus thoracis* between the 7th and 8th ribs, approximately 6 cm from the dorsal mid-line in a direction parallel to the plane of the cut surface of the side). The main results from this study are also included in Table 9.

Carcass Shape

The shape of the live meat animal has attracted considerable attention since the earliest attempts at selective breeding. Some of these live animal attributes were thought to be related to the value of the animal as a meat producer; others were purely aesthetic, and many in the former category have since been exposed as misconceptions, particularly in regard to tissue distribution and the confusion of fleshing with fatness. Similarly, the assumed importance of the variation in carcass shape, in terms of yield of meat, carcass composition and the relative proportions of different cuts, has often been inappropriate. The reasons behind the perpetuation of these misconceptions were due, in part, to a lack of objective data to evaluate the assumptions, and also in part to the complex nature of carcass shape itself, and the incomplete means of describing it.

Kempster *et al.* (1982*b*) have suggested that because correlations between visually assessed conformation and carcass composition have usually been low, scientists have tended to view conformation as an unimportant characteristic, in contrast to the meat industry which attaches importance to conformation as an indicator of commercial value. However, there has been a continuing search by scientists to measure carcass shape objectively, and it is probable that technological advancement in this area will lead to a more complete understanding and quantification of the variation in carcass shape and its relevance to carcass qualities.

The hypothesis which is the central element and the *raison d'être* for most studies of conformation has rarely been spelled out clearly. As stated by Dumont (1989) 'The relationships existing between conformation and composition have been studied for many years but mainly for practical purposes, and without sound biometrical basis. This situation can be explained to a great extent by the difficulties involved in the quantification of the variation of conformation'. But the practical difficulties should not be allowed to cloud the main objective which is based on the following rationale:

1. Variation in the external shape (conformation, topography) of carcasses is a reflection of the integrated variations in the shapes of individual tissues (bone, muscle, fat depots).
2. Bones vary both in their lengths (cranial–caudal axis in the majority of bones in the axial skeleton, proximal–distal axis in the limbs) and in planes perpendicular to their lengths. However, for the most part, skeletal variation is a small component of the overall variation in carcass *shape*, particularly in anatomical regions where bone structure is relatively simple and the mass of the overlying tissues is relatively large (e.g. the proximal pelvic limb). Skeletal variation (lengths) is directly responsible for variation in carcass *size* (the conceptual norm being carcass length), but variation in size can be accounted for by making various skeletal measurements. There-

fore, variation in carcass shape is primarily the result of variation in muscle and fat tissue deposition during growth.

3. Either (a) the effects of variation in the individual muscle and fat tissue (total volumes, distribution) on overall carcass shape cannot be identified and separated, and the proportions of each can only be estimated by additional measurements not directly related to shape (e.g fat cover), or (b) although both fat and lean deposition may have parallel effects on certain carcass dimensions, their effects in other regions may differ sufficiently to allow estimation of the proportions of each to be made by appropriate shape measurement.

It is clear that although that part of the hypothesis under point 2—the relatively invariant contribution of bone shape—may hold true, any variation in skeletal structure will be a source of error in predicting of lean or fat content. It is known that there are breed effects on bone shape, and there are certainly sex effects. It may be possible to remove some of the variation through measurement of bone thickness, such as the circumference of the carpus joint (Van der Peet, 1984).

The item requiring more research is that listed under 3(b)—can the separate effects of fat and lean deposition on overall shape be identified? It is interesting to note that we are probably more adept at distinguishing between fat and well-muscled examples of our own species, and there has long been the application of somatotyping in clinical studies. But there is some evidence of these effects in cattle. Using a set of monozygotic heifer triplets, fed different dietary energy levels from the approximate start of puberty, Fisher (unpublished) obtained carcass shape data in which the underlying skeleton was very similar in shape and size for all three animals but total dissectible fat ranged widely (31, 38 and 47%). Serial cross-sectional bandsawing of the carcasses enabled examination of the surface contours and individual tissue components to be made at defined anatomical planes. A simple index of 'roundness' of each cross-section was obtained by dividing the square root of its area by its perimeter. This shape index was practically identical in all three carcasses in the very lean region of the ischium (proximal pelvic limb), but the fatter carcasses had higher index values in the caudal rib and loin regions (Fig. 2).

Visually Assessed Conformation

Traditionally, conformation has been assessed visually and particular aspects of shape have been identified for appraisal (e.g. the degree of

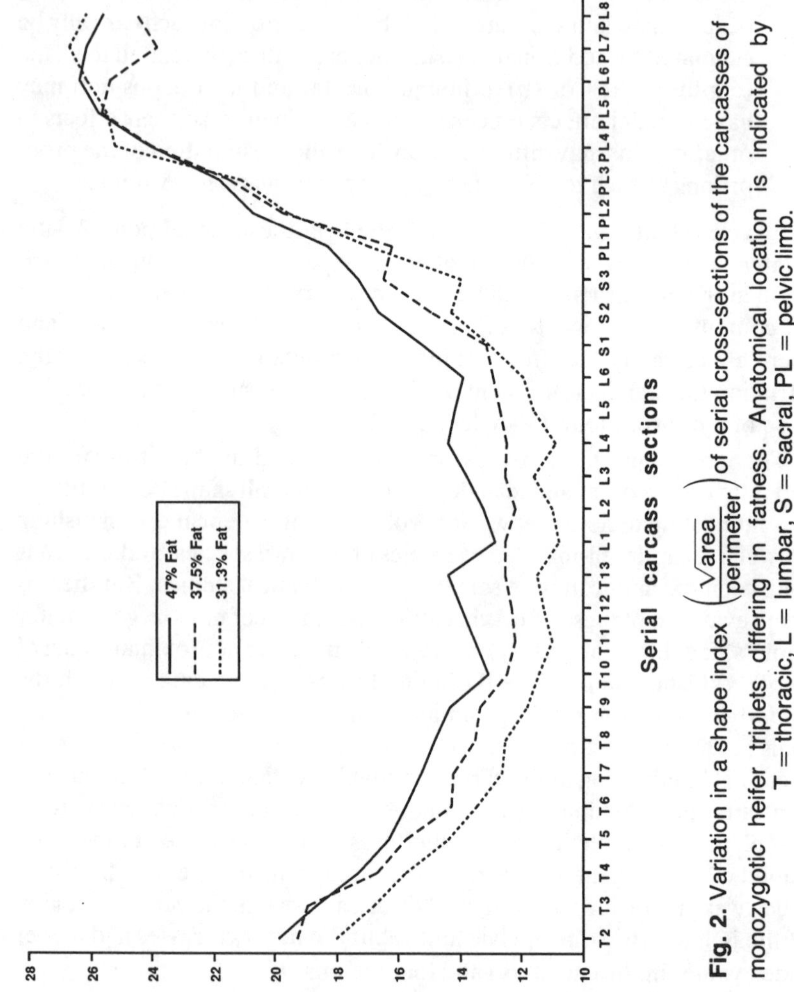

Fig. 2. Variation in a shape index $\left(\frac{\sqrt{\text{area}}}{\text{perimeter}}\right)$ of serial cross-sections of the carcasses of monozygotic heifer triplets differing in fatness. Anatomical location is indicated by T = thoracic, L = lumbar, S = sacral, PL = pelvic limb.

convexity or concavity of carcass profiles (Dumont *et al.*, 1975). This approach, which essentially aims to estimate the thickness of tissues relative to skeletal size, is highly dependent on an accurate estimate of fatness to provide further information on lean content. The reasoning behind this has been spelled out by Kempster *et al.* (1982*a*). Briefly, conformation and fatness are normally positively associated with each other but are related to lean proportion in opposite directions. Better conformation carcasses tend to have higher lean-to-bone ratios, and hence a higher lean percentage, than poorer conformation carcasses at the same fatness. But because increased fat deposition may improve conformation through increasing the bulk of tissue around the skeleton and produce a more convex profile, lean percentage may be depressed in some better conformation carcasses. This effect appears to be particularly evident in sheep carcasses, although the precise nature of the effects and their exaggerated manifestation in this species are not understood.

Conformation Assessment of Pig Carcasses

Even though carcass shape was accepted as an important predictor variable in some European countries (de Boer *et al.*, 1975), early British studies found no significant increase in the precision of predicting lean percentage or lean distribution when conformation score was added to P_2 and carcass weight (Kempster & Evans, 1979*b*). Later studies (Kempster & Evans, 1981) did show a marginal improvement using conformation in the prediction of carcass lean percentage, lean-to-bone ratio, and *m.longissimus* cross-sectional area. The data of Kempster and Evans (1981), presented in a slightly modified form by Kempster *et al.* (1982*a*), are shown in Table 10. However, de Boer (1984) has suggested that the British assessment of 'conformation' was not measuring the same attributes as the shape assessment under the 'type' category used in the original European Community Scheme.

Diestre and Kempster (1985) found that a score for leg conformation on a 5-point scale was a small, but significant, factor in the prediction of lean percentage when combined with P_2, P_2^2, and the minimum fat depth in the dorsal mid-line over *m.gluteus medius*.

What has clearly emerged is that the range of genetic types occurring in a population will indicate how likely conformation assessment will be in improving the prediction of lean percentage, and the incidence of the halothene gene is a key factor as the carriers have higher lean-to-bone ratios, and there is evidence of them having a different muscle distribution.

Table 10

The value of conformation scores in combination with carcass weight, fatness estimates, and breed in the prediction of carcass composition

Predictors	*Residual standard deviation*			
Reference: de Boer *et al.* (1975)		**PIGS**		
		% Lean		
		Factory 1		*Factory 2*
Sample standard deviation		4·42		3·34
Average backfat thickness (FT)		2·34		1·88
FT + 'type'		2·14		1·77
Reference: Kempster *et al.* (1982*a*)				
	% Lean	*Lean-to-bone ratio*	M.longissimus *area* (cm^2)	*% Total lean in ham+back*
Sample standard deviation	4·39	0·56	9·73	1·29
Carcass wt (W) + P_2	2·45	0·53	5·31	1·28
W + P_2 + conformation (C_5)[b]	2·44	0·51	5·22	1·28
W + P_2 + company (Cy)[c]	2·31	0·51	5·16	1·25
W + P_2 + C_5 + Cy	2·30	0·50	5·16	1·25
Reference: Diestre & Kempster (1985)				
			Lean %	
Sample standard deviation			4·42	
Carcass wt (W) + P_2			2·52	
W + P_2 + P_2^2			2·46	
W + P_2 + P_2^2 + loin fat thickness (LF)			2·37	
W + P_2 + P_2^2 + LF + conformation (C_5)[d]			2·31	
W + P_2 + P_2^2 + LF + company (Cy)[e]			2·17	
W + P_2 + P_2^2 + LF + C_5 + Cy			2·14	

Reference: Kempster & Harrington (1980)

	Saleable meat (SM) %	*SM-to-bone ratio*	M.longissimus *area* (cm^2)	*% Total SM higher-priced joints*
Sample standard deviation	1·83	0.36	10·9	1·05
Carcass wt (W)+fat class (FC_7)[f]	1.81	0·35	8·1	1·04
W+FC_7+conformation (C_6)[g]	1·68	0·30	7·5	0·98
W+FC_7+breed (B)[h]	1·45	0·27	6·5	0·92
W+FC_7+C_6+B	1·41	0·25	6·3	0·91

Reference: Colomer-Rocher *et al*. (1980)

		Saleable meat (kg)	*Muscle (kg)*	*Muscle-to-bone ratio*
Conformation (C_7)[i]		4·80	4·16	0·28
C_7+fatness (F)[j]		4·45	4·11	0·28
C_7+carcass wt (W)		2·06	2·66	0·26
C_7+F+W		1·43	1·82	0·26

Reference: Kempster (1986)

	16 Month dairy-bred cattle[k]			*Suckler-bred cattle (main trial)*[k]		
	% Lean	*Lean-to-bone ratio*	*% Total lean in higher-priced joints*	*% Lean*	*Lean-to-bone ratio*	*% Total lean in higher-priced joints*
Sample standard deviation	3·44	0·33	1·17	2·88	0·29	1·11
Carcass weight (W)+fat class (FC_7)[l]	2·46	0·32	1·08	2·41	0·29	1·07
W+FC_7+conformation (C_7)[l]	2·42	0·29	1·05	2·40	0·26	1·06
W+SF_e[m]	2·39	0·32	1·07	2·08	0·29	1·06
W+SF_e+conformation (C_{15})[n]	2·22	0·29	1·01	2·00	0·26	1·06
W+SF_e+breed (B)	2·09	0·27	0·96	1·91	0·27	0·97
W+SF_e+C_{15}+B	2.06	0·26	0·94	1·88	0·25	0·98

Table 10—*contd.*

SHEEP

Reference: Kempster *et al.* (1982*a*)

	% Lean	*Lean-to-bone ratio*	*M.longissimus depth (mm)*	*% Total lean in higher-priced cuts*
Trial 1				
Sample standard deviation	3·83	0·38	3·14	1·70
(1) Carcass wt (W)+fat class (FC_5)[o]	2·97	0·35	2·68	1·67
(2) W+FC_5+conformation (C_4)[p]	2·96	0·34	2·64	1·66
(3) W+FC_5+breed (B)[h]	2·88	0·33	2·61	1·62
(4) W+FC_5+B+C_4	2·88	0·33	2·59	1·61
Regression on C_4 (eqn (2))[q]	−0·52	0·15	1·04	0·43
Regression on C_4 (eqn (4))[q]	NS	0·12	0·83	0·40
(5) W+SF_e[m]	2·61	0·35	2.68	1·68
(6) W+SF_e+conformation (C_{15})[n]	2·61	0·34	2·60	1·65
(7) W+SF_e+B	2·56	0·33	2·62	1·62
(8) W+SF_e+C_{15}+B	2·56	0·32	2·56	1·60
Regression on C_{15} (eqn (6))[q]	NS	0·05	0·37	0·16
Regression on C_{15} (eqn (8))[q]	NS	0·04	0·35	0·14
Trial 2				
Sample standard deviation	4·36	0·39	3·22	1·80
(1) W+FC_5[o]	3·29	0·37	2·89	1·76
(2) W+FC_5+conformation (C_5)[b]	3.29	0·37	2·82	1·75
(3) W+FC_5+B	2·96	0·33	2.79	1·74
(4) W+FC_5+C_5+B	2·96	0·33	2·74	1·73
Regression on C_5 (eqn (2))[q]	NS	0·14	1·30	0·44
Regression on C_5 (eqn (4))[q]	NS	0·12	1·20	0·40

[a]Included some bones, and excluded lean in belly.
[b]Conformation assessed on 5-point scale.
[c]The breeding company supplying the pigs was fitted as a fixed effect in the model.
[d]Conformation of hind leg assessed on a 5-point scale.
[e]Breeding company effect estimated using common regression coefficient and different intercepts.
[f]Fatness assessed on a 7-point scale.
[g]Conformation assessed on a 6-point scale.
[h]Breed of sire fitted as a fixed main effect.
[i]Lateral hind limb conformation assessed on a 7-point scale.
[j]Dissectible total fat in hindquarter.
[k]Examples only: the data for two other groups (one each of dairy-bred and suckler-bred) showed similar trends.
[l]Fatness/conformation assessed on 7-point scales.
[m]Subcutaneous fat assessed to the nearest percentage unit.
[n]Conformation assessed on a 15-point scale.
[o]Fatness assessed on a 5-point scale.
[p]Conformation assessed on a 4-point scale.
[q]Values are partial regression coefficients.
NS signifies not significant.

Conformation Assessment of Lamb Carcasses

It has been argued that the importance and hence usefulness of conformation assessment in sheep is low because the variation in conformation itself is small in this species (Kempster *et al.*, 1982*b*). Even when progeny of Texel sires, having more extreme conformation than traditional British breeds, were included in two trials, the value of conformation at constant fatness was found to be minimal (Table 10). When fatness was estimated using a 5-point scale, conformation, assessed on a simple 4-point scale, was a significant factor in the prediction of percentage lean, lean-to-bone ratio, and *m.longissimus* depth. However, the partial regression coefficient for conformation score in the relation with lean percentage is negative, indicating that better conformation carcasses were fatter. When a more precise estimate of fatness was used (subcutaneous fat estimated to the nearest percentage point) there was no significant contribution from conformation score, even though this was assessed on a more discriminating 15-point scale. Lean-to-bone ratio, *m.longissimus* depth, and proportion of total lean in the higher-priced cuts were estimated more precisely using conformation score, but the reductions in the residual variance were small. Also, the association between conformation score and lean-to-bone ratio is weakened through breed effects. There were examples of breed differences in lean-to-bone ratio which did not correlate with conformation score in the study by Kempster *et al.* (1981*a*), and it was suggested that differences in bone structure may be responsible.

Fisher and Bayntun (1981) compared four morphologically contrasting sheep breeds (Southdown, Hampshire, Cheviot and Scottish Blackface, ranked in order of decreasing conformation score) in the relations between the muscle weight and the weight and dimensions of bones in the pelvic limb. They found that breed differences in the weight of lean per unit bone length were not maintained in the expression of weight of lean per unit bone weight: relative to the breed having the highest former ratio (Southdown), the Hampshire and Cheviot had lower, and the Scottish Blackface higher, than expected muscle-to-bone weight ratios. Moreover, variation in bone density did not explain these differences, and the implication was that differences in bone shape were responsible; some breeds have thinner bones relative to length than others ('true breed' effects), but there is also an effect of stage of maturity. Young (1989) showed that the major factor involved in the change of bone weight relative to length was bone shape, and not density. Relative to their length, limb bones become thicker at their mid-lengths as they grow, but volume decreases ('maturity' effects), the hypothesis being that the proportion of length comprising the relatively massive epiphyses decreases during growth. In the study by Fisher and Bayntun (1981), bone density data indicated that the rate of bone development was faster in the Southdown and Scottish Blackface breeds than in the Hampshire and Cheviot and the 'maturity' effects on bone shape may at least partly explain the observed changes in the ratios studied.

Conformation Assessment of Beef Carcasses

Using data from 805 commercially deboned and fat-trimmed carcasses and 500 fully dissected half carcasses, in total comprising 13 sire breeds, Kempster and Harrington (1980) examined the correlations between conformation score (5, 6, or 7-point scales) and carcass composition within fat class or at constant dissected subcutaneous fat percentage. Correlations were low to moderate, conformation rarely accounting for more than 30% of the variation in the compositional traits. The overall standard deviation of saleable meat percentage in the 805 carcasses was 1·83, and the correlation with conformation score, pooled within fat class, was 0·38. Corresponding values, pooled within beef unit (source) and breed, were 1·45 and 0·15, respectively. The correlations with saleable meat-to-bone ratio were higher, both overall and within breed (0·57 and 0·39, respectively). The residual standard deviations of saleable meat percentage, saleable meat-to-bone ratio, higher-priced cuts percentage and *m.longissimus* area—using combinations of predictors, are given in Table 10.

Bone measurements (weights, lengths, circumferences and densities of the femur, humerus and tibia) did not significantly improve the prediction of lean percentage in the analysis by Kempster (1986) when added to an equation containing estimated subcutaneous fat percentage, eye-muscle area, and conformation assessed on a 15-point scale as predictors. Bone density did not improve the prediction of lean-to-bone ratio, but bone weights and measurements, particularly of the femur, did improve prediction precision. The addition of breed identity to the equations including bone weights and dimensions did not markedly improve the precision of prediction, indicating that the main differences in these parameters occurred between breeds rather than within breeds.

Objective Measures of Conformation

Early attempts to quantify carcass conformation objectively were based either on ratios of weights to linear dimensions, the simplest and perhaps most useful being hot carcass weight divided by carcass length (Yeates, 1952), or on the use of several dimensions which essentially gave a crude index of shape. Investigations of the repeatability (between sides) of linear dimensions, changes during carcass cooling, and their relations with carcass composition and tissue distribution, many of them in the 1950s (e.g. Bodwell, 1959), showed that such measurements were poor predictors of carcass composition. Kempster *et al.* (1982*a*) state that in a mixture of carcass types where the standard deviation of lean percentage at equal weight is 3·0, the use of carcass length, depth or width would not be expected to lower the residual standard deviation below 2·5.

In later studies, more sophisticated measuring techniques or a more complex methodology using simpler techniques, have produced more encouraging results. Bass *et al.* (1981) constructed reference lines over the dorsal profile of the pelvic limb on hanging beef sides from ten sire breeds using skeletal structures as reference points. Deviations of the carcass profile in the lateral view from the reference line were measured to facilitate estimation of the proportional area, within that determined by the reference points, occupied by the carcass profile. This area represented an integrated thickness of the mass of flesh lying posterior to the femur and tibia in the lateral view, and an additional simple linear thickness in the medio-lateral plane was also recorded. The precision of the objective measurements in the prediction of hindquarter composition was compared with that obtained using two visual assessment schemes. At the same hindquarter weight and fatness (dissected value), the conformation scores improved the precision of prediction of trimmed, saleable meat, dissected muscle and dissected bone weights only slightly (maximum additional percentage variation accounted for = 3·3). In contrast, the objective measurements, particularly the lateral area, increased the variation explained by more than ten percentage points, and whereas sire breed effects (pooled within breed) reduced the precision of visual score to non-significance in the prediction of muscle weight, the objective measurements were still significant variables. Interestingly though, the visual scores were more precise predictors of saleable meat-to-bone or lean-to-bone ratios, and the medio-lateral hindquarter width was more precise than the lateral area in the prediction of this ratio.

A different approach was used by Fisher (1975) in a study of steers of one breed (Hereford). Photographic negatives of the dorsal and lateral views of 30 carcasses were obtained and profile areas measured from projected images on a specially designed imaging table. Relations between fore- and hindquarter areas in the two views, were adjusted to the cubic dimension, and the weights and volumes of tissues were computed. Correlations between weights of fat depots and areas were considerably higher for the dorsal areas than the lateral areas (0·63 and 0·80 versus 0·19 and 0·60 for subcutaneous and intermuscular fat, repsectively). Differences were much reduced in the relation with dissected lean weight, but the largest correlation (0·71) was obtained from the lateral area of the hindquarter. In multiple regression analysis, the sample standard deviation of total dissectible fat weight (4·4) was reduced to residual values of 2·35 using side weight, dorsal area and carcass length as predictors, and 2·32 using side weight and a mean visual score for fatness assessed by six assessors, independently, on a 7-point scale.

Dumont (1989) examined in detail the conformation of the ham (proximal pelvic limb) in 65 pigs comprising entire and castrated males, females, and four breeds including the highly-muscled Pietrain. Photographic diapositives of the medial and dorsal views were projected and 'transverse apparent diameters' (TAD) were constructed on the traced profiles at 26 equidistant points and perpendicular to the line joining the calcaneal tuber to the cranial edge of the symphysis pubis. Multivariate analysis of centred data techniques were used to identify interrelations of the most discriminating widths in relation to muscle-to-bone ratio. The results showed that those TAD values which varied most, and were thus the best indicators of conformation differences, were not necessarily those which correlated best with muscle-to-bone ratio. The multivariate analysis showed that, in the dorsal view, the first two axes defining the first plane of the projected variables contrasted TAD values on the sloping medial surface of the ham with the remainder (first axis) and contrasted the proximal and distal TAD values along the second axis. In the medial view, muscle-to-bone ratio is a major discriminant variable mainly along the first axis and contrasted TAD values in the penultimate distal quartile of total length with those near the extremities. Selection of multiple regression variables to predict muscle-to-bone ratio based on these observations yielded a best equation which included a ratio of TAD values from the medial view, a ratio from the dorsal view, and an additional dorsal value and accounted for 84·6% of the variation.

Techniques which would enable the three-dimensional shape of carcasses to be measured have been developed over the years. Sterophotogrammetry was used by Brinks *et al.* (1964) to estimate the proportions of wholesale cuts in live cattle, and a cheaper system which also estimates surface contours is the Moiré method, applied to carcasses by Speight *et al.* (1974). Neither method has been applied in detail to relate carcass shape to composition. More recently, there has been interest in video image analysis (VIA) to estimate composition. Although this technique does not directly provide data on three-dimensional shape, information obtained from a number of angles around the carcass axis does allow aspects of shape to be measured with a high degree of accuracy and, unlike the photographic methods, data are generated almost instantaneously and so there are substantial advantages to be gained if used in a commercial classification system.

Danish work on the development of a VIA-based 'classification centre' for beef was reported by Sorensen (1984). The approach adopted was determined by the relations between carcass composition and yield of saleable meat (and hence carcass value) in young bulls. It was argued that the percentage of bone and the lean-to-bone ratio were the most important compositional characteristics in this type of cattle (at standard weight). This would be true in a population where level of fat was low and relatively invariant, but this philosophy also accepts the limitations inherent in the use of saleable meat as the ultimate end-point: fat proportion in saleable meat varies—that from fatter animals contains a higher proportion of fat than that from leaner animals, and it is an inconsistent product. If more and more trimming of fat becomes standard practice as a result of consumer demand for practically fat-free meat at the point of retail sale, then lean meat percentage will emerge as the characteristic most highly correlated to carcass value. This Danish work was at its infancy when reported (Sorensen, 1984) but promising correlations were obtained from VIA determined measurements of thigh widths in relation to carcass length and weight, and bone proportion, or lean-to-bone ratio ($r = 0{\cdot}64$ and $0{\cdot}59$, respectively). Folllowing the initial investigations, the Danish approach has been to combine the VIA data with tissue depth recorded by a light reflectance probe. The measurements obtained from this beef classification centre include VIA coordinates of the carcass contour viewed from the lateral aspect; VIA distribution of grey tone values on the carcass surface using green light illumination; probe fat depth over a rump site and fat and muscle depth in the loin; and carcass weight (Sorensen *et al.*, 1988). A total of 2948 carcasses were processed, covering

Table 11
Results from the Danish Beef Classification Centre (BCC)

	Saleable meat %	*Fat trim %*	*Bone %*	*EUROP*[a]		*Muscularity index*
				Conformation	*Fatness*	
Sample standard deviation	2·1	2·1	2·1	1.38	1·03	10
Simple or multiple correlation coefficients from						
Carcass weight (W)	0·40	0·17	0·47	0·53*	0·44	—
W+EUROP scores	0·74	0·75	0·77	—	—	—
BCC equations	0·73	0·77	0·79	0·84*	0·67	—
Residual standard deviation from						
BCC equations	1·5	1·4	1.3	1·0*	0·6	5·5

[a]EUROP scales: data relate to 5-point scales for fatness, and for conformation except where indicated by * (15 subclasses).
Data of Sorensen *et al.* (1988).

the 5 EUROP conformation and fat classes, and 389 of these were commercially dissected into saleable meat, fat trim and bone. The main results are shown in Table 11. The correlations between visual scores, in combination with carcass weight, and carcass composition were very similar to those obtained using the classification centre equations, but the authors claimed that repeatability was better using the objective measurements. Also, correlations between fat depths measured by probe and the visually assessed fat class were considerably higher ($r = 0{\cdot}45–0{\cdot}55$) than the corresponding correlations between VIA fat areas and fat class ($r = 0{\cdot}17–0{\cdot}25$).

The details of the practical operation of the beef classification centre have been reported by Petersen *et al.* (1989). They report that in addition to calculating carcass composition and estimating EUROP grades, the system also has algorithms to calculate a muscularity index (weights of six hindquarter muscles and area of *m.longissimus dorsi* both corrected for carcass weight), and a value index (based on carcass composition, muscularity and 'EUROP classification of fat/lean colour'). The precision in estimating the muscularity index reported by Petersen *et al.* (1989) is included in Table 11, but no data are provided for the value index.

In the apparatus designed for grading pig carcasses in the Federal Republic of Germany, video imaging was used to measure conformation parameters of the ham (Sack, 1983). This apparatus known as the SKG, was replaced by the SKG II which used a pneumatically operated mechanical probe to provide the same information on carcass dimensions and shape. According to Casteels *et al.* (1984), the SKG II provides greater accuracy and speed of measurement, and also provides information on the percentage of expensive primal cuts.

A development in France, reported in 1985, utilised mechanical probes to estimate conformation of beef carcasses (Anon., 1985). This apparatus, known as the Colavaug-B01 (a derivative of the inventor's name Auge and the manufacturers Lavaur), is automatic in operation and positions a beef side within a rigid frame and adjusts the positions of the measuring probes according to the carcass length determined by mechanical sensors detecting the os pubis. Motorised mechanical sensor probes then make contact with the carcass at four positions cranio–caudally from the middle of the ribs to the proximal pelvic limb. Each probe consists of two sensors—one fixed and one telescopic. The fixed locates a skeletal point, and the telescopic is set to make contact with the carcass surface at a set distance from the bone. The actual extension of the telescopic section, which is determined by tissue resistance, is recorded. In addition, the width of the thorax is measured by a horizontally moving arm. These measurements plus carcass weight are used to estimate EUROP conformation sub class (+,=,− about each main class). Visual assessment of conformation was also undertaken by three official classifiers, a mean calculated and compared with the Colavaug estimated subclass. Over 85% of the individual classifiers scores were within one subclass of each other, and 87% of the Colavaug estimates were within one subclass of the average visual score. There was a tendency for the Colavaug to underestimate the conformation of the highest scoring carcasses (E+ . . . E−) and overestimate the conformation class in the middle of the range (R− . . . O+) when compared with the average visual scores. These results are interesting in that this apparatus, by measuring the thickness of soft tissues overlying bones, provides such a high correlation with visually assessed conformation, but of course, there are no data relating the Colavaug values to actual carcass composition or yields.

Electrical Methods

Equipment known by its acronym EMME (electronic meat measuring equipment) was developed by the EMME Corporation, Phoenix, Arizona, USA to estimate the lean content of live pigs. The theoretical basis of the method, in simple terms, is that conductivity of electricity through lean tissue exceeds that through fat tissues about twenty fold, and the conductivity of a sample object placed within a glass fibre tunnel surrounded by a solenoid interferes with the induced magnetic field. Eddy currents are induced within the body water of an animal or carcass, and impedance to current flow in the system results in irreversible energy loss as heat. This energy loss is detected in the coil as an index of conductive mass of the subject. In addition to this conductive effect, there is a second component which defines the flow of current in a subject, namely its capacitance which is related to cell membranes. A more complete description of the theoretical basis of the method is given by Kuei *et al.* (1989). In the original EMME machine, both conductive and dielectric properties are represented in a single number 'the EMME number' which is purported to be proportional to lean mass.

A version of the EMME SA-1 was modified to provide appropriate signal levels from a beef quarter in the study by Koch and Varnadore (1976). Using 66 beef carcasses, the EMME numbers so generated were correlated with the weights of primal cuts, which were not trimmed to a

Table 12
Results of some studies using electrical methods to estimate carcass composition

Predictors	*Parameters from prediction equations*					
Reference: Koch & Varnadore (1976)	*Residual standard deviation*					
	Forequarter lean cuts (kg)		*Hindquarter lean cuts (kg)*		*Side lean cuts (kg)*	
Sample standard deviation	4·7		4·0		8·4	
Weight of quarter/side (W)	1·37		2·10		3·26	
EMME number	3·35		2·46		5·25	
W+EMME	0·98		1·38		2·04	
On a sample of 19 forequarters						
	Standard partial regression coefficents (b) and associated R^2 values					
	Primal rib-untrimmed		*Primal rib-constant trim*		*Fat-free rib*	
	b	R^2	b	R^2	b	R^2
Forequarter wt (FW)	0·88	0·78	0·84	0·71	0·62	0·38
EMME number	0·59	0·35	0·71	0·51	0·75	0·57
FW+EMME	0·76+0·20	0·81	0·65+0·38	0·82	0·32+0·59	0·64
Reference: Jones & Haworth (1983)	*Residual standard deviation*					
	Group 1: half-carcass wt 30–33·9 kg		*Group 2: half-carcass wt 34–37·9 kg*		*Combined groups*	
	Commercial joint wt (kg)					
Sample standard deviation	1·08		1·03		—	
Half-carcass wt	1·18		1·25		1·23	
EMME number	1·24		1·29		1.41	
Best probe fat depth	0·91		1·04		1·53	
	Carcass lean (g/kg)					
Sample standard deviation	38·2		34·7		—	
Half carcass wt	38·2		34·9		36·4	
EMME number	37·1		34·3		35·9	
Best probe fat depth	24·7		25·2		25·8	

Reference: Kuei *et al*. (1989)

	Sample standard deviation	R^2	Residual standard deviation
(a) Weight of fat-standardised lean (kg)			
Carcass (*H*100, *L*2, *Tem*, *HCW*)[a]	4·9	0·91	1·47
Ham (*H*45, *Tem*, *A*0–45, *L*1)	0·8	0·91	0·25
Loin (*A*75-*E*, *L*2, *Tem*, *H*15)	0·7	0·80	0·29
Shoulder (*H*100, *H*75, *H*135, *Tem*)	0·7	0·89	0·22
(b) Percentage of fat-standarised lean			
Carcass (*H*100, *HCW*, *L*2, *Tem*)	4·7	0·82	1·96
Ham (*H*110, *HCW*, *L*2, *Tem*)	4·7[b]	0·76	2·32
Loin (*H*120, *HCW*, *L*2, *H*150)	4·6[b]	0·46	3·35
Shoulder (*H*100, *HCW*, *Tem*, *L*2)	4·5[b]	0·69	2·52

Reference: Jenkins *et al*. (1988)

Sample standard deviation (kg) = 1·63

	Regression models (Y = *weight of fat-free mass, kg*)				
	1	*2*	*3*	*4*	*5*
Regression coefficients					
Carcass wt (kg)	0·59***	0·65***	0·69***	0·39***	0·53***
Fat depth 12th rib (cm)		−2·37**			
Fat depth 4th sacral (cm)			−0·80***		−0·65***
Length^2/resistance (cm^2/ohms)				0·35***	0·24**
Residual standard deviation (kg)	0·50	0·44	0·39	0·42	0·35

**, $p < 0{\cdot}01$. ***, $p < 0{\cdot}001$.
[a]For explanation of all variable abbreviations, see text. [b]Values not given in paper; estimated from RSD and R^2.

uniform fat cover, but from which external fat in excess of 1·27 cm thick was removed. This rather poorly defined end-point, as the authors admit, limits the conclusions that could be derived from this study. Even so, significant improvements in the precision of prediction over that provided by the weight of the quarters was obtained adding EMME numbers as shown in Table 12. In a subsample of 19 forequarters, three levels of fat variation were obtained: the primal rib cut weight (no fat trim); the rib trimmed to approximately 0·8 cm surface fat; and fat-free lean from the rib estimated by dissecting the 9-10-11th rib slice and its chemical composition. The results, also included in Table 12, show that the partial regression coefficients for forequarter weight decrease in size whilst those for EMME number increase in size as variation in fatness reduces. These results show that the EMME number was more closely related to lean weight than to total quarter weight, but the R^2 value was only moderately high at 0·64 for fat-free lean.

The EMME technique was used to evaluate pig carcass composition by Jones and Haworth (1983). Fifty-seven commercial pig carcasses were studied within two weight groups, 30–33·9 and 34–37·9 kg. Over all weights the EMME number accounted for 0·34 of the variation in commercial joint weight (all joints excluding the belly, trimmed to 6 mm subcutaneous fat) and 0·26 of the variation in dissectible lean weight obtained from these joints. Corresponding R^2 values using carcass weight as predictor were 0·82 and 0·44, respectively. Within a weight group, EMME correlations were much lower (Table 12). Using stepwise multiple regression techniques, EMME number entered at step 5 in the estimation of lean weight in joints, and at step 4 in the estimation of carcass lean proportion. In both cases the more important variables were fat and lean depths measured by probe, and inclusion of the EMME number reduced the RSD only marginally. The authors concluded that EMME was not a useful prediction method based on these results, and carcass weight in combination with fat and lean depths, as measured by a probe, were better predictors of carcass composition.

More recent equipment has produced much more encouraging results. Forrest *et al*. (1988) used a medical electromagnetic scanner—a Dj Medical Corporation HA-2 Electromagnetic Scanner (TOBEC—acronym for total body electrical conductivity) to predict weight of total dissected carcass lean. Carcass length and temperature (carcasses were measured as dehaired uneviscerated whole bodies, as eviscerated warm sides and as chilled sides) were also used in the prediction equation which gave an R^2 value of 0·913 with an RSD of 0·69 (presumably kg—not stated).

Although other techniques were included in this study (ultrasonic scanning, tissue depths measured by automatic light reflectance probe), different subsamples of a total of 412 pigs were used to test each piece of equipment, and direct comparisons are not possible.

In a later study, Kuei *et al.* (1989) used the HA-2 to calculate values for conductivity and capacitance measured at 64 points on the carcasses of 49 gilts and 63 castrated male pigs ranging from 85–140 kg live weight. Conductivity was recorded as a curve of the summed magnetic and electric field intensities against position of the carcass within the coil. Data recorded from the curve were defined by setting the X-axis to range from 0 to 150, where 0 = starting point of curve (hind foot) and maximum curve height = 100 (entire carcass in the field). The variables used (Table 12) were: A = area under specific portions of the curve; H = curve height at specific point; E = last reading (shoulder region remaining in field). In addition, carcass length from the cranial edge of the first rib to the cranial edge of the aitch bone ($L2$), length from hind foot to most cranial point ($L1$), carcass temperature (*Tem*) and hot carcass weight (HCW) were recorded. After jointing and dissection, lean samples were chemically analysed for lipid and fat-standardised (10% lipid) primal cuts and carcass were calculated.

The results in Table 12 show that the predictions of the weights of standardised lean content in the total carcass and the three primal cuts (shoulder, ham, and loin) had a high degree of precision, but temperature was a significant variable in all equations. Temperature varied between 31 and 40°C in this trial, but the authors argue that in a commercial slaughter line the throughput is of a constant rate and temperature measurement may not be required. In the prediction of percentage composition of the carcass and cuts, the R^2 values were lower than in the prediction of weights of lean. This is to be expected for the joint predictions, as weight of joint was not known and the variation in fat and bone would tend to lead to increased error, but the prediction equation for total carcass percentage lean explained a higher proportion of the variation as it contained the relevant carcass weight as a predictor variable.

Resistive impedance to current flow through lamb carcasses was measured by Jenkins *et al.* (1988) following earlier work reported on human subjects. A current of 800 mA at 50 kHz was applied through electrodes on the muscle complex of a thoracic limb and on the extensor muscle complex of a pelvic limb, both placed at a standard 1·5 cm proximal to the carpal/tarsal articulations. Detecting electrodes were placed on corresponding positions on the other two limbs, and resistive impedance was measured on a tetrapolar impedance plethysmograph (Model BIA-101, RJL Systems Inc., Detroit, USA). Forty crossbred ram lambs were used, and in addition to the resistive impedance measurement, fat depths were measured over the 4th sacral vertebra on the dorsal mid-line and over *m.longissimus thoracis* 0·75 of its width from the dorsal mid-line at the 12th rib. Fat-free mass was used as the end-point and was defined as the sum of carcass water and protein (N×6·25), estimated from proximate analysis of ground whole carcass tissues (including kidney fat and skeletal tissue). The coefficients of variation of the fat-free mass and (carcass length)^2/resistance were very similar (10·6 and 10·3%, respectively). Although there was not a great variation in carcass weight (coefficient of variation 13%), weight alone accounted for 91% of the variation in fat-free mass, and was a highly significant predictor in multiple regression equations (Table 12). In conjunction with carcass weight, resistive impedance was not as good a predictor as the best fat depth (over sacrum), but a model which included both fat depth and resistance accounted for the highest proportion of variation in fat-free mass.

Computerised Tomography

It is possible to produce computerised cross-sectional images (tomographs) from 128×128 or 256×256 matrices of data generated by X-ray absorption or nuclear magnetic resonance. Both approaches were pioneered in human medicine studies, and the principles of operation are described in Chapter 6 (Allen, 1990).

Because these techniques require sophisticated and expensive machinery, including advanced computing facilities, and because they are usually located in hospitals and centres for human medical research, there are few studies which have investigated their potential in farm animals, and even fewer applied to carcass measurement. Allen and Leymaster (1985) used a pig carcass to estimate machine error (effects of 'noise' in the generation of electronic signals by the detectors logging the attenuation of radiation energy in a Siemens Somatom 2 whole-body X-ray computer tomograph). Twenty-four combinations of instrument setting were used and were representative of combinations of different slice thickness (2); number of projections (effectively the number of pathways of radiation, varying from half-degree intervals and a full 360° rotation to one degree intervals and a 240° rotation) (3); voltage (2); and current (3). The slice scanned was just caudal to the last rib. The results showed that considerable differences between repeat scans occurred when the instrumental setting varied if the end-point used was individual pixel value, less so when the frequency distribution of pixel values based on classes of varying attenuation units (Hounsfield Units) were used. If the mean pixel value of a carcass image was defined as the end-point, there was very little effect. The crucial question to be answered if this technique is to be standardised and equations formulated to predict carcass composition is therefore which form of analytical data should be used to relate to carcass dissection values. The authors suggest that individual pixel values will be unlikely, and class frequency data when the class interval is 2 Hounsfield Units or less will be too detailed. In studies on live pigs, Allen and Vangen (1984) found that areas of subcutaneous fat and *longissimus* muscle, which can be obtained from the computerised tomography (CT) scan using appropriate software, were more highly correlated with weights of dissected fat and lean than pixel means obtained from the same defined areas.

Sehested and Vangen (1989) compared, as predictors of dissected lean proportion, 'absorption values' (= mean pixel value, as defined above?) obtained from computerised tomography X-ray scans at six locations along the lengths of the carcasses of 128 pigs, with combinations of backfat thicknesses, longissimus area, carcass weight and percent ham and loin. Stepwise multiple regression techniques were used for both types of predictor variables, and the models chosen were those that included the highest number of significant covariates but with a restriction that not more than 20 should be used. However, the large number included using CT data probably means that the true residual standard deviations are higher than those indicated (Table 13). The results show clearly that CT variables were much more precise indicators of dissected composition than the simpler tissue depth and weight variables. In addition, there were no significant biases due to sex, breed or carcass weight in this experiment.

Some preliminary work has employed nuclear magnetic resonance imaging in carcass evaluation. Groenveld *et al.* (1984) showed that both X-ray and nuclear magnetic resonance (NMR) imaging were able to discriminate adequately between the three main carcass tissues of lean, fat and bone in pig carcasses. They also showed that simplified and shorter (non-imaging) NMR procedures could be adopted to estimate the proportions of constituent parts in a two component mixture (fat and lean) from studies *in vitro*. These estimates were very precise, but there is no evidence of the application of this technique to entire carcasses or half carcasses.

Table 13
Relative precision of prediction of pig carcass composition from computerised tomography (model 1) and simpler carcass measurements (model 2) (data from Sehested & Vangen, 1989)

Dependent variable	*Sample standard deviation*	*Residual standard deviation*	
		Model 1[a]	*Model 2*[b]
Lean, %	4·75	1·02	1·92
Lean, kg	2·27	0·40	0·73

[a]Model 1—Absorption values at six cross-sectional CT scans used as independent variables.
[b]Model 2—Backfat thicknesses over the dorsal mid-line (4) and at last rib 7 cm from mid-line. Percentage ham and loin, *longissimus* area and carcass weight used as independent variables.

Partial Dissection

A method of evaluation which is practically inapplicable to commercial classification of carcasses but which has much to offer the researcher in terms of precision and ease of application, is partial dissection or sample joint dissection. Different approaches have been used during the last 40 years, the predictor portion varying from individual muscles to large primal joints, the dependent data being weights of individual tissues or chemical components obtained from proximate analysis, and the use of data in different statistical models has also varied. The latter was analysed in some depth by Williams (1976), who compared the precision of equations which included the weight of the (predicted) tissue in the joint (1); 1 plus side weight (2), and 2 plus the weight of the joint itself (3). Almost without exception, precision increased as the progression was made from 1 to 2 to 3. The same conclusions were reached by Harrington and King (1963) who reanalysed the data of Callow (1962) and showed that his conclusion, that the weight of tissue in a joint was not a reliable guide to its weight in the total carcass, was unjustified when the controlling influences of joint and carcass weights were taken into account. Williams (1976) provided evidence that, for most joints, a multiple regression using absolute weights as in 3, above, was preferable to simple regression using the percent tissue in a joint to predict the percent of the same tissue in the side.

Kempster and Jones (1977) also addressed this problem in a study of the prediction of lean content of 753 steer carcasses from 17 breed type×feeding system groups. They concluded that the question of whether it is better to use weights or percentages is a controversial one, because although weights are preferred on statistical grounds, percentages are often more convenient to use. In their study, with the exception of the shin joint, there was little difference in precision between the use of weights or percentages. The examples from the studies by Williams (1976) and Kempster and Jones (1977) are shown in Table 14 for the absolute weight of lean in the side. This was calculated from the RSD in percentage terms in both studies, by dividing by side weight. However, Kempster and Jones (1977) used both the percentage lean in the joint *and* side weight as independent variables, whereas percentage lean in joint was the single predictor used by Williams (1976). Nevertheless, there was little difference between these two types of equation in the prediction of lean content in the study by Kempster and Jones (1977). The results in Table 14 show that there was an advantage, in terms of precision, gained from using absolute weights rather than percentages for ten out of the twelve joints in the study by Williams (1976), but only for five out of the eleven joints in the study by Kempster and Jones (1977). This may be due, in part, to the difference between the animals in the two studies, those in the study by Williams (1976) being more variable in side weight and weight of muscle, and generally fatter. What clearly does emerge from these studies is that there can be a marked advantage in using absolute weights for some joints, the size of the advantage varying with the characteristics of the carcass sample, but if there are advantages in using percentage values for some joints, these advantages are relatively small. Furthermore, Williams (1976) showed that the advantage of using absolute weights was even more marked in the prediction of fat and, particularly, bone.

Kempster and Jones (1977) noted that the addition of joint weight to the equation containing weight of tissue in joint and side weight as predictors, significantly increased the precision of prediction and accuracy (breed/feeding system bias) for all joints except those from the distal portions of the limbs (leg and shin). They argued that it was the low fat content of these joints which accounted for this observation, as the addition of joint weight effectively adds information on fat content (bone, where present being closely related to lean weight), and this added information improves the prediction for all other joints. Evans and Kempster (1979) used percentage lean in joints to predict carcass percentage lean in

Table 14
The relative precision of using absolute weight of lean in a sample joint, or percentage lean in the joint, to predict weight of lean in the side—beef studies

Reference: Williams (1976)	*Residual standard deviation of lean (kg)*	
	Using absolute weights	*Using %*
Shin	2·77	3·77
Leg	3·25	3·61
Middle rib	2·06	2·22
Round	1·73	1·90
Brisket	1·65	1·80
Loin	2·22	2·33
Flank	1·89	1·95
Steak piece	2·35	2·38
Neck	2·24	2·27
Foreloin	1·98	2·00
Forerib	1·53	1·51
Rump	2·11	2·02
Reference: Kempster & Jones (1977)[a]		
Shin	3·15	3·79
Leg	3·43	3·58
Wing rib	2·05	2·06
Top piece	1·48	1·60
Coast	1·44	1·46
Sirloin	1·85	1·85
Thin flank	2·14	2·25
Pony	1·57	1·55
Clod and sticking	2·08	2·03
Forerib	1·82	1·79
Rump	1·90	1·86

[a]For definition of joints, see Cuthbertson *et al.* (1972).

pigs, and found that the square root transformation of the fat thickness P_2 ($P_2^{0\cdot5}$), when added to the equation as an additional independent variate, significantly improved the precision of estimation for most joints.

This approach can be carried one step further by including the weight of dissected fat (subcutaneous, intermuscular, or both) in the sample joint in the prediction equation. This information is usually available at the end of the dissection process, or could be with little additional effort, and it is surprising that so little use has been made of this extra information. The results in Table 15 show that, in a study of composition of 66

Table 15
The comparative precision (residual standard deviations, kg lean) of equations based on sample joint data and including joint weight and/or various fat depots as predictors

Reference: Fisher (unpublished)

Sample standard deviation = 6·18 kg lean

Joint[a]	Predictor variables						
	X_1+X_2	$X_1+X_2+X_3$	$X_1+X_2+X_4$	$X_1+X_2+X_5$	$X_1+X_2+X_3+X_4$	$X_1+X_2+X_3+X_5$	$X_1+X_2+X_3+X_4+X_5$
Distal thoracic limb	1·93	NS	1·81	NS	1·76	NS	1·54
Proximal thoracic limb	1·33	1·23	1·20	1·27	1·14	NS	NS
Neck and thorax	1·15	NS	0·97	0·99	NS	NS	NS
Lumbar	1·84	1·74	1·59	1·78	1·58	NS	NS
Distal pelvic limb	2·00	NS	1·94	NS	1·94	NS	1·87
Proximal pelvic limb	1·26	1·20	1·12	1·17	1·11	1·16	1·06
Abdominal	2·26	1·44	1·74	1·42	1·34	1·38	NS

[a] For definitions, see Williams & Bergstrom (1980).
X_1 = Side weight.
X_2 = Weight of lean in joint.
X_3 = Weight of joint.
X_4 = Weight of subcutaneous fat in joint.
X_5 = Weight of intermuscular fat in joint.
NS = Not significant.

Hereford×Friesian steers and heifers, the substitution of joint weight by subcutaneous fat weight in the equations including side weight and lean weight in joint, improved the precision of total carcass lean estimation in every case except that where the abdominal joint is used as the predictor. For that joint, the weight of intermuscular fat was an improved substitution for joint weight. However, when used in addition to joint weight, subcutaneous fat was a better predictor than intermuscular fat for every joint. For the distal thoracic limb, the distal pelvic limb and the proximal pelvic limb joints, the most precise equations included both subcutaneous and intermuscular fat. For the other joints, the intermuscular fat was a non-significant variable in the equation containing both fat depots or, in the case of the neck and thorax, joint weight was non-significant.

The high precision achieved using sample joint dissections has been well demonstrated, and examples of this degree of precision compared with that from other methods of prediction have been given by Kempster *et al*. (1982*a*). Typically, for samples whose standard deviation of carcass lean percentage is about 3·0, visual scores (beef and sheep) and fat thickness measurements have residual standard deviation (RSD) values around 2·5 (fat thickness 2·0–2·5 for pigs), whereas RSD values of less than 1·5 are achieved using the lean content of some sample joints.

The choice of which joint to use must depend on the anticipated precision and cost. Generally, the cheaper and smaller joints (e.g. leg of beef, breast of lamb) are less precise and also are subject to a greater degree of bias in predicted composition among different groups of animals. Cook *et al*. (1983) estimated the cost of dissecting each sample joint as a proportion of the total cost of dissecting a complete half-carcass and, using the correlation between the lean content of the joints and total side lean, calculated which joint was best to use for cattle, sheep and pigs in situations where a proportion of the carcasses was fully dissected (double sampling), given a total budget for completing the evaluation. Based on set costs for each joint in studies conducted within the Meat and Livestock Commission of Great Britain, the best joint for bacon-weight pigs was the ham, for beef the coast (brisket), closely followed by fore-rib and thin flank, and for sheep the breast or best end of neck in the prediction of total lean percentage.

CONCLUSIONS

The key objectives set out at the beginning of this chapter were choice of end-point, accuracy and precision, and the need for objectivity. It can be argued that practicability and the balance between costs and precision are equally important. But they are not equally weighted considerations in, on the one hand, the commercial situation in which carcass description is obtained on the abattoir line and, on the other, as tools to provide compositional data in scientific studies or the evaluation of breeding stock. Practicability is more important in the former, where line speeds and other situation-specific considerations impose economic acceptance levels, dictated by time and labour costs. In animal production studies, there are examples of the dilemma between low-cost methods of evaluation which have been used on numerous carcasses and high capital cost methods used to provide a restricted number of observations. The research application, however, differs from the commercial in that there is usually more flexibility in terms of time and attention to detail.

The methods reviewed are those of greatest current interest, and the list is not exhaustive. For example, carcass specific gravity has been used in many studies, and continues to be a valuable tool in studies of the composition of human subjects, and there are many examples where methods developed for use in human medicine have been applied to animal research. There are limitations of the specific gravity method, most notably the variation in density of individual tissues (Miles, 1976), but of course other techniques have their own endemic limitations. However, this chapter has been directed at evolving technologies, or has attempted to assess those widely used, and the impractibility of other techniques has been the basis for their lack of inclusion. Other approaches (e.g. gamma radiation, neutron activation analysis) are recognised as too unexplored to be evaluated constructively, although applications to in-vivo measurement of body composition of farm livestock have been attempted (e.g. East *et al.*, 1984).

The biological differences between the three species—pigs, cattle, and sheep, in terms of the distribution of carcass tissues, are reflected in the success that the different techniques have achieved. For pigs, measurement of subcutaneous fat depths and, to a lesser extent, underlying muscle depths, has proved to be successful, and the use of probes for this purpose is likely to continue and develop. This approach is not likely to be as fruitful in cattle, and sheep also present a different problem. In these species, objective measures of total fatness, or muscle mass, are the goal, and its achievement is the only really progressive step. Fat-to-lean estimation is possible using the velocity of ultrasound technique, and some methods based on electrical properties seem promising. Computerised tomography is very effective but is limited by other considerations. For research and breeding development purposes, sample joint dissection is probably the best current option, but careful consideration needs to be given to the question of costs versus numbers if bias in prediction is to be avoided.

In the future, the targets may change. If reduction in fatness proceeds (already apparent in pigs, possibly accelerated via new technologies in cattle and sheep, such as multiple ovulation and embryo transfer, and cloning techniques), then the determination of muscle-to-bone ratio may assume even greater importance. It is difficult to predict whether this will be quantified more precisely and accurately by carcass shape measurement or by invasive techniques in future years.

Related to this changing profile of carcass composition is the requirement to determine the amount of intramuscular, or marbling fat. At present, some meats (e.g. ground or minced) are categorised and labelled, at the point of retail sale, according to fat content, the source of which may be fatty tissue processed with lean meat. In some products, these levels are already so low that it is not added fat but the fat within the muscle which is the main determinant of lipid content. Quantification of intramuscular fat may therefore become more important, and because of its potential effects in influencing meat quality, it gains even greater emphasis. The techniques which potentially can measure intramuscular fat include light reflectance probes (but modified to achieve this objective by utilising optimal wavelengths); nuclear magnetic resonance, (claimed by Cross *et al.* (1989) to hold much promise); ultrasonic detection (velocity of ultrasound, scattering and attenuation); and near-infrared reflectance analysis.

There are many factors, operating in the long sequel of events leading from fundamental science and quantification in the laboratory to commercial manufacture, which influence the course and impetus of the development of a measurement technique. Adequate funding, links with potential manufacturers and the support and cooperation of the meat industry are vital. In many cases, the decision of statutory bodies will determine the fate of a particular technique (e.g. the criteria imposed in the European Community pig grading scheme). In this respect, there are large differences between countries in their commitment to new technology which, to some extent, reflect their national industry structure, particularly in regard to number of abattoirs, their ownership, and throughput. Exemplary developments have been achieved in Denmark, where high unit costs of, for example, the pig classification centre, are apparently justified by high throughput and increased profitability. The integrated structure of the meat industry in Denmark, where producers and abattoirs are cooperatively involved, is udoubtedly instrumental in facilitating these innovations.

Nations which export large volumes of their home production have been at the forefront in developing carcass classification or grading, for reasons explained by Kempster *et al.* (1982*a*). But increasing discriminations by consumers, particularly in regard to fatness, must be a concern which demands attention at all levels of meat marketing, whether it be export grading, national classification, or local selling.

REFERENCES

Abraham, H. C., Murphey, C. E., Cross, H. R., Smith, G. C. & Franks, W. J. (1980). Factors affecting beef carcass cutability: an evaluation of the USDA yield grades for beef. *J. Anim. Sci.*, **50,** 841–51.

Adolph, E. F. & Heggeness, F. W. (1971). Age changes in body water and fat in fetal and infant mammals. *Growth,* **35,** 55–63.

Allen, P. (1990). New approaches to measuring body composition in live meat animals. In *Reducing Fat in Meat Animals,* eds J. D. Wood & A. V. Fisher. Elsevier Applied Science Publishers Ltd, London, pp. 00–00.

Allen, P. & Leymaster, K. A. (1985). Machine error in X-ray computer tomography and its relevance to prediction of in vivo body composition. *Livest. Prod. Sci.,* **13,** 383–98.

Allen, P. & Vangen, O. (1984). X-ray tomography of pigs. Some preliminary results. In *In Vivo Measurement of Body Composition in Meat Animals,* ed. D. Lister. Elsevier Applied Science Publishers, London, pp. 52–66.

Anon. (1985). Un classement objectif et automatique des carcasses de gros bovins sur leur conformation le COLAVAUG. *Viandes et Produits carnes,* **6,** 122–5.

Bass, J. J., Johnson, D. L., Colomer-Rocher, F. & Binks, G. (1981). Prediction of carcass composition from carcass conformation in cattle. *J. Agric. Sci., Camb.,* **97,** 37–44.

Bass, J. J., Woods, E. G. & Greville, E. (1982). Prediction of beef carcass composition by tissue depth measurements taken over the 11th rib. *Livest. Prod. Sci.,* **9,** 337–48.

Berg, R. T., Jones, S. D. M., Price, M. A., Fukuhara, R., Butterfield, R. M. & Hardin, R. T. (1979). Patterns of carcass fat deposition in heifers, steers and bulls. *Can. J. Anim. Sci.,* **59,** 359–66.

Bodwell, C. E. (1959). The Use of Linear Measurements in Evaluating Beef Carcasses. MSc thesis, King's College, Cambridge.

Branscheid, W., Sack, E., Kallweit, E., Höreth, R. & Baulein, U. (1989). Non-invasive methods in pig grading: a reliable possibility for calibration of grading devices. *Proc. 35th Int. Congr. Meat Sci. and Technol., Copenhagen,* pp. 239–43.

Brinks, J. S., Clarke, R. T., Kieffer, N. M. & Urick, J. J. (1964). Predicting wholesale cuts of beef from linear measurements obtained by photogrammetry. *J. Anim. Sci.*, **23**, 365–74.

Butler-Hogg, B. W. (1984). The growth of Clun and Southdown sheep: body composition and the partitioning of total body fat. *Anim. Prod.*, **39**, 405–11.

Butler-Hogg, B. W. & Whelehan, O. P. (1986). Carcass quality of dairy sheep. *Anim. Prod.*, **42**, 461 (abstr.).

Butler-Hogg, B. W. & Wood, J. D. (1982). The partition of body fat in British Friesian and Jersey steers. *Anim. Prod.*, **35**, 253–62.

Butterfield, R. M., Thompson, J. M. & Reddacliff, K. J. (1985). Changes in body composition relative to weight and maturity of Australian Dorset Horn rams and wethers. 3. Fat partitioning. *Anim. Prod.*, **40**, 129–34.

Callow, E. H. (1962). The relationship between the weight of a tissue in a single joint and the total weight of tissue in a side of beef. *Anim. Prod.*, **4**, 37.

Casteels, M., Verbeke, R. & Matthaus, R. (1984). Tests involved in Belgian approval procedures for carcass classifying equipment: Instrument supported carcass classification. *Die Fleischerei*, November 1984, 16 pp.

Chadwick, J. P. & Kempster, A. J. (1983). The estimation of beef carcass composition from subcutaneous fat measurements taken on the intact side using different probing instruments. *J. Agric. Sci., Camb.*, **101**, 241–8.

Chadwick, J. P., Kempster, A. J. & Homer, D. L. M. (1986). A comparison of the Hennessy Lamb Probe, Ruler Probe and visual fat scores for use in sheep carcass classification. *Anim. Prod.*, **42**, 445 (abstr.).

Coleman, M. E., Rhee, K. S. & Cross, H. R. (1988). Sensory and cooking properties of beef steaks and roasts cooked with and without external fat. *J. Food Sci.*, **53**, 34–61.

Colomer-Rocher, F., Bass, J. J. & Johnson, D. L. (1980). Beef carcass conformation and some relationships with carcass composition and muscle dimensions. *J. Agric. Sci., Camb.*, **94**, 697–708.

Cook, G. L., Jones, D. W. & Kempster, A. J. (1983). A note on a simple criterion for choosing among sample joints for use in double sampling. *Anim. Prod.*, **36**, 493–5.

Cook, G. L., Chadwick, J. P. & Kempster, A. J. (1989). An assessment of carcass probes for use in Great Britain for the EC Pig Carcass Grading Scheme. *Anim. Prod.*, **48**, 427–34.

Cross, H. R., Gilliland, D. A., Durland, P. R. & Seideman, S. (1983). Beef carcass evaluation by use of a video image analysis system. *J. Anim. Sci.*, **57**, 908–17.

Cross, H. R., Whittaker, D. & Savell, J. W. (1989). The objective measurement of value in meat animals. In The *Automated Measurement of Beef*, eds L. E. Brownlie, W. J. A. Hall & S. V. Fabiansson. Australian Meat and Livestock Corp., Sydney, pp. 1–13.

Cuthbertson, A., Harrington, G. & Smith, R. J. (1972). Tissue separation—to assess beef and lamb variation. *Proc. Brit. Soc. Anim. Prod.*, **1**, 113–22.

De Boer, H. (1984). Classification and grading—principles, definitions and implications. In *Carcass Evaluation in Beef and Pork: Opportunities and Constraints, Satellite Symposium*. EAAP, The Hague, The Netherlands, pp. 9–20.

De Boer, H., Bergstrom, P. L., Jensen, A. A. M. & Nijeboer, H. (1975). *Carcass*

Measurements and Visual Assessments as Predictors of Lean Meat Content, with Reference to the EEC Classification and Grading System. 26th Annual Meeting of EAAP, Warsaw, Poland, 21 pp.

Department of Health and Social Security. (1984). *Diet and Cardiovascular Disease.* Report on Health and Social Subjects No. 28. Committee on Medical Aspects of Food Policy. HMSO, London, 32 pp.

Diestre, A. & Kempster, A. J. (1985). The estimation of pig carcass composition from different measurements with special reference to classification and grading. *Anim. Prod.,* **41,** 383–91.

Diestre, A., Gispert, M. & Oliver, M. A. (1989). The evaluation of automatic probes in Spain for the new scheme for pig carcass grading according to the EC regulations. *Anim. Prod.,* **48,** 443–8.

Dumont, B. L. (1989). Estimation of muscle/bone ratio of ham from conformation measurements. *Proc. 35th Int. Congr. Meat Sci. and Technol., Copenhagen,* pp. 233–8.

Dumont, B. L., Le Guelte, P. & Sornay, J. (1975). The judgement of fleshiness in carcasses of cattle. [Eng. version] INRA/ITEB mimeograph. 26 pp.

East, B. W., Preston, T. & Robertson, I. (1984). The potential of *in vivo* neutron activation analysis for body composition measurements in the agricultural sciences. In *In Vivo Measurement of Body Composition in Meat Animals,* ed. D. Lister. Elsevier Applied Science Publishers, London, pp. 134–8.

Evans, D. G. & Kempster, A. J. (1979). A comparison of different predictors of the lean content of pig carcasses. 2. Predictors for use in population studies and experiments. *Anim. Prod.,* **28,** 97–108.

Fisher, A. V. (1975). The profile area of beef carcasses and its relationship to carcass composition. *Anim. Prod.,* **20,** 355–61.

Fisher, A. V. & Bayntun, J. A. (1981). Differences between sheep breeds in the growth of the limb bones and the development of conformation. *Anim. Prod.,* **32,** 376–7 (abstr.).

Fisher, A. V. & Bayntun, J. A. (1984). The effect of breed type on the relative variability of different fat depots in cattle. *Anim. Prod.,* **38,** 543 (abstr.).

Fisher, A. V., Wood, J. D., Stevens, G. & Robelin, J. (1983). The relationship between carcass fatness and the lipid and protein content of beef. *Proc. 29th Europ. Meet. Meat Res. Workers,* Salsomaggiore, Italy, Vol 1, pp. 48–54.

Fisher, A. V., Broadbent, J. S., Coutts, C., Kay, R. M. & Rigby, I. (1988). Differences in the relation between carcass visual scores and composition in heifers and steers. *Anim. Prod.,* **46,** 497 (abstr.).

Forrest, J. C., Kuei, C. M., Orcutt, M. W., Schinkel, A. P., Stouffer, J. R. & Judge, M. D. (1988). Electromagnetic scanning, ultrasonic imaging, and electronic probing for estimation of pork carcass composition. *Proc. 34th Int. Congr. Meat Sci. and Technol., Brisbane,* pp. 31–3.

Fortin, A., Sim, D. W. & Talbot, S. (1980). Ultrasonic measurements of back fat thickness at different locations and positions on the warm pork carcase and comparisons of ruler and ultrasonic procedures. *Can. J. Anim. Sci.,* **60,** 635–41.

Fortin, A., Jones, S. D. M. & Haworth, C. R. (1984). Pork carcass grading: a comparison of the New Zealand Hennessy Grading Probe and the Danish Fat-O-Meater. *Meat Sci.,* **10,** 131–44.

Fortin, A., Wood, J. D. & Whelehan, O. P. (1987). Breed and sex effects on the development, distribution of muscle, fat and bone, and the partition of fat in pigs. *J. Agric. Sci., Camb.*, **108,** 141–53.
Fredeen, H. T. & Weiss, G. M. (1981). Comparison of techniques for evaluating lean content of hog carcasses. *Can. J. Anim. Sci.*, **61,** 319–33.
Groeneveld, E., Kallweit, E., Henning, M. & Pfau, A. (1984). Evaluation of body composition of live animals by X-ray and nuclear magnetic resonance computed tomography. In *In Vivo Measurement of Body Composition in Meat Animals,* ed. D. Lister. Elsevier Applied Science Publishers, London, pp. 84–8.
Hammond, J. (1932). *Growth and Development of Mutton Qualities in the Sheep*. Oliver and Boyd, Edinburgh and London.
Hammond, J. (1933). The anatomy of pigs in relation to market requirements. *Pig Breed.*, A. **13,** 18–25.
Harries, J. M., Williams, D. R. & Pomeroy, R. W. (1975). Prediction of comparative retail value of beef carcasses. *Anim. Prod.*, **21,** 127–37.
Harrington, G. & King, J. W. B. (1963). A note on the prediction of muscular tissue weight in sides of beef. *Anim. Prod.*, **5,** 327.
Hök, P. (1988). New technology in slaughter and processing pigs. *Proc. 34th Int. Congr. Meat Sci. and Technol., Brisbane,* pp. 631–4.
Huxley, J. S. (1932). *Problems of Relative Growth*. Methuen and Co., London.
Jenkins, T. G., Leymaster, K. A. & Turbington, L. M. (1988). Estimation of fat-free soft tissue in lamb carcasses by use of carcass and resistive impedance measurements. *J. Anim. Sci.*, **66,** 2174–9.
Johnson, D. D., Savell, J. W., Smith, G. C. & Weatherspoon, L. (1984). Prediction of pork belly composition using various measurements of the carcass or belly. *J. Anim. Sci.*, **58,** 611–18.
Johnson, E. R. & Ball, B. (1988). An evaluation of rump P8 and twelfth rib fat thickness measuring sites for estimating saleable beef yield in export and local carcasses. *Proc. 34th Int. Congr. Meat Sci. and Technol., Brisbane,* pp. 71–3.
Johnson, E. R. & Vidyadaran, M. K. (1981). An evaluation of different sites for measuring fat thickness in the beef carcass to determine carcass fatness. *Aust. J. Agric. Res.*, **32,** 999–1007.
Jones, S. D. M. (1982). The accumulation and distribution of fat in ewe and ram lambs. *Can. J. Anim. Sci.*, **62,** 381–6.
Jones, S. D. M. & Haworth, C. R. (1982). The measurement of subcutaneous fat thickness in cold beef carcasses with an automatic probe. *Can. J. Anim. Sci.*, **62,** 645–8.
Jones, S. D. M. & Haworth, C. R. (1983). The electronic prediction of commercial yield and lean content in pig carcasses. *Anim. Prod.*, **37,** 33–40.
Jones, S. D. M., Tong, A. K. W., Martin, A. H. & Robertson, W. M. (1986). The effect of ribbing site on fat thickness measurements and the prediction of beef carcass composition. *Can. J. Anim. Sci.*, **66,** 541–5.
Jones, S. D. M., Tong, A. K. W. & Robertson, W. M. (1988). The prediction of beef carcass lean content. *Proc. 34th Int. Congr. Meat Sci. and Technol., Brisbane,* pp. 47–8.
Kempster, A. J. (1981). Fat partition and distribution in the carcasses of cattle, sheep and pigs: a review. *Meat Sci.*, **5,** 83–98.

Kempster, A. J. (1984). Cost-benefit analysis of *in vivo* estimates of body composition in meat animals. In *In Vivo Measurement of Body Composition in Meat Animals*, ed. D. Lister. Elsevier Applied Science Publishers, London, pp. 191–203.

Kempster, A. J. (1986). Estimation of the carcass composition of different cattle breeds and crosses from conformation assessments adjusted for fatness. *J. Agric. Sci., Camb.*, **106**, 239–54.

Kempster, A. J. & Evans, D. G. (1979*a*). The effects of genotype, sex and feeding regimen on pig carcass development. 2. Tissue weight distribution and fat partition between depots. *J. Agric. Sci., Camb.*, **93**, 349–58.

Kempster, A. J. & Evans, D. G. (1979*b*). A comparison of different predictors of the lean content of pig carcasses. 1. Predictors for use in commercial classification and grading. *Anim. Prod.*, **28**, 87–96.

Kempster, A. J. & Evans, D. G. (1981). The value of shape as a predictor of carcass composition in pigs from different breeding companies. *Anim. Prod.*, **33**, 313–18.

Kempster, A. J. & Harrington, G. (1980). The value of 'fat-corrected' conformation as an indicator of beef carcass composition within and between breeds. *Livest. Prod. Sci.*, **7**, 361–72.

Kempster, A. J. & Jones, D. W. (1977). Relationships between the lean content of joints and overall lean content in steer carcasses of different breeds and crosses. *J. Agric. Sci., Camb.*, **88**, 193–201.

Kempster, A. J., Avis, P. R. D. & Smith, R. J. (1976). Fat distribution in steer carcasses of different breeds and crosses. 2. Distribution between joints. *Anim. Prod.*, **23**, 223–32.

Kempster, A. J., Jones, D. W. & Cuthbertson, A. (1979). A comparison of the Danish MFA, Ulster and Optical probes for use in pig carcass classification and grading. *Meat Sci.*, **3**, 109–20.

Kempster, A. J., Croston, D. & Jones, D. W. (1981*a*). Value of conformation as an indicator of sheep carcass composition within and between breeds. *Anim. Prod.*, **33**, 39–49.

Kempster, A. J., Chadwick, J. P., Jones, D. W. & Cuthbertson, A. (1981*b*). An evaluation of the Hennessy and Chong Fat Depth Indicator and the Ulster Probe for use in pig carcass classification and grading. *Anim. Prod.*,**33**, 319–24.

Kempster, A. J., Cuthbertson, A. & Harrington, G. (1982*a*). *Carcass Evaluation in Livestock Breeding, Production and Marketing*. Granada, London.

Kempster, A. J., Cuthbertson, A. & Harrington, G. (1982*b*). The relationship between conformation and the yield and distribution of lean meat in the carcasses of British pigs, cattle and sheep: A review. *Meat Sci.*, **6**, 37–53.

Kempster, A. J., Chadwick, J. P. & Jones, D. W. (1985). An evaluation of the Hennessy Grading Probe and the SFK Fat-O-Meater for use in pig carcass classification and grading. *Anim. Prod.*, **40**, 323–9.

Kempster, A. J., Cook, G. L. & Grantley-Smith, M. (1986*a*). National estimates of the body composition of British cattle, sheep and pigs with special reference to trends in fatness. A review. *Meat Sci.*, **17**, 107–38.

Kempster, A. J., Chadwick, J. P. & Charles, D. D. (1986*b*). Estimation of the carcass composition of different cattle breeds and crosses from fatness measurements and visual assessments. *J. Agric. Sci., Camb.*, **106,** 223–37.

Kempster, A. J., Jones, D. W. & Wolf, B. T. (1986*c*). A comparison of alternative methods for predicting the carcass composition of cross-bred lambs of different breeds and crosses. *Meat Sci.*, **18,** 89–110.

Kempster, A. J., Chadwick, J. P., Cue, R. I. & Grantley-Smith, M. (1986*d*). The estimation of sheep carcass composition from fat and muscle thickness measurements taken by probes. *Meat Sci.*, **16,** 113–26.

Kirton, A. H. (1982). *Lamb Carcass Grading*. Aglink FPP 490 (2nd revise), Ministry of Agriculture and Fisheries, Wellington, 4 pp.

Kirton, A. H., Woods, E. G. & Duganzich, D. M. (1984). Predicting the fatness of lamb carcasses from carcass wall thickness measured by ruler or by a total depth indicator (TDI) probe. *Livest. Prod. Sci.*, **11,** 185–94.

Kirton, A. H., Duganzich, D. M., Feist, C. L., Bennett, G. L. & Woods, E. G. (1985). Prediction of lamb carcass composition from GR and carcass weight. *Proc. NZ Soc. Anim. Prod.*, **45,** 63.

Kirton, A. M., Feist, C. L., Duganzich, D. M., Jordan, R. B., O'Donnell, K. P. & Woods, E. G. (1987). Use of the Hennessy Grading Probe (GP) for predicting the meat, fat and bone yields of beef carcasses. *Meat Sci.*, **20,** 51–63.

Koch, R. M. & Varnadore, W. L. (1976). Use of electronic meat measuring equipment to measure cutout yield of beef carcasses. *J. Anim. Sci.*, **43,** 108–13.

Kuei, C. H., Forrest, J. C., Orcutt, M. W., Judge, M. D. & Schinkel, A. P. (1989). Electromagnetic scanning to estimate composition and weight of pork primal cuts and carcasses. *Proc. 35th Int. Congr. Meat Sci. and Technol., Copenhagen,* pp. 249–56.

Kutsley, J. A., Murphey, C. E., Smith, G. C., Savell, J. W., Stiffler, D. M. & Terrell, R. N. (1982). Use of the Hennessy and Chong fat depth indicator for predicting fatness of beef carcasses. *J. Anim. Sci.*, **55,** 565–71.

Lewis, R. W., Brungardt, V. H. & Bray, R. W. (1964). Influence of subcutaneous fat contours in estimating trimmable fat and retail yield of heifer carcasses. *J. Anim. Sci.*, **23,** 1203 (abstr.).

Lush, J. L. (1926). Practical methods of estimating the proportions of fat and bone in cattle slaughtered in commercial packing plants. *J. Agric. Res.*, **32,** 727–54.

Luthje, H., Olsen, E. V. & Busk, H. (1988). Effects of Introducing the Classification Centre for Pig Carcass Grading on Payment to Pig Producers. Danish Meat Research Institute mimeograph, 7 pp.

MacNeil, M. D., (1983). Choice of a prediction equation and the use of the selected equation in subsequent experimentation. *J. Anim. Sci.*, **57,** 1328–36.

Martin, A. H., Fredeen, H. T., Weiss, G. M. & Carson, R. B. (1972). Distribution and composition of porcine carcass fat. *J. Anim. Sci.*, **35,** 534–41.

Masoro, E. J. (1968). *Physiological Chemistry of Lipids in Mammals.* W. B. Saunders Company, Philadelphia.

Matzke, P., Peschke, W., Averdunk, G., Blendl, H., Sauerer, G., Gunter, I. & Huber, I. (1986). Untersuchungen zur apparativen Klassifizierung von Schweinehälften durch die Meßsysteme FOM und SKG II. *Fleischwirtsch,* **66,** 391–7.

McMeekan, C. P. (1940). Growth and development in the pig, with special refer-

ence to carcass quality characters. I. *J. Agric. Sci.*, **30,** 276–343.
McMeekan, C. P. (1941). Growth and development in the pig, with special reference to carcass quality characters. Pt IV. The use of sample joints and of carcass measurements as indices of the composition of the bacon pig. *J. Agric. Sci.*, **31,** 1–49.
Meat and Livestock Commission (1983). *Beef prices Related to Carcase Classification*. 83/300 2M 12/83, MLC, Bletchley.
Meat and Livestock Commission (1986). *Briefing Notes on the EEC Pig Carcase Grading Scheme.* 96 1M 4/86 (23/73) (mimeograph).
Miles, C. A. (1976). Chemical composition of carcasses and sample joints: specific gravity determinations. In *Criteria and Methods for Assessment of Carcass and Meat Characteristics in Beef Production Experiments* (eds A. V. Fisher, J. C. Tayler, H. de Boer & D. H. van Adrichem Boogaert) Commission of the European Communities, Brussels, EUR 5489. pp. 253–62.
Miles, C. A., Fursey, G. A. J., Pomeroy, R. W., Harries, J. M., Stouffer, J. R., Williams, L. & Scruton, G. (1970). Displaying the soft tissue components of living animals and carcasses by two dimensional ultrasonic scanning. Meat Research Institute Record Memorandum, Langford, Bristol.
Miles, C. A., Fursey, G. A. J. & York, R. W. R. (1984). New equipment for measuring the speed of ultrasound and its application in the estimation of body composition of farm livestock. In *In Vivo Measurement of Body Composition in Meat Animals*, ed. D. Lister. Elsevier Applied Science Publishers, London, pp. 93–105.
Miles, C. A., Fisher, A. V., Fursey, G. A. J. & Page, S. J. (1987). Estimating beef carcass composition using the speed of ultrasound. *Meat Sci.*, **21,** 175–88.
Miles, C. A., Fursey, G. A. J., Page, S. J. & Fisher, A. V. (1990). Progress towards using the speed of ultrasound for beef leanness classification. *Meat Sci.*, **28,** 119–30.
Murphey, C. E., Johnson, D. D., Smith, G. C., Abraham, H. C. & Cross, H. R. (1985). Effects of sex-related differences in external fat deposition on subjective carcass fatness evaluations—steer versus heifer. *J. Anim. Sci.*, **60,** 666–74.
Newman, P. B. (1984). The use of video image analysis for quantitative measurement of visible fat and lean in meat: Part 1—boneless fresh and cured meats. *Meat Sci.*, **10,** 87–100.
Newman, P. B. (1987). The use of video image analysis for quantitative measurement of visible fat and lean in meat: Part 4—application of image analysis measurement techniques to minced meats. *Meat Sci.*, **19,** 139–50.
Newman, P. B. & Wood, J. D. (1989). New techniques for assessment of pig carcasses—video and ultrasonic systems. In *New Techniques in Pig Carcass Evaluation*, ed. J. F. O'Grady. Pudoc, Wageningen, pp. 37–51.
Pedersen, O. K. & Busk, H. (1982). Development of automatic equipment for grading of pig carcasses in Denmark. *Livest. Prod. Sci.*, **9,** 675–86.
Petersen, F., Klastrup, S., Sorensen, S. E. & Madsen, N. T. (1989). Beef classification centre. *Proc. 35th Int. Congr. Meat Sci. and Technol., Copenhagen,* pp. 49–52.
Phillips, D., Herrod, W. & Schafer, R. J. (1987). The measurement of subcutaneous fat depth on hot beef carcasses with the Hennessy Grading Probe. *Aust. J. Exp. Agric.*, **27,** 335–8.

Pond, C. N. (1984). Physiological and ecological importance of energy storage in the evolution of lactation: evidence for a common pattern of anatomical organisation of adipose tissue in mammals. *Symp. Zool. Soc. Lond.*, No. 51, 1–32.

Richmond, R. J. & Berg, R. T. (1971). Fat distribution in swine as influenced by live weight, breed, sex and ration. *Can. J. Anim. Sci.*, **51**, 523–31.

Robelin, J. (1986). Growth of adipose tissues in cattle; weight distribution, chemical composition and cellularity: a review. *Livest. Prod. Sci.*, **14**, 349–64.

Rook, A. J., Ellis, M., Whittemore, C. T. & Phillips, P. (1987). Relationships between whole-body chemical composition, physically dissected carcass parts and backfat measurements in pigs. *Anim. Prod.*, **44**, 263–73.

Roy, G. & Dumont, B. L. (1975). Méthode de jugement descriptif de l'état d'engraissement des carcasses de bovins adultes. *Rev. de Med. Vét.*, **126**, 387–400.

Sack, E. (1983). Apparative Klassifizierung von Schweinehälften. *Fleischwirtsch.*, **63**, 27–42.

Sehested, E. & Vangen, O. (1989). Computer tomography, a non-destructive method of carcass evaluation. In *New Techniques in Pig Carcass Evaluation*, ed. J. F. O'Grady. Pudoc, Wageningen, pp. 98–102.

Smith, W. C. & Pearson, G. (1985). Relationships between chemical composition and physically separated tissues of the pig carcass. *Meat Sci.*, **14**, 29–41.

Sorensen, S. E. (1984). Possibilities for application of video image analysis in beef carcass classification. In *In Vivo Measurement of Body Composition in Meat Animals* ed. D. Lister. Elsevier Applied Science Publishers, London, pp. 113–22.

Sorensen, S. E., Klastrup, S. & Petersen, F. (1988). Classification of bovine carcasses by means of video image analysis and reflectance probe measurements. *Proc. 34th Int. Congr. Meat Sci. and Technol., Brisbane,* pp. 635–8.

Speight, B. S., Miles, C. A. & Moledina, K. (1974). Recording carcass shape by a moiré method. *Med. and Biol. Engng.*, **8**, 221–6.

Sterrenburg, P. (1989). *Rugspekdikte als voorspeller van het E G—vleespercentage in hot varkenskarkas, in relatie met enkele uitwendige karkasmaten.* Rapport B-315 (23 pp.) of the Instituut voor veeteeltkundig Onderzoek 'Schoonord', The Netherlands.

Terry, C. A., Savell, J. W., Recio, H. A. & Cross, H. R. (1989). Using ultrasound technology to predict pork carcass composition. *J. Anim. Sci.*, **67**, 1279–84.

Thompson, J. M., Atkins, K. D. & Gilmour, A. R. (1979). Carcass characteristics of heavyweight crossbred lambs. II. Carcass composition and partitioning of fat. *Aust. J. Agric. Res.*, **30**, 1207–14.

Timon, V. M. & Bichard, M. (1965). Quantitative estimates of lamb carcass composition. 1. Sample joints. *Anim. Prod.*, **7**, 173–81.

Truscott, T. G., Wood, J. D. & Macfie, H. J. H. (1983). Fat deposition in Hereford and Friesian steers. 1. Body composition and partitioning of fat between depots. *J. Agric. Sci., Camb.*, **100**, 257–70.

Ulyatt, M. J. & Barton, R. A. (1963). A comparison of the chemical and dissectible carcass composition of New Zealand Romney Marsh ewes. *J. Agric. Sci.*, **60**, 285–9.

Van der Peet, G. F. V. (1984). *Skeletvariatie bij rundeve i.v.m. classificatie*. I. V. O. Rapport B-246 (43 pp.).

Wallace, M. A., Stouffer, J. R. & Westervelt, R. G. (1977). Relationships of ultrasonic and carcass measurements with retail yield in beef cattle. *Livest. Prod. Sci.*, **4,** 153–64.

Walstra, P. (1989). Automated grading probes for pigs currently in use in Europe, their accuracy, costs and ease of use. In *New techniques in Pig Carcass Evaluation*, ed. J. F. O'Grady. Purdoc, Wageningen, pp. 16–27.

Wassenberg, R. L., Allen, D. M. & Kemp, K. E. (1986). Video image analysis prediction of total kilograms and percent primal lean and fat yield of beef carcasses. *J. Anim. Sci.*, **62,** 1609–16.

Williams, D. R. (1969). The visual description of carcasses. *J. Agric. Sci., Camb.*, **73,** 495–9.

Willliams, D. R. (1976). Beef carcass weights, sample joints and measurements as predictors of composition. *Wld Rev. Anim. Prod.*, **12,** 13–31.

Williams, D. R. & Bergstrom, P. (1980). *Anatomical Jointing, Tissue Separation and Weight Recording. EEC Standard Method for Beef.* Commission of the European Communities, Brussels, EUR 6878 EN.

Wood, J. D. (1984). Fat deposition and the quality of fat tissue in meat animals. In *Fats in Animal Nutrition*, ed. J. Wiseman. Butterworths, London, pp. 407–35.

Wood, J. D. (1990). Consequences for meat quality of reducing carcass fatness. In *Reducing Fat in Meat Animals,* eds J. D. Wood & A. V. Fisher. Elsevier Applied Science Publishers Ltd, London, pp. 334–97.

Wood, J. D., MacFie, H. J. H., Pomeroy, R. W. & Twinn, D. J. (1980). Carcass composition in four sheep breeds: the importance of type of breed and stage of maturity. *Anim. Prod.*, **30,** 135–52.

Yeates, N. T. M. (1952). The quantitative definition of cattle carcasses. *Aust. J. Agric. Res.*, **3,** 68–94.

Young, M. J. (1989). The influence of changes in tissue shape on muscle: bone ratio in growing sheep. *Anim. Prod.*, **48,** 635 (abstr.).

Chapter 8

Consequences for Meat Quality of Reducing Carcass Fatness

J. D. WOOD
*Department of Meat Animal Science, University of Bristol, Langford, Bristol, UK**

INTRODUCTION

It is now well accepted that consumers in most countries prefer to purchase meat with lower levels of fat than formerly. The main reason is the possible association between high levels of saturated fat and heart disease. However, there is a strong body of opinion which maintains that fat in meat contributes to eating quality and that reducing fat to too low levels will reduce satisfaction at a time when consumer interest in the quality of food in general is increasing. The dilemma is discussed in this chapter—what is the evidence for the role of fat in meat quality and what are the consequences of reducing fat?

The question is more complex than first appears. In practice, reductions in fatness are achieved on farms by changing breeds, sexes, production systems, weights and ages and each of these may exert effects on meat quality independently of fatness. Meat quality can also be affected by the way animals are handled before slaughter and carcasses processed after slaughter. In the latter case there is good evidence that carcass fatness modifies the effects exerted.

The aim of this chapter is therefore to unravel some of the direct and indirect associations between carcass fatness and meat quality.

*Formerly Carcass and Abattoir Department of the AFRC Institute of Food Research—Bristol Laboratory, Langford, Bristol, UK.

DEFINITIONS OF MEAT QUALITY

In this chapter the main emphasis will be on eating quality, usually determined by taste panels and defined in terms of tenderness (texture), juiciness and flavour. Other attributes contribute to the overall appreciation of meat when eaten but these three are the most important, with tenderness being the most important of all (Asghar & Pearson, 1980).

Eating quality is usually assessed in taste panel tests in which the meat is prepared, cooked and presented in carefully controlled ways and standard scoring systems are used. Useful information can also be obtained in consumer tests in which people cook and eat the meat themselves and compare with past experience. Often the results of trained- and consumer-panel tests are in broad agreement although the former are considered more accurate (Wood *et al.*, 1986; Medeiros *et al.*, 1987). Objective tests are valuable although only tenderness can be reliably assessed this way. Measurements made using the Warner–Bratzler shear press or the Instron materials testing instrument correlate quite well with taste panel assessments of tenderness (Rhodes *et al.*, 1972; Harris, 1976). Moreover, results can be readily compared between experiments.

Visual appearance is an important aspect of meat quality especially at the point of sale. The ratio of lean to fat influences consumer choice as do colour and wetness (drip loss). Finally, presentational characteristics, including the firmness of tissues and cohesiveness between them, are important to consumers and butchers.

ROLE OF FAT IN MEAT QUALITY

It is recognised that fat plays a part in the eating quality of various foods: arguments therefore surround the size of the effect rather than the existence of it. In general, fat affects all three aspects of eating quality, i.e. tenderness, juiciness and flavour.

Tenderness in muscle is influenced by the amount and type of connective tissue, which provides 'background toughness', and the contractile state of the muscle fibres which provide 'myofibrillar toughness' and which can have an overriding effect (Marsh & Leet, 1966; Moller *et al.*, 1981). Thus rapid post-mortem chilling of beef and lamb carcasses induces shortening and toughening of the muscle fibres if the temperature falls to a critical level before the onset of rigor mortis.

It has been suggested that fat affects myofibrillar toughness by providing insulation against the effects of rapid cooling on muscle fibres. Subcutaneous and intramuscular fat have both been implicated (Cross *et al.*, 1972; Smith *et al.*, 1976). Slower cooling of the muscles in heavy fat carcasses could therefore explain the commonly held view that such muscles are more tender (Dikeman, 1987).

Toughness due to connective tissue arises mainly in old animals and that due to cold shortening in carcasses which have been rapidly cooled. When neither of these factors operate, there may be an effect of fat due to its lower resistance to shearing compared with the muscle fibres. This dilution effect of fat has been observed in various foods which have a background protein structure. For example, Emmons *et al.* (1980) found that the force required to cut cheese was increased as the proportion of non-fat protein matrix increased. High-fat cheeses were considered by taste panellists to be less elastic, softer and smoother than low-fat cheeses. A more attractive texture is also found in biscuits in which the gluten structure is more 'interrupted' by fat (Griffiths, 1985). In meat, intramuscular (marbling) fat could have a similar role.

A recent consideration of the physical basis of meat texture concluded that the breakdown which begins as meat is chewed always occurs initially between fibre bundles in the perimysial connective tissue (Purslow, 1985). This is where intramuscular fat is located and it may be that fat infiltration interrupts the bonding between fibre bundles, allowing fracture to occur more easily. Presumably, fat inclusion in the perimysium would also reduce the force required to break the connective tissue itself.

Juiciness is often associated with tenderness but can be separately assessed by taste panellists. It depends on the amount of liquid released during mastication both from the food and saliva (Harris, 1976; Asghar & Pearson, 1980). Fat may affect saliva production through controlling the ease with which the meat is chewed or by introducing flavour compounds which stimulate saliva flow (Blumer, 1963). The absence of fat in meat which has lost a high proportion of water during cooking will cause it to be registered as 'dry' by taste panellists.

The flavour of meat during cooking and eating arises from both water soluble and lipid soluble components in the tissues and interactions between them (Asghar & Pearson, 1980; Mottram & Edwards, 1983). Mottram and Edwards (1983) demonstrated that the characteristic flavour of beef could be produced when triglycerides were extracted, leaving only the phospholipids. This suggests that the amount of lipid required for flavour perception is very low indeed.

Variability in the fat content of meat is not normally associated with changes in colour, which are determined more by variations in the post-mortem biochemistry of muscle. Rapid glycolysis and a low pH immediately *post mortem* cause pig muscle to become pale soft and exudative (PSE); and slow glycolysis (high pH) causes dark firm and dry (DFD) meat in pigs and dark cutting beef (DCB). The firmness of the fat tissue itself is determined partly by the concentrations of the gross constituents, e.g. lipid but mainly by variation in fatty acid composition, particularly the balance between saturated and unsaturated fatty acids which have high and low melting points respectively (Wood, 1984).

ANATOMICAL AND GROWTH ASPECTS OF FAT DEPOSITION RELEVANT TO MEAT QUALITY

Carcass Fat

Fat tissue, containing lipid, water and collagen, is concentrated in specific parts of the body, termed depots (the word depot indicates the physiological role of fat as a source of energy when feed is not available). In the red meat animals, the major depots of the carcass are subcutaneous (under the skin and overlying superficial muscles), intermuscular (between muscles) and intramuscular (within muscles). The relative proportions of subcutaneous and intermuscular fat (the carcass fat tissues) depend on the species. In pigs the proportion of subcutaneous fat is high, around 70% of total body fat and the proportion of intermuscular fat is low, about 20%. In cattle and sheep on the other hand the proportions are more similar, 20–40% of total body fat being found in these two major depots. These are broad conclusions since the proportions of all depots are affected by weight and breed. As weight increases in all species the proportion of subcutaneous fat increases more than intermuscular because it has a higher 'relative growth rate'. Also, the proportion of subcutaneous fat is higher in beef breeds of cattle and meat breeds of sheep than dairy types of both species (Wood *et al.*, 1980; Truscott *et al.*, 1983).

Both subcutaneous and intermuscular fat, if they remain with the meat during cooking, will contribute to tenderness, juiciness and flavour for the reasons discussed in the previous section. However, it is the intramuscular fat which has the greatest influence on the eating quality of lean meat (Murphy & Carlin, 1961).

Intramuscular Fat

As in the other fat depots, lipid in muscle tissue is present in groups of identifiable fat cells generally considered to be similar in origin to connective tissue cells (fibroblasts). They are situated mainly in the perimysium (connective tissue sheath surrounding fibre bundles) and also in the endomysium (around myofibrils) (Fjelkner-Modig, 1985) and are typically associated with areas of greatest blood supply (Moody & Cassens, 1968). Lipid is also present in muscle as intracellular lipid droplets (at higher concentrations in red than white fibres) and as integral constituents of cell membranes (mainly phospholipids).

The major component of marbling fat is triglyceride, located in groups of fat cells in the perimysium of muscle. Triglycerides can be extracted with a solvent such as diethyl ether and indeed most published studies on muscle lipid have used this standard procedure. If analysis of total lipid is required, i.e. including phospholipids in cell membranes, a more polar solvent is used such as chloroform:methanol (2:1) which will extract about 20–30% more total lipid than diethyl ether depending on the fat content of the sample (Sahasrabudhe & Smallbone, 1983). Other workers have combined diethyl ether extraction with acid hydrolysis which also produces values some 20–30% higher than diethyl ether extraction alone.

The concentration of marbling fat increases as carcass weight and fatness increase. However, the rate of increase of lipid in muscle is less than that in the dissectible fat depots as shown by the results in Table 1. Fat depots and muscle from sheep covering a wide age range were extracted with appropriate solvents and the weights of lipid classes regressed against carcass weight. Phospholipid had a slower rate of growth than triglyceride or total lipid and all classes increased more rapidly in the fat depots than in muscle. Evidence for low relative growth rates of intramuscular fat compared with the carcass fat depots in pigs was presented by Davies and Pryor (1977). Growth coefficients in the log–log regressions of fat depot weight on total fat weight were 1·01, 0·97 and 0·91 for subcutaneous, intermuscular and intramuscular fat respectively.

Changes in muscle lipid associated with increasing fatness in cattle of commercial slaughter weights have been described by Fisher *et al.* (1983) (Table 2). The animals were of mixed sex, from various British and French breeds and had average values for dissectible fat of 16·8% of side weight. The values for muscle lipid are similar to those obtained by Johnson (1987) in Australian beef and dairy breeds and by Callow (1948) in cattle, sheep and pigs of much higher average levels of carcass fat than in the more recent studies. Comparative figures for the red meat species have also been presented by Kempster *et al.* (1986*a*) (Fig. 1). There is evidence here of faster rates of accumulation of muscle lipid relative to dissectible fat and higher absolute levels in cattle and sheep compared with pigs. The concentration of muscle lipid in the average UK pig carcass in 1984 was also estimated by Kempster *et al.* (1986*a*) to be less in pigs (5·0%) than in cattle (6·3%) and sheep (6·2%). These values are high compared with the range 1·0–1·8% reported by Kay *et al.* (1981) in the dissected carcass muscle of red deer raised on a Scottish hill farm but

Table 1

Growth of lipid classes in fat depots of Romney sheep from pre-natal stage to maturity. Values are the growth coefficient, *b*, in the equation $\log y = \log a + b \log x$, where *y* is lipid class and *x* is carcass weight

	Subcutaneous[a]	*Intermuscular*[a]	*Intramuscular*[b]
Total lipid	2·69	2·01	1·24
Triglyceride	2·70	2·05	1·45
Phospholipid	1·99	1·19	0·95

[a] In forequarter.
[b] In total carcass.
Broad & Davies (1980, 1981).

Table 2

Concentrations of ether-extractable lipid in dissectible muscle, edible meat and the carcass as influenced by the concentration of dissectible fatty tissue in 166 beef animals of commercial slaughter weights

	Dissectible fat (% of side weight)		
	10	*20*	*30*
Lipid (%) in:			
Muscle	2·3	4·9	7·5
Edible meat	8·1	18·1	28·0
Carcass	9·3	20·1	30·8

Fisher *et al.* (1983).

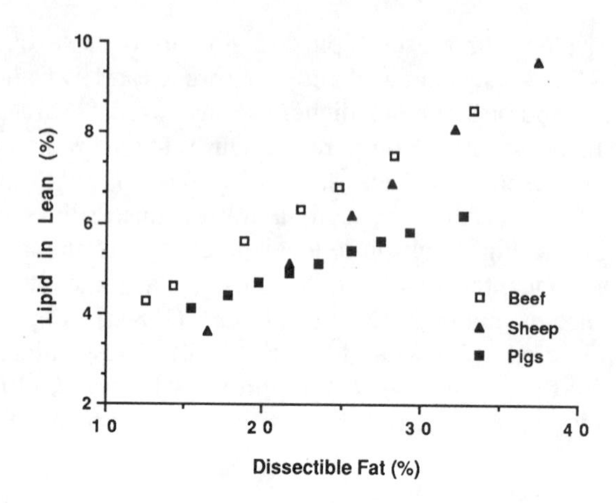

Fig. 1. Ether-extractable lipid in dissected lean tissue plotted against dissectible fat as a percentage of carcass weight for UK beef, lamb and pig carcasses in 1984. Carcass weight includes head and feet in pigs but not in cattle and sheep (Kempster *et al.*, 1986*a*).

probably similar to lipid concentrations normally found in poultry muscle (Moran, 1986).

The lipid concentration in dissected muscles from the carcass is greater than in cores taken from the centre of individual muscles because these include no trace of subcutaneous or intermuscular fat. The most common muscle analysed is *m.longissimus,* being the heaviest muscle in the carcass and the most popular as meat. In Large White × Landrace pigs, Davies and Pryor (1977) observed that the ether-extractable lipid content of dissected *m.longissimus* was 4·2% compared with 5·4% for the musculature as a whole. However, the concentration in cores of *m.longissimus* is typically 1–2% in pigs of the white breeds (Fjelkner-Modig, 1985; Barton-Gade, 1987) and in very lean pigs declines to 0·5% or less (Wood *et al.*, 1986). In beef animals containing 20% dissectible fat in the carcass, Fisher *et al.* (1983) found that the concentration of ether-extractable lipid in cores of *m.longissimus* was 2·1%, which contrasts with the value of 4·9% in dissected muscle (Table 2).

Breed Effects on Intramuscular Fat

Early American work showed that pigs of the Duroc breed had higher concentrations of muscle lipid than other breeds (Hiner *et al.*, 1965;

Jensen *et al.*, 1967). Some published data on breed effects in pigs are shown in Table 3 which confirm the results of these early studies and show that the dark skinned breeds (Duroc and Hampshire) tend to have higher concentrations than the white skinned breeds (e.g. Large White and Landrace). Not all papers quote fat thickness but where this is given it is clear that high muscle lipid concentrations in the Duroc breed are independent of carcass fatness. For example in the study of McGloughlin *et al.* (1988) the three breeds had similar fat thickness measurements (16 mm P_2, i.e. fat and skin thickness 65 mm from the dorsal mid-line at the level of the last rib). In the experiment of Edwards *et al.* (1990) a significantly

Table 3
Breed effects on the concentration of lipid in *m.longissimus* of pigs

Breed	*Pure or cross bred*	*Lipid (%)*	*Carcass weight (approx); extraction method; reference; country*
Duroc	P	7·44	76 kg; ether extract;
Yorkshire	P	4·32	Hiner *et al.* (1965); USA
Duroc	P	7·03	72 kg; ether extract;
Yorkshire	P	3·69	Jensen *et al.* (1967);
Hampshire	P	4·36	USA
Poland China	P	4·32	
Spotted White	P	4·27	
Hampshire	P	2·00	73·5 kg; ether extract
Swedish Landrace	P	1·33	with acid hydrolysis;
Swedish Yorkshire	P	1·70	Fjelkner-Modig (1985); Sweden
Hampshire	P	1·94	80 kg; ether extract with
Swiss Large White	P	1·36	acid hydrolysis;
Swiss Landrace	P	1·16	Schworer *et al.* (1986);
			Switzerland
Large White	C	1·31	70 kg; ether extract with
Duroc	C	1·73	acid hydrolysis;
Hampshire	C	1·31	Barton-Gade (1987); Denmark
Duroc	C	1·38	62 kg; ether extract;
Large White	C	1·04	Edwards *et al.* (1989); UK
Duroc	C	2·91	67 kg; petroleum ether
Large White and			extract; McGloughlin *et al.*
Landrace	C	2·04	(1988); Ireland

higher regression constant was found in Durocs for the relationship between muscle lipid and P_2 fat thickness. However, part of the reason for the very high muscle lipid concentrations in the early American studies is greater overall fatness. For example, in the study of Jensen *et al.* (1967), average backfat thickness was 33–38 mm (37 mm in Durocs) and in the work of Hiner *et al.* (1965) it was higher than this (probably 50 mm). In the former study, the subjective marbling score (scale 1–28) was 13·4 in Durocs and approximately 8·5 in the other breeds.

Although it appears that cores of *m.longissimus* were evaluated in all the studies quoted in Table 3, small differences in sample preparation can have large effects on absolute values for lipid. In a recent study (the main results are given by Wood *et al.*, 1988) the concentration of ether-extractable lipid in cores of *m.longissimus* was on average one-third of that in the total muscle of the 'steak', i.e. including *mm multifidi,* the small muscles attached to the spinous processes of the vertebrae, and their associated intermuscular fat (Table 4, see also Fig. 2). Breed effects were apparent whichever procedure was used, pure Duroc entire males having higher lipid concentrations than pure Landrace or Duroc × Landrace. Other factors which affect lipid concentrations include the site of sampling. For example higher values are found in the lumbar than the thoracic portion of *m.longissimus* (Lawrie, 1961).

Evidence that fat accumulation within muscle is under different physiological control from that in the rest of the carcass in pigs was provided by Duniec *et al.* (1961). The heritabilities of marbling fat percentage and fat tissue content were high (0·50 and 0·69 respectively) although

Table 4
The concentration of ether-extractable lipid in *m.longissimus* and muscle steak comprising *m.longissimus, multifidi* and associated intermuscular fat. Chops from last rib region

	Lipid (%)	
	m.longissimus	*Steak*
Duroc	1·60	3·99
Landrace	0·76	2·24
Duroc×Landrace	0·91	2·82
Standard error of difference and significance of breed effect	0·104***	0·326***

*** $P < 0{\cdot}001$.
From Wood *et al.* (1988).

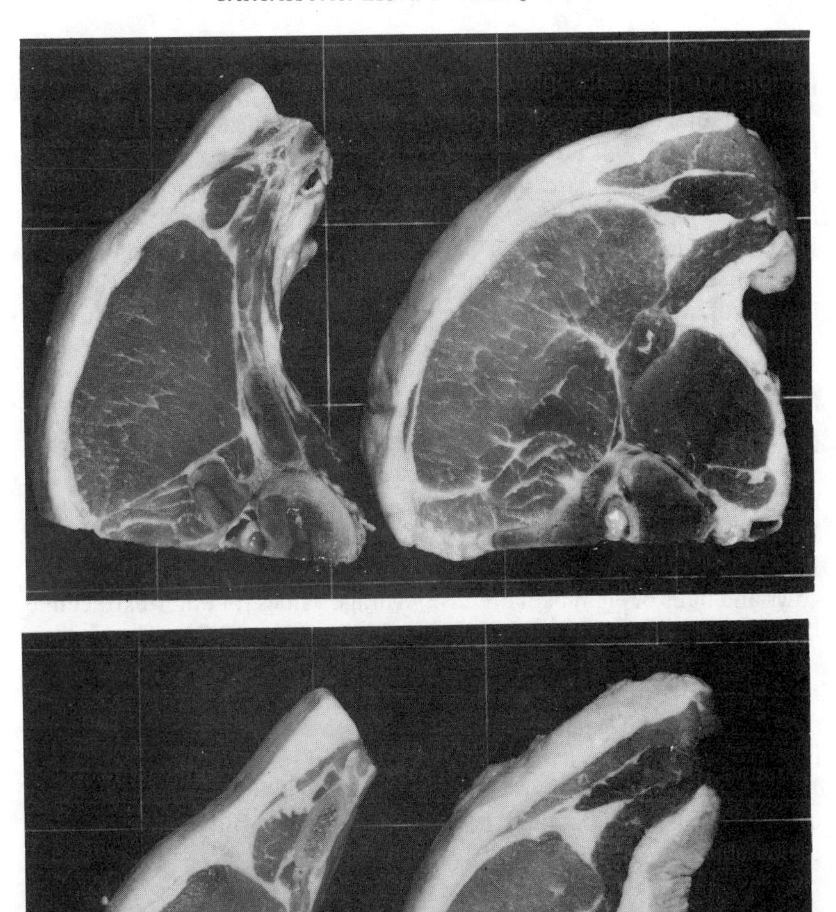

Fig. 2. Thoracic (left) and lumbar (right) chops from Duroc-sired pigs containing different amounts of marbling fat. The concentration of ether-extractable lipid in *m.longissimus* cores from the lumbar chop was top: 0·53, bottom: 1·41

the genetic correlation between them was very low (0·11). This suggests that selection for high marbling fat in lean carcasses would be successful.

Breed effects on muscle lipid are less obvious in cattle and sheep but have been reported. For example Fisher *et al.* (1983) found that at 22% dissectible fat in the carcass, the concentration of ether-extractable lipid in cores of *m. longissimus* was 3·44, 2·25 and 1·51% in Jersey, Friesian and Hereford steers, respectively. This was a surprising result in view of the belief in some quarters that muscle from beef breeds is 'more heavily marbled'. Callow (1961) also reported lower lipid concentrations in the total dissected muscle of Hereford steers compared with Shorthorns (Friesians were similar to Herefords) although Johnson (1987) found no real differences between Angus, Hereford, Friesian or Charolais cross steers compared at the same carcass fatness. At 30% carcass fat, the values for ether-extractable lipid in dissected muscle were 5·94, 5·45, 5·74 and 6·07 in Angus, Hereford, Friesian and Charolais crosses respectively. Much lower values than these were found in 12-month old bulls by Liboriussen *et al.* (1977). The animals were crosses between Danish Red cows and bulls of eight beef breeds. Average values for ether-extractable lipid in *m. longissimus* ranged from 1·41 in Chianina to 2·06 in Hereford.

Major differences in muscle lipid between Japanese cattle breeds were reported by Zembayashi *et al.* (1988). At similar concentrations of dissectible carcass fat (approx. 30%), the concentration of ether-extractable lipid in *m. longissimus* at the 6th thoracic vertebra was on average 17% in Japanese Black steers, 8·6% in Shorthorns and 9·6% in Holsteins. An example of 'heavily marbled' beef from a Japanese Black steer is shown in Fig. 3.

EFFECTS OF FAT ON MEAT QUALITY

Visual and Presentational Aspects of Quality

In pigs, the amount of fat affects the appearance of meat, its firmness and the cohesiveness between tissues, all of which are aspects of meat quality important to consumers and meat traders (Meat and Livestock Commission, 1983). The effects of carcass fatness on visual and presentational aspects were investigated in a study of 300 pigs covering the commercial range of fatness in the UK by Kempster *et al.* (1986*b*) and Wood *et al.* (1986). Carcasses, average weight 58 kg, were selected in ten abattoirs to fall into three fatness groups: lean (8 mm P_2 fat thickness), medium (12

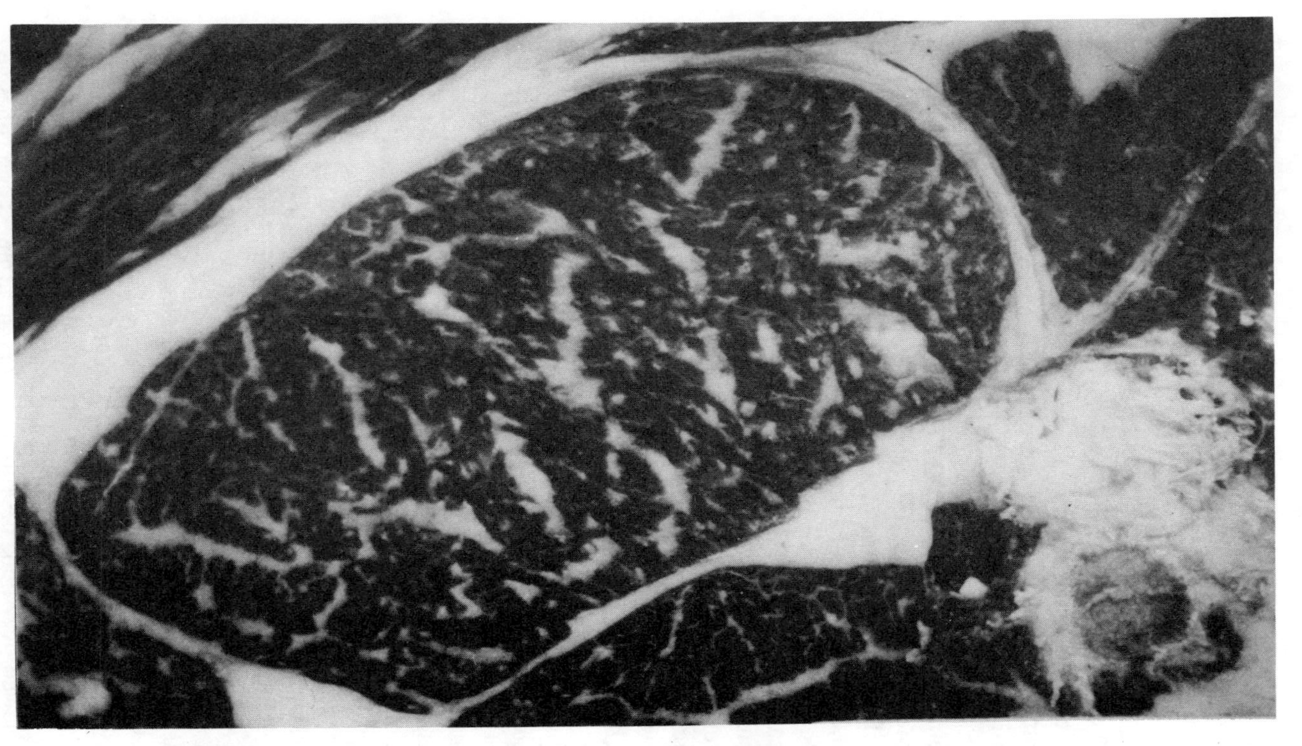

Fig. 3. Section through loin at level of 6th thoracic vertebra in a Japanese Black steer, live weight 466 kg, age 756 days. The concentration of dissectible fat in the carcass was 30·3% and the concentration of ether extractable lipid in *m.longissimus* 22·9%. Photograph courtesy of M. Zembayashi, Kyoto, Japan.

mm P_2) and fat (16 mm P_2). In each abattoir, six pigs from five production units were selected, an entire male and a female for each of the fatness groups. In this way the effects of production factors on quality were standardised across the groups. At the Meat and Livestock Commission's Meat Technology Centre, a panel of 45 butchers prepared the carcasses for retail sale as joints and cuts of meat and commented on their presentational characteristics. Also, a panel of consumers (500 families) commented on the eating characteristics of loin chops and shoulder and leg joints. At the Institute of Food Research, Bristol, trained taste panellists gave scores for visual characteristics and eating quality and objective measurements were also made. The results (Table 5) show clear effects of fat thickness on meat quality. Butchers and consumers were critical of the high level of fatness in cuts of meat from carcasses with 16 mm P_2 and butchers, but not consumers, thought the leanest loins (8 mm P_2) were too lean. The viewing panel detected these differences in fatness and also in the appearance of fat tissue. Butchers considered the 8 mm group had fat which was too soft and exhibited excessive tissue separation, findings which were confirmed in the objective meat quality tests. Butchers also considered that a higher percentage of the leanest loins had muscle which was too wet and a difference in wetness was detected (narrowly) by the viewing panel. This was possibly due to a greater incidence of PSE muscle in the 8 mm group as judged by the higher values for reflectance (i.e. greater paleness) and drip loss in *m.longissimus*. However, the viewing panel detected no difference in muscle colour and there was no evidence of PSE in the leg or hand (forequarter) joints. The higher mean score for overall attractiveness given by the viewing panel to loins from the 8 mm group disguised an important effect of carcass weight (indicated by INT (interaction) in Table 5). There was a larger difference between lean and fat loins in heavy carcasses (60–70 kg) (mean scores 1·66 and −0·19) than in light carcasses (45–55 kg) (mean scores 0·54 and 0·27). This shows that the individual aspects of quality that were unattractive in the leanest group (soft fat, fat separation, wetness) were particularly unattractive at light carcass weights, counteracting the panel's major concern, which was excessive fatness in the 16 mm group.

The marked effect of the amount of carcass fat on the appearance and handling characteristics of fat tissue is explained by changes in its chemical composition. Results in Fig. 4 (Wood *et al.*, 1989) show that as P_2 fat thickness increased from 5 to 20 mm, the concentrations of water, collagen and $C_{18:2}$ (linoleic acid) declined and that of lipid increased. Previous work (reviewed by Wood, 1984) has shown that firmness and cohesive-

Table 5
Visual and presentational aspects of meat quality in loin joints and chops from pig carcasses (58 kg) differing in fatness. One hundred carcasses in each fatness group

	P_2 fat thickness (mm)			*Significance of difference between fatness groups*
	8	*12*	*16*	
Butcher panel[a]				
Much too fat	0	4	39	***
Much too lean	15	1	0	***
Fat soft or very soft	32	15	6	***
Excessive tissue separation	46	18	11	***
Muscle too wet	40	23	18	***
Carcass measurements[a]				
Reflectance	47·3	45·7	44·9	*
Drip loss	4·8	3·8	3·0	***
Consumer panel[a]				
Much too fat	5	—	24	***
Much too lean	0	—	0	NS
Objective tests[b]				
Firmness of fat (g)	432	637	913	***
Fat separation (%)	52	23	4	***
Water in *m.longissimus* (%)	75·7	75·6	75·4	**
Trained viewing panel[b]				
Fatness	−0·7	—	1·1	***
Appearance of fat	−0·4	—	−0·1	***
Colour of muscle	0·04	—	−0·03	NS
Wetness of muscle	−0·5	—	−0·2	***
Overall attractiveness	1·1	—	0·04	INT

[a] Tests conducted by Meat and Livestock Commission. Figures for panels are percentage of samples in a given category. Reflectance given in EEL units; drip loss expressed as percent muscle weight (both in *m.longissimus*) (Kempster *et al.*, 1986*b*).
[b] Tests conducted at Institute of Food Research Bristol. Scores of viewing panel as follows: fatness −2 (much too lean) to +2 (much too fat); appearance of fat −4 (extremely unacceptable) to 0 (satisfactory); colour of muscle −2 (much too pale) to +2 (much too dark); wetness of muscle −2 (very wet) to +2 (very dry); and overall attractiveness −7 (extremely unattractive) to +7 (extremely attractive). INT indicates interaction with carcass weight—see text (Wood *et al.*, 1986).
*$P < 0·05$, **$P < 0·01$, ***$P < 0·001$.
NS, not significant.

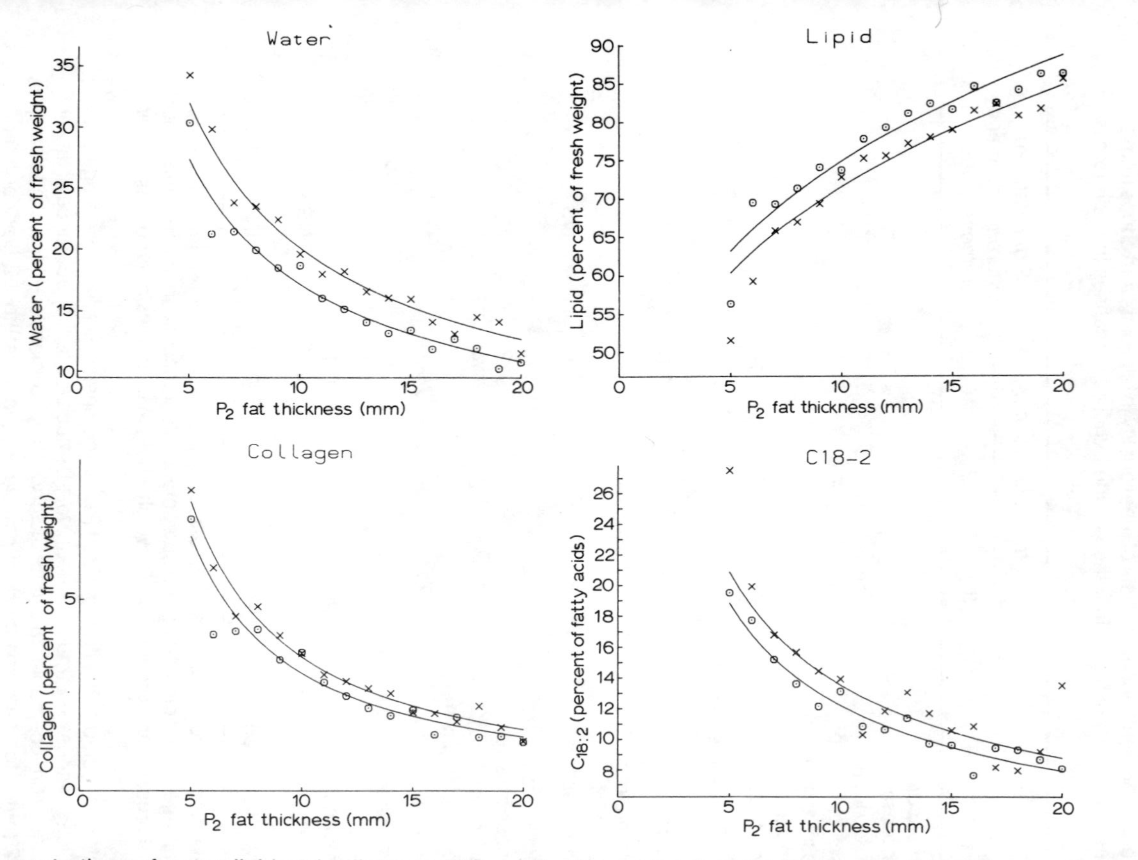

Fig. 4. Concentrations of water, lipid and collagen and $C_{18:2}$ (linoleic acid) in backfat from last rib (both layers combined) in pigs. Values for entire males (x−x) and females (o−o) plotted against P_2 fat thickness (mm). Log–log curves have been fitted to the data.

ness in fat tissue are related most strongly to the concentrations of $C_{18:0}$ (stearic acid) and $C_{18:2}$. In the present case, the concentration of $C_{18:2}$ varied much more than that of $C_{18:0}$ and provided the best prediction of firmness, measured subjectively and objectively (correlations around 0·75).

Linoleic acid is a major constituent of plant oils and dietary fat sources such as soya oil. Increasing the concentration of these in the pig's diet increases the concentration of $C_{18:2}$ in carcass and muscle fat causing a reduction in fat tissue firmness (Wood, 1984). The effect is even more marked in lean pigs with thin fat containing inherently higher levels of $C_{18:2}$ (Fig. 4).

The fatty acids of fat tissues from cattle and sheep are more saturated on average than those from pigs, due mainly to a lower concentration of $C_{18:2}$ and a higher concentration of $C_{18:0}$ (Wood, 1984). Tissues from carcasses with low levels of fat contain relatively high concentrations of saturated fatty acids and low concentrations of unsaturated fatty acids leading to a higher melting point of extracted lipid (Bensadoun & Reid, 1965; Leat, 1975; Wood, 1984). Consequently, undeveloped fat tissues in very lean ruminant carcasses are not excessively soft as in pigs. However, the concentrations of water and lipid are affected by the amount of fat in a similar way in all three species so the fat tissues of young cattle (Truscott, 1980) and emaciated cattle (Vickery, 1977) are less attractive through being wetter. Lipid content also affects the firmness of muscle which is an important component of the USDA quality grading scheme for beef (Campion *et al.*, 1975).

Eating Quality

Pigs

There is a large amount of published information on the effects of the fat content of the carcass and tissues on the tenderness, juiciness, flavour and overall appeal of cooked pigmeat. Some studies come to the conclusion that there is an important positive effect of fatness, particularly of marbling fat (muscle lipid) (e.g. Batcher & Dawson, 1960; Murphy & Carlin, 1961) whilst others conclude that the effect is small or non-existent (e.g. Hiner *et al.*, 1965; Rhodes, 1970). Some of the confusion is due to differences in the range and the absolute levels of fatness investigated. Where the range is narrow and the meat of average to high fat content, a smaller effect is observed than when a wide range is studied, especially

Table 6
Marbling fat (total lipid) and eating quality in grilled pork loin steaks

Muscle lipid (%)		*No. of loins*	*Taste panel scores*[a]			*Shear force value*
Average	*Range*		*Flavour*	*Tenderness*	*Overall acceptability*	
0·86	< 1·0	14	0·8	0·6	0·0	100
1·24	1·00–1·49	74	1·6	1·7	1·2	86
1·73	1·50–1·99	36	1·7	1·9	1·4	78
2·37	2·00–2·49	29	1·9	2·2	1·9	79
2·76	2·50–2·99	12	2·5	2·7	2·3	76
3·94	> 3·0	33	2·3	2·7	2·3	69

[a] Scores −5 (poor) to +5 (ideal).
Bejerholm & Barton-Gade (1986).

encompassing low levels. In the latter case, more meaningful statements about 'optimum' levels of marbling fat can be made (Kirkegaard *et al.*, 1979; Bejerholm, 1984; Bejerholm & Barton-Gade, 1986; DeVol *et al.*, 1988). Danish research has consistently concluded that 'about 2% intramuscular fat (total lipid) is necessary for good taste characteristics' (e.g. Bejerholm & Barton-Gade, 1986) and the evidence for this is given in Table 6. Loin steaks from the rib-end of *m.longissimus* were taken from 232 pigs of mixed breeds. Of these, 198 had normal water holding capacity and were divided into groups with different total lipid concentrations (ether extract with acid hydrolysis, approximately 1·2 × ether extract). Steaks were fried on a griddle to a central temperature of 65°C and presented to a 9-member taste panel. Shear force measurements were made with a Karl Frank testing instrument. The results (Table 6) show that toughness measured instrumentally decreased throughout the range of lipid. Flavour, tenderness and overall acceptability as assessed by taste panellists peaked at 2·5–3·0% lipid. This conclusion is similar to that of DeVol *et al.* (1988) in their (American) study of roasted chops from 120 carcasses, viz. 'the data suggest a threshold value of 2·5–3·0% fat below which chops are significantly tougher but above which there is little effect on tenderness'. Muscle lipid in this study was extracted using chloroform: methanol (2 : 1) which gives results similar to those obtained in the Danish work. This value for lipid is apparently close to the average for modern commercial pigs in the USA according to the results of DeVol *et al.* (1988) but is higher than average figures in the UK, Denmark and other European countries (Table 3). In pigs of average carcass weight (70 kg), values of 1·0% ether-extractable lipid are commonly observed in European white breeds, corresponding to 1·2% total lipid. This is close to the recently published figure of 1·31% total lipid in Danish Large White and Hampshire crossbred pigs (Barton-Gade, 1987) suggesting that the marbling fat levels identified as ideal for eating quality in Denmark are almost double those normally found.

Comparison of present UK values for muscle lipid with those published by Rhodes (1970) shows a marked recent reduction associated partly with the reduction in carcass fatness since then. It was possible in 1970 to claim for pigs of 70 kg carcass weight within the range 11–42 mm P_2 and 1·0–3·9% ether-extractable lipid that 'the results do not support the contention that fatness in the carcass is associated with tenderness in the lean meat'. However, recent research on lighter (58 kg carcass weight) pigs within the range 5–20 mm P_2 and 0·25–2·25% ether-extractable lipid

Table 7
Eating quality of loin chops taken from 58 kg pig carcasses differing in fat thickness (same pigs as Table 5)

	P_2 fat thickness (mm)		*Significance of difference between fatness groups*
	8	*16*	
Consumer panel[a]			
Extremely or very tender	35	37	**
Slightly to extremely tough	18	12	
Extremely or very juicy	16	23	***
Dry	16	9	
Excellent or very good flavour	35	35	NS
Trained taste panel[b]			
Tenderness	1·0	1·1	NS
Juiciness	1·1	1·3	**
Flavour	1·5	1·7	NS
Overall acceptability	0·7	1·0	NS

[a] Tests conducted by Meat and Livestock Commission involving 500 families. Figures are percentage of samples in a given category (Kempster *et al.*, 1986*b*).
[b] Tests conducted at Institute of Food Research, Bristol. Scores for tenderness, flavour and overall acceptability on a scale −7 (very low) to +7 (very high). Juiciness scale 0 (dry) to 4 (extremely juicy) (Wood *et al.*, 1986).
** $P < 0·01$, *** $P < 0·001$.

detected significant differences in juiciness and tenderness between 'lean' (average 8 mm P_2, 0·55% lipid) and 'fat' (average 16 mm P_2, 0·96% lipid) carcasses. The results from this study are shown in Table 7—the pigs are those also described in Table 5. The Meat and Livestock Commission consumer panel (300 families) cooked and ate the meat as part of an ordinary meal and compared eating quality with their normal purchases. The taste panel at the Institute of Food Research, Bristol (12 members) assessed the eating quality of the lean meat following grilling and preparation under controlled 'laboratory' conditions. Both panels detected significantly drier (less juicy) meat in the lean group and the consumer panel detected less tender meat. The results show that in these carcasses, 0·5% extractable lipid was too low for acceptable eating quality, levels closer to 1% being necessary. The actual level will depend on a range of factors such as carcass weight, processing conditions, etc. In a recent large-scale trial involving meat-type and white-type sires, meat with 0·8% extractable lipid in *m. longissimus* was highly acceptable (Meat and Livestock Commission, 1989).

Cattle and Sheep

The effects of the fat content of beef on its eating quality have been debated for many years, especially in the USA where the amount of marbling fat in the exposed loin is still the major determinant of meat quality as judged by the USDA quality grade. In the UK and other countries, the tendency of graders to give lower scores for 'under finished' carcasses rests mainly in the belief that eating satisfaction will be lower.

The debate mainly concerns the effects of fat in steaks or roasts, i.e. in fresh meat. In hamburgers, a form in which beef is increasingly consumed in many countries, the evidence is clearly that the difference in fat content between fat and high-fat products (perhaps 10–15% vs 25–30%) is important for eating satisfaction, taste panellists and consumers giving higher scores for juiciness and tenderness to the high-fat products (Cross *et al.*, 1980; Kregel *et al.*, 1986).

In *m. longissimus* loin steaks or roast loins, commonly tasted in fresh meat studies, the average values and variability of lipid content are lower than in hamburgers, ranging from 1–4% in European studies (e.g. Liboriussen *et al.*, 1977; Fisher *et al.*, 1983) to 3–7% in US studies where carcass fat levels are somewhat higher (e.g. Campion *et al.*, 1975; Seideman *et al.*, 1987). Within these ranges, numerous studies have been conducted, some showing important positive effects of lipid on tenderness and juiciness of the cooked meat and others showing no significant effects. On average the literature shows small positive effects and the conclusions of Blumer (1963) are still appropriate: he concluded from a review, mainly of American work, that 5% of the variation in panel tenderness and 16% of the variation in juiciness could be ascribed to marbling fat in beef. There was a wide range of values around these averages.

A more recent comprehensive study which reaches conclusions consistent with those of Blumer (1963) is summarised in Table 8. In loin steaks from 496 crossbred steers with mean USDA quality grade 'average choice', corresponding to 5·4% ether-extractable lipid in *m. longissimus* (standard deviation 2·0%), there were positive although low correlations

Table 8
Correlations between USDA quality grade, marbling fat (*m. longissimus* 12th rib) and eating quality in 496 steer carcasses

		Taste panel scores			
	Warner–Bratzler shear force	*Tenderness*	*Juiciness*	*Acceptability*	*USDA quality grade score*
Marbling score	−0·13	0·22	0·29	0·24	0·87
Lipid (%)	−0·15	0·22	0·32	0·27	0·72

Campion *et al.* (1975).

between muscle lipid (expressed subjectively as marbling score or objectively as ether-extractable lipid) and eating quality measurements. The range in marbling score corresponded to a range in extractable lipid in *m. longissimus* of 2·7% ('traces') to 14·7% ('abundant'). Ranking according to marbling score or lipid produced no consistent trend in shear force measurements or taste panel scores. The authors concluded that eating quality was acceptable when muscle lipid was at least 2·9%. Only 14 out of 496 samples fell below this.

Other US workers have maintained that a value of approximately 3% ether-extractable lipid ('minimum-slight' marbling score) is necessary to provide tender juicy beef (Tatum, 1981; Dikeman, 1987). However, in a Danish study, high levels of tenderness, juiciness and flavour were found in grilled steaks averaging only 1·6% extractable lipid (Liboriussen *et al.*, 1977). In these young bulls of various breeds, average taste panel scores for the three aspects of eating quality lay between 7 and 8 (scoring range 1–10), indicating a high level of acceptability. A lipid concentration of 1·6% corresponds to a USDA marbling score of 'practically devoid' (Savell *et al.*, 1986) which is invariably associated with very low scores for eating quality in US studies (Smith *et al.*, 1984). It is possible that production and processing differences between European and US beef systems are responsible for these very different conclusions.

The relationships between the concentration of *m.longissimus* lipid and eating quality scores reported for lamb are of a similar magnitude to those for beef, i.e. small positive correlations have been found for tenderness, juiciness and flavour (Carpenter & King, 1965; Smith *et al.*, 1970; Purchas *et al.*, 1979). The report of Carpenter and King (1965) suggests that levels of ether-extractable lipid above 6% such as are found in beef at marbling scores in the 'moderate to abundant' range (Savell *et al.*, 1986) are less likely in lamb. As with beef, a concentration of 3% ether-extractable lipid in *m.longissimus* has been suggested as necessary for acceptable eating quality in lamb (Dikeman, 1987).

Poultry

Poultry is regarded as a low-fat meat although concentrations of ether-extractable lipid in the whole carcass (including skin) are not much lower than in the red meat species (Wood, 1989). Between individual muscles, concentrations vary typically from below 1% in the breast to over 6% in the thigh (Salmon, 1979) and in dissected carcass muscle as a whole, values similar to those in the red meat species are found (i.e. 3–6%) (Moran, 1986).

Lack of juiciness in breast meat is associated with low lipid levels (< 1%) and attempts are made commercially to overcome this by injecting oils into the muscle. Larmond and Moran (1983) assessed the eating quality of breast meat from Small White turkeys of two finish grades (fatness levels in the carcass) with or without injected coconut oil (to 3% of muscle weight). There was no clear effect of finish grade itself on the tenderness, juiciness or flavour of roasted meat but oil injection greatly increased taste panel scores for these characteristics.

EFFECTS OF PRODUCTION, PRE-SLAUGHTER AND PROCESSING FACTORS ON MEAT QUALITY: DIRECT EFFECTS AND EFFECTS OF FATNESS

In this section the aim is to determine whether production or processing changes directly affect meat quality or whether they do so via changes in fatness. In the case of processing factors such as chilling rate and ageing, the interactions with fatness are of interest.

Production Factors

Breed

It is often claimed that particular breeds have outstanding meat quality. However, the conclusion from the large body of scientific data on the topic is that the separate role of breed is small except when unusual genes are introduced, e.g. the double muscling gene in cattle, the halothane gene in pigs and genes for marbling fat, also in pigs. The general conclusion, in all the red meat species, is that breed effects are actually effects of fatness, meat from the fatter breeds having slightly better eating quality, when all other factors are controlled, as outlined in the previous section.

Results of a major beef breed study are given in Table 9 (the pooled results in Table 8 were from the same study). Seven sire breeds were crossed on Hereford × Angus cows and the calves (approximately 70 steers in each breed group) finished in a feedlot after weaning at 200 days. Hereford and Angus carcasses were the fattest as judged by fat thickness measurements although marbling fat scores and muscle lipid were highest in Jerseys. Jerseys also had the lowest shear force measurements and (with South Devons) the highest taste panel scores for tenderness and other aspects of eating quality. High tenderness scores in Jerseys were also found by Cole *et al.* (1964) and Moore and Bass (1978). These results support the view that muscle lipid rather than carcass fat is important for eating quality.

Examination of the breed means in Table 9 shows a high degree of correlation between lipid content and tenderness. Pooling of the breeds reduced the correlations to about 0·2 (Table 8). This supports the view that breed effects are mainly connected with the fat content of meat. However, connective tissue plays a role in the extra tenderness of cattle

Table 9
Effects of beef breed on fatness and eating quality of loin steaks

	Hereford	*Angus*	*Jersey*	*South Devon*	*Limousin*	*Simmental*	*Charolais*	*Pooled SD*
USDA quality grade[a]	9·5	10·1	9·9	9·7	8·7	9·4	9·7	1·3
Marbling score[b]	10·8	12·3	14·1	11·5	9·3	10·6	11·4	3·2
Lipid in *m.longissimus*[c] (%)	5·5	6·4	7·2	5·4	3·9	4·7	5·0	2·0
Fat thickness (mm)[c]	15·2	15·5	12·7	13·2	9·9	10·2	10·2	3·8
Warner–Bratzler shear force (kg/cm^2)	2·5	2·5	2·3	2·4	2·7	2·7	2·5	0·5
Taste panel scores[d]								
Tenderness	7·3	7·4	7·5	7·5	7·0	6·9	7·4	0·8
Juiciness	7·0	7·1	7·3	7·2	7·0	7·1	7·1	0·6
Flavour	7·5	7·5	7·6	7·5	7·5	7·5	7·5	0·4
Overall acceptability	7·3	7·2	7·4	7·4	7·1	7·1	7·3	0·6

[a] Av. Good = 8; Av. Choice = 11; Av. Prime = 14.
[b] Scale 1 (devoid) to 27 (abundant).
[c] 12th rib.
[d] Scale 1 (extremely undesirable) to 9 (extremely desirable).
Campion *et al.* (1975).

having the double muscling gene. Here, low concentrations of connective tissue in muscle are associated with higher tenderness (Boccard, 1978). Robertson *et al.* (1986) also showed that higher concentrations of collagen as found in water buffalo (*Bubalus bubalis*) produced higher peak force measurements of toughness when compared with Brahman cross cattle (*Bos indicus*) of the same age.

Most studies of sheep breeds have shown that the separate effects of breed on eating quality are small although there is a positive effect of fatness, possibly connected with the slower cooling and smaller chance of cold shortening of fatter carcasses (Dransfield *et al.*, 1979). Lamb carcasses are particularly prone to cold shortening if chilling or freezing is too rapid (Meat Industry Research Institute of New Zealand, 1985). Several studies with pigs have also shown only small effects of breed on meat quality when carcass fatness is similar. For example in the study of Wood *et al.* (1979), a 'traditional' breed, Gloucester Old Spot, had similar eating quality to two 'modern' hybrid genotypes. In pigs, though, two specific genetic effects have important effects on meat quality. Firstly the halothane gene, a recessive gene conferring stress sensitivity, causes some undesirable effects on meat quality including pale, soft, exudative (PSE) muscle which, because of a low water holding capacity, is dry and tough when cooked (Bejerholm, 1984). The halothane gene also confers a low appetite and a leaner carcass which explains why breeds such as the Pietrain (with a high incidence of the halothane gene) are leaner and have poorer meat quality than breeds such as the Large White which have a low incidence (Webb *et al.*, 1982). This negative association between meat quality and fatness caused by different incidences of the halothane gene should not be taken to be a general effect in pigs, applicable across a range of genotypes.

The second specific genetic effect in pigs involves levels of marbling fat which are unconnected with carcass fatness. The Duroc is an example of a breed with high concentrations of muscle lipid in relation to fat thickness compared with other breeds. Renewed interest in the Duroc in the UK and other countries in recent years is partly connected with the possibility that eating quality could be raised through high levels of marbling fat (above 0·5%), whilst keeping the more obvious subcutaneous and intermuscular fat deposits at low levels.

Early American work found that concentrations of ether-extractable lipid in Durocs were around 7% compared with 4% in other breeds (Table 3). This was associated with slightly more tender, juicy meat.

More recent Danish research, using pigs much leaner than this, indicated that total lipid values of 4% as found in some Duroc crosses, produced significantly higher taste panel scores than concentrations of < 1% as found in some Yorkshire pigs (Bejerholm & Barton-Gade, 1986, Table 6). However, other reports show that average lipid values in modern commercial crossbred stock are lower, between 1–2% in Duroc-sired pigs and about 1% in progeny from White sires (Large White, Yorkshire or Landrace, Table 3). Some recent studies of eating quality comparing Duroc- and

Table 10
Lipid (marbling fat) concentrations in *m.longissimus* and measures of eating quality in recent European studies comparing Durocs with other breeds as terminal sires (60–70 kg carcasses)

	Lipid (%) (method of analysis)[a]	*Eating quality (method of analysis)*[b]			
Denmark (Barton-Gade, 1987)[c]					
	(Method 2)	*(Shear force measurements)*			
Large White	1·31[e]	91·2[e]			
Duroc	1·73[d]	82·6[d]			
Hampshire	1·31[e]	78·2[d]			
Ireland (McGloughlin *et al.*, 1988)[f]					
	(Method 1)	*(Taste panel scores)*			
		J	*Fl*	*T*	*Oa*
Duroc	2·9	1·2	3·9	4·5	5·1
Large White/Landrace	2·0	1·2	3·6	4·5	4·8
	***	NS	*	NS	NS
UK (Edwards *et al.*, 1990)[f]					
	(Method 2)	*(Taste panel scores)*			
		J	*Fl*	*T*	*Oa*
Duroc	1·8	1·3	2·1	−0·3	1·0
Large White	1·4	1·3	2·0	0·0	1·1
	***	NS	NS	NS	NS

[a] Method 1 involves ether extraction alone; method 2 also involves acid hydrolysis. Method 2 produces values approx. 1·2×method 1.
[b] In Danish study, tenderness was measured objectively (as in Table 6). In Irish study, taste panellists used following scales: J, juiciness 0–4; Fl, flavour 0–5; T, tenderness 0–7; Oa, overall acceptability 0–7. UK taste panellists scored Fl, T and Oa −7 to +7 and J 0 to 4.
[c,d,e] Different superscripts within a column indicate significant differences ($P < 0{\cdot}05$).
[f] *$P < 0{\cdot}05$); ***$P < 0{\cdot}001$.

White-sired pigs are summarised in Table 10. Higher values for muscle lipid, whether extractable or total, were found in Durocs but no consistent effect on eating quality was seen.

Sex

Studies with cattle, sheep and pigs have shown small but commercially important effects of sex on eating quality. Tenderness (in beef and lamb) and flavour (in pigmeat) are mainly affected, meat from entire males tending to produce lower scores in comparative tests. There is good evidence that differences in fatness are not involved, the concentration of connective tissue (tenderness) and differences in steroid metabolism (flavour) being more important.

In cattle, the advantages of the bull over the steer in growth rate, feed conversion efficiency and lean meat yield are well documented. However, the evidence of greater toughness in meat from bulls is also clear (Field, 1971; Arthaud *et al.*, 1977). Data from six pairs of bull–steer twins collected at the Institute of Food Research, Bristol are shown in Table 11. Four muscles were cooked in typical ways and presented to taste panellists for scoring on 0–100 scales. Only one score was significantly different between bulls and steers, juiciness being lower in roast *m.longissimus* from bulls, but in all cases scores for steers were the same

Table 11
Composition and eating quality of muscles from six pairs of twin bulls (*B*) and steers (*S*) 400 days of age

	Composition of fresh muscle				*Taste panel scores*[b]					
Muscle[a]	*Lipid (%)*		*Collagen (%)*		*Tenderness*		*Juiciness*		*Flavour*	
	B	*S*	*B*	*S*	*B*	*S*	*B*	*S*	*B*	*S*
LD	1·9*	2·5	0·7	0·6	61	65	30***	38	41	43
SS	1·0	1·8	0·8	0·7	56	62	50	52	39	41
PM	0·7	0·8	0·5	0·4	59	59	42	42	49	49
G	1·8**	3·2	0·4	0·4	65	67	—	—	46	48

[a] Muscle and cooking method: Ld, *m.longissimus dorsi* (roast); SS, *m.supraspinatus* (casserole); PM, *m.psoas major* (grill); G, *m.gastrocnemius* (mince).
[b] Scale 0 (poor) to 100 (ideal).
* $P < 0{\cdot}05$; ** $P < 0{\cdot}01$; *** $P < 0{\cdot}001$.
Dransfield *et al.* (1984).

or higher. Muscles from bulls had lower lipid concentrations and tended to have higher concentrations of collagen.

In the study of Dransfield *et al.*, (1984) it was concluded that larger differences in tenderness between bulls and steers are found when the meat is produced under commercial conditions in which processing factors, especially the rate of carcass chilling, are relatively uncontrolled compared with research studies. Several workers have questioned whether bull beef is tougher because the carcasses, with lower levels of fat cover, would be more prone to cold shortening. However, after the application of electrical stimulation (Crouse *et al.*, 1983; Riley *et al.*, 1983) and extended ageing/conditioning treatments (Crouse *et al.*, 1983; Johnson *et al.*, 1988) bull beef was still relatively tough, showing that greater myofibrillar toughness because of cold shortening is not the cause. Similarly, Arthaud *et al.* (1977) found greater toughness in *m. longissimus* muscles from bulls than steers when both had similar marbling scores and quality grades.

Other workers (e.g. Boccard *et al.*, 1979) have shown that the concentration of collagen is higher in muscles from bulls and suggested that this explains the sex difference in toughness. A difference in the type of collagen has also been found, bulls having a lower concentration of heat-soluble collagen (Boccard *et al.*, 1979; Hinch & Thwaites, 1982; Gerrard *et al.*, 1987). Although sometimes small, this difference would also lead to greater toughness in bulls.

A similar effect of sex (castration) on tenderness has been observed in sheep. Thus muscles from rams are slightly tougher than those from wethers even in the absence of cold shortening (Field, 1971; Purchas *et al.*, 1979; Kirton *et al.*, 1983). Greater toughness has also been explained on the basis of higher concentrations of collagen. The question of ram taint has also been raised in connection with increased use of entire male lambs. In a New Zealand study, taste panellists were able to detect a higher incidence of 'foreign' flavours in meat from old rams (2+ years of age) although consumers gave similar scores to rams and ewes aged 12 months or older (Kirton *et al.*, 1983). Butler-Hogg *et al.* (1984) also detected a slightly higher incidence of foreign flavours in meat from young ram lambs (18 weeks of age) compared with ewes. The incidence was 12·6% overall of which 8·1% was in meat from rams.

In pigs there is no evidence that meat from entire males is tougher than that from females and often the opposite effect has been observed (e.g. Mottram *et al.*, 1982). Some recent results from the Institute of Food Research, Bristol are shown in Table 12. Carcasses were selected in ten

Table 12
Eating quality of grilled *m.longissimus* in 48 entire male and 48 female pigs

	Entire male	*Female*	
Fat tissue			
Pork odour	1·1	1·2	NS
Abnormal odour	−7·0	−7·0	NS
Muscle tissue			
Tenderness	1·2	0·9	NS
Juiciness	1·2	1·2	NS
Flavour (liking)	1·5	1·7	NS
Pork flavour	0·8	0·7	NS
Abnormal flavour	− 5·7	−6·0	NS
Overall acceptability	0·8	1·0	NS

Taste panel scores all −7 to +7 except juiciness 0 to 4.
NS, not significant.
Wood *et al.* (1986).

abattoirs from individual producer groups to have the same average weight (58 kg) and fat thickness (12 mm P_2). Tenderness of grilled chops was slightly higher in entires although the sex difference was not significant.

The major question surrounding the use of entire male pigs for meat production relates to boar taint. Higher concentrations of androstenone and skatole are found in backfat from entires (Mottram *et al.*, 1982; Mortensen *et al.*, 1986) but this has not always led to lower scores for odour or flavour from taste panellists (Malmfors & Lundstrom, 1983). In tests conducted at Bristol, taste panellists and consumers have consistently made no distinction between meat from the different sexes (Rhodes, 1972; Mottram *et al.*, 1982; Wood *et al.*, 1986). As shown in Table 12, taste panellists in a recent study gave marginally lower scores for flavour liking to chops from entires compared with females but scores for abnormal odours and flavours were equally low in both sexes.

It is possible that taint levels are lower in UK pigs because of lower carcass weights and ages and a different breed mix compared with the rest of Europe. Taint levels are higher in the Pietrain and Landrace breeds than the Large White which is more numerous in the UK (Malmfors & Lundstrom, 1983). A recent study found higher concentrations of androstenone in Duroc than Landrace breeds with Duroc × Landrace being intermediate (Wood *et al.*, 1988).

The evidence that the incidence of unfavourable odours and flavours in pigmeat is correlated more closely with the concentration of skatole rather than androstenone suggests that dietary effects rather than sex effects are critical in taint development (e.g. Mortensen *et al.*, 1986). However, the concentration of skatole, as well as androstenone, seems to be higher in fat tissue from entires. In a Danish study, elimination of the 5% or so of entire male carcasses with greater than 0·25 ppm skatole in backfat produced similar consumer responses as were found in meat from castrates (Mortensen *et al.*, 1986).

In pigs, the chemical composition and therefore the quality of backfat is affected by sex. Wood *et al.* (1989) found that within the range 5–20 mm P_2 fat thickness, the composition of backfat in entire males was similar to that in females some 15% lower in P_2 (Fig. 4). In both sexes, composition changed with fat thickness but at the same P_2 value, backfat in entire males was wetter and contained a higher concentration of linoleic acid ($C_{18:2}$).

Production System and Diet

Eating quality can be influenced by changes in the production system which cause animals to differ in age, weight, fatness and the composition of fat tissues at slaughter. The effects on the amount and composition of fat tissue seem most important in determining tenderness.

Several American papers have compared eating quality in beef finished on grain (high energy) or forage (low energy) diets. Most of these (six examples are given in Table 13) show that grain-finished animals grow more rapidly and produce heavier and fatter carcasses at slaughter. The meat is usually more marbled and more tender.

It has been suggested that the main reason for greater tenderness in grain-finished beef is slower post-mortem cooling in fatter, heavier carcasses, leading to a reduced likelihood of cold shortening and the added possibility of pre-rigor conditioning (increased activity of proteolytic enzymes) (Tatum, 1981; Dikeman, 1987). This cannot be the only explanation since in reports which have studied the interactions, the difference in tenderness between production systems was greater than that between chilling rates. In the study of Lochner *et al.* (1980) for example, the tenderness scores for forage-finished carcasses were 2·6 and 2·9 for rapid and slow chilling rates respectively compared with scores of 4·1 and 3·8 in grain-finished carcasses.

The beneficial effect of high levels of feeding on the eating quality of

Table 13
Effects of forage (F) or grain (G) feeding on weight, fatness and tenderness in beef

Reference	*Carcass weight (kg)*		*Fat thickness (mm)*[g]		m.longissimus *lipid (%)*[h]		*Tenderness*[i]		*Toughness (kg)*[j]	
	F	*G*	*F*	*G*	*F*	*G*	*F*	*G*	*F*	*G*
Bowling *et al.* (1977)[a]	216	229	4·1	8·4	—	—	5·5	6·1	7·6	5·6
Lochner *et al.* (1980)[b]	240	319	5	26	4·6	10·1	2·7	3·9	8·1	6·4
Schroeder *et al.* (1980)[c]	182	309	2·5	12·7	2·1	3·9	3·7	5·4	6·4	3·5
Aberle *et al* (1981)[d]	199	287	1·8	9·7	—	—	2·7	4·9	5·1	3·3
Lee & Ashmore (1985)[e]	237	293	3·6	12·6	1·7	5·1	5·0	5·3	8·2	7·3
Medeiros *et al.* (1987)[f]	235	336	3·8	10·9	1·4	3·0	5·9	8·4	11·4	9·3

[a] Average of slow and rapid chilling treatments.
[b] Assuming 0·56 killing-out proportion. Average of slow and rapid chilling treatments.
[c] Averages for 3F and 2G treatments.
[d] After 230 (F) or 210 (G) days on feed.
[e] Average of slow and rapid chilling treatments.
[f] Steers 17 (F) or 24 months (G) of age.
[g] 12th rib.
[h] Extractable lipid.
[i] Taste panel scores 1–8 except [b] 1–7 and [f] 1–13.
[j] Warner–Bratzler peak force measurements in most cases.

beef could be due directly to higher levels of marbling fat. An alternative explanation was offered by Aberle *et al.* (1981) who suggested that differences in the type of collagen could be involved, more rapid protein synthesis and turnover leading to a higher proportion of young, heat-soluble collagen. Another possibility is that the type of fat deposited during grain feeding is responsible for improved eating quality. Several reports have shown that in ruminants, grain feeding leeds to higher concentrations of unsaturated fatty acids, particularly oleic acid ($C_{18:1}$) and lower concentrations of saturated fatty acids, particularly stearic acid ($C_{18:0}$) compared with forage or grass feeding (Wood, 1984). In studies by Dryden and Marchello (1970) and Westerling and Hedrick (1979), significant positive correlations were found between the concentration of $C_{18:1}$ in extracted lipid of *m.longissimus* and its tenderness and flavour.

There is also evidence for greater tenderness and overall eating satisfaction in lamb which has been finished on high energy diets rather than on pasture (e.g. Rhodes, 1971; Kemp *et al.*, 1981). The tendency for light lean lamb carcasses to cold-shorten if chilling is inappropriate has been well documented (Smith *et al.*, 1976) although for lamb as for beef the beneficial effects of high level feeding are greater than its role in simply providing a thick layer of insulation on the carcass.

In pigs, there is evidence that *ad libitum* feeding, which results in high growth rates and a younger age at slaughter, produces more tender meat than low-level or restricted feeding scales (Meat and Livestock Commission, 1989). This effect is greater than the increased concentration of marbling fat would suggest. Pig fat tissues are also very amenable to changes in fatty acid composition as the result of feeding diets containing different fat sources (Wood, 1984). Linoleic acid ($C_{18:2}$) shows the greatest variability between diets and consequently in pig tissues. This greatly affects the firmness of pig backfat because of the low melting point of linoleic acid, a characteristic which might also be expected to influence juiciness and tenderness during cooking. However, when fatty acid composition has been changed over a wide range using different diets, no consistent effects on tenderness, juiciness or flavour of the cooked meat have been observed in US (West & Myer, 1987) or UK studies (Edwards *et al.*, 1990).

β-Agonists and Somatotropins

Recent debate has surrounded the effects on eating quality of β-agonists and somatotropins used as growth promoters. β-agonists reduce carcass fatness in cattle, sheep and pigs but appear to consistently reduce tenderness by an amount which is greater than expected from the reduction in carcass or intramuscular fat (Tarrant, 1987; Warriss *et al.*, 1990). Some recent results for pigs are shown in Table 14. Fat thickness was reduced by a commercially significant amount as the result of feeding salbutamol even in these extremely lean pigs although muscle lipid was unaffected. There was an increase in instrumentally-measured toughness which could be associated with changes in muscle histology. The reduction in haem pigment and collagen concentrations in *m.longissimus* suggested an increase in the proportion of white glycolytic fibres which might be expected to increase shear force. It has also been suggested that toughness could be increased through an increased degree of cross-linking in collagen and a reduction in the activity of protease enzymes (Tarrant, 1987).

Table 14
Effect on carcass composition and meat quality in pigs of feeding the β-agonist salbutamol (3 ppm) between weaning and 85 kg live weight

	Control	*Salbutamol*	
Live weight (kg)	84·7	84·3	NS
Carcass weight (kg)	64·1	65·7	**
P_2 fat thickness (mm)	10·5	8·7	***
LD depth (mm)	60·0	64·0	***
LD lipid (%)	0·81	0·86	NS
LD collagen (%)	0·38	0·33	***
LD haem pigments (mg/g)	0·82	0·69	***
LD toughness[a] (kg)	4·11	5·04	***

[a] First yield force in *m.longissimus dorsi* (LD).
** $P < 0{\cdot}01$, *** $P < 0{\cdot}001$.
Warriss *et al.* (1990).

Large reductions in carcass fatness in pigs have also been observed following treatment with porcine somatotropin (PST) (Campbell *et al.*, 1988). In early studies this had no deleterious effect on tenderness although in more recent work treatment with 100 μg PST/kg body weight/day between 25 and 55 kg live weight increased toughness over controls (7·88 and 6·73 kg shear force respectively) (Solomon *et al.*, 1988). In this case there was no effect of treatment on the proportions of different muscle fibre types.

Age

As meat animals age, they also increase in weight and fatness which tends to increase the eating quality of the lean meat as already described. Apart from this, age effects are due to changes in the amount and particularly the type of collagen in muscles. The degree and strength of cross-linking between collagen molecules increases with age which reduces the solubility of collagen at a particular cooking temperature and leads to increased toughness (Cross *et al.*, 1973, 1984; Liboriussen *et al.*, 1977). However, the consensus view from the many papers on this subject is that toughening due to collagen is only significant when extremes of age are compared and is not important within the normal commercial range of ages. These interrelationships are of most interest in beef, less so in lambs and of little interest in pigs which show limited variation in age at slaughter.

Results from an American study involving bulls and steers slaughtered at 12, 15, 18 and 24 months following consumption of a high-energy diet are shown in Table 15. Subjective and objective measures of tenderness were unaffected by age in contrast with the effects of sex—meat from bulls was tougher at all ages.

Collagen plays an important role in determining the tenderness of individual muscles. In a study of 18 muscles from beef animals (Dransfield, 1977), the concentration of collagen was low in tender muscles such as *m. longissimus* and high in tough muscles such as *m. extensor carpi radialis* (collagen as percent of fat-free dry matter 2·76 and 5·22 respectively). Proportions of heat-soluble collagen did not differ so widely between muscles, total collagen being the chemical component most highly correlated with toughness.

Pre-slaughter Factors

The degree of stress experienced between the production unit and slaughter can affect meat quality, particularly the colour and water holding capacity of muscle and also eating quality. In both cattle (Dransfield, 1981; Warriss *et al.*, 1984) and pigs (Dransfield *et al.*, 1985; Warriss & Brown, 1985), excessive fighting and mixing deplete muscle glycogen leading to meat with a high ultimate pH and dark colour. In general, this is more tender and juicy than normal meat after cooking because of a higher water holding capacity. However, flavour is adversely affected (Dransfield, 1981). Detailed examinations of the relationships between ultimate pH and eating quality parameters have been made by Bouton *et*

Table 15

Composition and tenderness of *m.longissimus* muscles from bulls (*B*) and steers (*S*) at four ages

	Age (months)							
	12		*15*		*18*		*24*	
	B	*S*	*B*	*S*	*B*	*S*	*B*	*S*
Carcass weight (kg)	185	186	272	240	275	261	300	302
Extractable lipid (%)	3·2	4·6*	4·1	7·4**	4·3	7·6**	5·0	7·9**
Myoglobin (mg/g)	1·71	1·75	2·54	2·26	2·87	2·44	4·44	4·01
Toughness (kg)	7·9	6·1**	7·3	6·0*	7·0	5·7*	7·5	5·9**
Tenderness (1–10)	6·5	7·6**	6·7	7·6*	7·1	7·7	6·3	7·4**

Effects of sex indicated (* $P < 0{\cdot}05$; ** $P < 0{\cdot}01$; *** $P < 0{\cdot}001$). Effects of age significant for carcass weight, extractable lipid and myoglobin ($P < 0{\cdot}001$) but not significant for tenderness, however measured.
Arthaud *et al.* (1977).

al. (1973) in beef and Dransfield *et al.* (1985) in pigs. In the case of tenderness, both reports show quadratic relationships, tenderness decreasing in the pH range 5·4–6·0 and increasing markedly between 6·0 and 7·0. This helps to explain why, although the majority of reports observe more tender meat as the result of stress, others find tougher meat. In a recent Canadian study for example (Jones *et al.*, 1988), steers more frequently mixed and transported longer distances had tougher meat than controls (shear force values 8·51 and 6·01 kg respectively). Bouton *et al.* (1973) showed that muscles within the ultimate pH range 5·4–6·0 could have experienced cold shortening in the pre-rigor state whereas those in the range 6·0–7·0 would have been unlikely to have done so. Also, water holding capacity is increased much more above than below pH 6·0.

The meat quality changes which occur as the result of pre-slaughter stress are largely independent of carcass and muscle fatness. However, there are some indirect associations. Entire male cattle are leaner than castrates and more likely to produce meat with a high ultimate pH (Warriss & Brown, 1985; Brown *et al.*, 1989). Some leaner breeds of cattle and pigs may also be more prone to a certain level of stress. For example Limousin × Friesian bulls containing 67% lean in the side had higher ultimate pH values and darker meat than Friesian bulls containing 63% lean following the same transportation and mixing procedures (Fisher, personal communication). In pigs, lean genotypes which have inherited the halothane gene from one parent (heterozygotes) are more susceptible to poor pre-slaughter treatments and more likely to produce poor quality meat than non-carriers (Barton-Gade, 1984). However, unlike cattle, such meat is likely to be pale, soft and exudative (PSE) rather than dark. Homozygotes for the halothane gene produce PSE meat under all pre-slaughter conditions. For the opposite reasons to dark

Table 16
Eating quality of pale, soft, exudative (PSE) and non-PSE pork chops. Taste panel scores −5 to +5

	PSE	*Normal*	
Tenderness	−0·7	1·7	***
Juiciness	1·1	2·0	***
Flavour	0·9	1·7	***
Overall acceptability	−0·5	1·3	***

*** $P < 0{\cdot}001$.
Bejerholm (1984).

cutting meat, i.e. because of a low water holding capacity, PSE muscle is relatively tough and dry (Table 16).

Post-slaughter Processing Factors

The eating quality of meat, particularly tenderness, can be altered by changing the rate of carcass chilling and by varying the time and temperature of conditioning (ageing). Differences in carcass fatness can modify both of these effects.

Rate of Chilling: Cattle and Sheep

If muscles are cooled too rapidly so that their temperature falls to about 10°C whilst they are still capable of contraction, severe shortening, termed cold shortening occurs and the meat is tough when cooked (Locker & Hagyard, 1963). The effect also occurs in carcasses, in which muscles such as *m.longissimus* which cool rapidly because of their superficial location and lack of surrounding muscle bulk, are particularly susceptible (Marsh *et al.*, 1968; Marsh, 1981). Lamb carcasses are most at risk because of their small size (e.g. Table 17) but cold shortening is also a significant problem in beef. It can also occur in pigs (James *et al.*, 1983) and poultry (Wakefield *et al.*, 1989). It is generally thought for all species that a muscle pH above 6·0 is critical since it indicates sufficient available energy for contraction (Bendall *et al.*, 1976).

Cold shortening has become a serious commercial problem in recent years because of recognition of the cost savings possible in terms of reduced evaporative weight loss when the heat is removed quickly from carcasses. An EEC directive requiring that carcasses must not be butchered or transported above a temperature of 7°C has also encouraged the use of very rapid chilling systems in European countries (James & Bailey, 1986).

The effects of rapid chilling on cold shortening in key muscles such as *m.longissimus* and *m.semimembranosus* can be prevented or lessened by hanging the carcass so the muscles are under restraint during chilling. Hanging from the pelvis is beneficial in both beef (Newbold & Harris, 1972) and pig carcasses (Moller & Vestergaard, 1986). Electrical stimulation of the carcass also protects against cold shortening by depleting muscle glycogen, lowering muscle pH and thereby removing the energy available for contraction (Bendall *et al.*, 1976). Bendall *et al.* (1976) showed that the optimum current for beef carcasses was 700V 25 Hz

Table 17
Effects of chilling rate and conditioning on tenderness of lamb loin roasts. Tenderness measured objectively using Instron materials testing instrument or subjectively by taste panel

	Chilling rate[a]	*Not conditioned*[b]	*Conditioned*[c]	
Total work (J)	Rapid	0·33	0·25	**
	Slow	0·18	0·14	NS
		***	**	
First break (kgf)	Rapid	11·85	7·97	*
	Slow	3·95	2·48	*
		***	**	
Final compression (kgf)	Rapid	10·31	7·67	NS
	Slow	2·91	2·34	NS
		***	**	
Tenderness score[d]	Rapid	−1·56	0·85	**
	Slow	2·86	4·42	**
		***	***	

[a] Deep leg temperature at 10 h *post mortem*: rapid 1·5°C; slow 16·5°C.
[b] 24 h (rapid chill) or 48 h (slow chill) after slaughter.
[c] 6 days (rapid chill) or 7 days (slow chill) at 0°C after slaughter.
[d] Panel score −7 (very tough) to +7 (very tender).
* $P < 0{\cdot}05$, ** $P < 0{\cdot}01$, *** $P < 0{\cdot}001$.
Taylor *et al.* (1972).

applied within 1 h of slaughter although lower voltages applied immediately after slaughter are also effective (e.g. 50V 60 Hz, Koh *et al.*, 1987).

The tenderising effect of electrical stimulation has been explained on the basis of muscle energy depletion and prevention of cold shortening, actual fibre rupture so as to introduce breaks in the tissue to assist conditioning and enhanced activity of proteolytic enzymes as pH falls (Marsh, 1981). However, the beneficial effects have not always been observed in practice. Several studies conducted in New Zealand have shown that it is necessary to apply electrical stimulation as part of a strictly controlled 'accelerated conditioning specification' in order for the procedure to be routinely effective (Meat Industry Research Institute of New Zealand, 1985).

The amount of fat in the carcass, particularly on the outside (subcutaneous) and the weight of the carcass, are important in controlling heat loss during chilling. James and Bailey (1986) estimated that for 140 kg sides cooling in air at 0°C and 0·5 m/sec, cooling times of the fattest and leanest carcasses could be respectively 20% above or below the average. Several authors have suggested that the perceived beneficial effect of fatness on tenderness is due to its insulating role against the effects of rapid chilling or cold shortening. Both Tatum (1981) and Dikeman (1987), in reviews of published work, concluded that 7·6 mm fat at the 12th rib position was necessary to prevent cold shortening in beef. A key paper which supports this insulating role for fat in the enhancement of tenderness in lamb carcasses is that of Smith *et al.* (1976) (Table 18). Forty crossbred lambs were selected live to have carcasses of thick, intermediate or thin levels of finish. They were all approximately the same age (5–9 months) but as usually happens in studies of this kind, the thick finish group were considerably heavier (muscle mass as well as fat cover being important in heat retention). All carcasses were placed in a chiller at 0°C within 40 min of slaughter. The central temperature of *m.longissimus* fell below 10°C at 2 h *post mortem* in the thin finish group and was below 1°C at 6 h (Table 18). The differences in muscle temperature were

Table 18
Effects of finish (subcutaneous fat thickness) on tenderness of lamb muscles after rapid chilling

	Finish group		
	Thick	*Intermediate*	*Thin*
Carcass weight (kg)	28·6[a]	21·5[b]	16·8[c]
Fat thickness 12th rib (mm)	7·1[a]	3·3[b]	1·1[c]
m.longissimus temperature (°C)[d]	3·6[a]	0·9[a]	0·7[a]
Tenderness[e] *m.longissimus*[a]	6·2[a]	6·0[a]	5·0[b]
m.semimembranosus	5·5[a]	5·2[a]	4·0[b]
Toughness (kg)[f] *m.longissimus*	4·6[a]	6·1[b]	7·5[c]
m.semimembranosus	5·8[a]	7·2[b]	7·9[b]

[a,b,c]Means in a row with different superscripts are significantly different ($P < 0{\cdot}05$).
[d]Centre of muscle at 13th rib 6 h *post mortem.*
[e]Taste panel scores 1 (very tough) to 8 (very tender).
[f]Warner–Bratzler shear press.
Smith *et al.* (1976).

associated with differences in taste panel scores for tenderness and objective measures of toughness with particularly tough meat being produced in the lambs with least fat cover. The importance of fat cover was confirmed by the greater toughness of muscles from the sides of lambs in the intermediate group from which subcutaneous fat had been trimmed. For example, the taste panel score for *m. longissimus* tenderness was 6·0 in the untrimmed and 5·0 in the trimmed sides ($P < 0{\cdot}05$).

In a study of beef carcasses, Koohmaraie *et al.* (1988) could find no such effect of fat cover on toughness after rapid cooling. After slaughter, carcasses were split, one side remaining intact and the other trimmed of subcutaneous fat. Sides were then either conditioned at a high temperature (26°C) for 6 h or chilled at 0°C for 24 h. Muscles in the rapidly chilled sides, particularly those from which fat had been removed, cooled very rapidly so that toughening would have been expected. However, no effect of fat cover or rapid chilling on toughness was observed.

Other workers have observed significant differences in tenderness between muscles from lean and fat carcasses which could not be explained by differences in the rate of cooling. Lochner *et al.* (1980) and Lee & Ashmore (1985) both found lower tenderness/higher toughness in *longissimus* muscles from lean carcasses which were slowly cooled so as to avoid cold shortening than from fatter carcasses either slowly or rapidly chilled. Both reports suggested that the system of production was critical, lean carcasses having been produced by feeding forage (i.e. low energy) rather than grain-based (i.e. high energy) diets (see Table 13).

Table 19
Effects of carcass fat thickness and low voltage electrical stimulation on toughness in beef *m. longissimus* muscles

Unstimulated controls			*Stimulated*[a]		
Fat depth[b] *(mm)*	*Toughness*[c]	*Carcass wt (kg)*	*Fat depth*[b] *(mm)*	*Toughness*[c]	*Carcass wt (kg)*
2·4	10·5	184	2·7	4·5	190
4·8	11·2	201	5·5	5·1	183
12·7	10·7	242	13·0	6·5	240
16·0	11·2	254	18·0	6·1	248

[a] 45 V peak for 40 s, nostril–rectum configuration.
[b] Above *m. longissimus* last rib.
[c] Warner–Bratzler initial yield force (kg).
Powell *et al.* (1986).

The results of a large scale Australian study also question the insulating role of fat in preventing cold shortening (Table 19). Carcasses of different fatness were selected in four abattoirs and chilled rapidly following low-voltage electrical stimulation. Toughness measurements made using a Warner–Bratzler shear press (the initial yield force taken as a measure mainly of myofibrillar toughness) were much lower in stimulated carcasses than in unstimulated controls showing that cold shortening had occurred in the controls. However, there was no effect of fat cover on toughness in either group.

There is therefore good evidence that the amount of subcutaneous fat affects the cooling rate of muscles and is a major factor in the degree of cold shortening. The effect of fat is important when it has prevented muscle temperatures falling to a critical level (< 10°C) whilst they are still able to contract (indicated by a pH > 6·0). There is clearly a limit to this insulating effect and in some circumstances chilling has been so rapid that the effect has not been observed.

Rate of Chilling: Pigs

It was formerly believed that cold shortening could not occur in pig muscles because of the fast rate of pH fall *post mortem* compared with beef and the generally thicker layer of subcutaneous fat. However, the effect has now been demonstrated in isolated muscles cooled rapidly (Dransfield & Lockyer, 1985) and in carcasses chilled in air at the very low temperatures (−20 to −30°C) of modern tunnel chillers (James *et al.*, 1983; Moller *et al.*, 1987). Problems of toughness in pork arising in Danish factories have been explained on the basis of cold shortening (Barton-Gade *et al.*, 1987).

The deleterious effects of rapid chilling on toughness in pigs are less than in lamb or beef and have not always been observed despite chilling at temperatures of −30°C or below (e.g. Crenwelge *et al.*, 1984). Indeed, rapid chilling has been recommended as a way of avoiding the combination of low pH and high temperature which causes pale, soft, exudative (PSE) pork. Muscles with a fast rate of pH fall are unlikely to cold shorten even under severe chilling conditions (Honikel, 1986).

Electrical stimulation can improve the tenderness of rapidly chilled pork by depleting the energy source for muscle contraction (Gigiel & James, 1984). Pelvic suspension has also been successfully employed to prevent muscles from shortening despite very rapid chilling at −25°C for 65 min within 90 min of slaughter (Moller & Vestergaard, 1986). The

Table 20
Effects of rapid chilling and carcass suspension method (conventional or pelvic) on toughness in pork. Values are kg/cm^2 peak force

		Chilling rate[a]	
		Fast	*Slow*
m.longissimus:	conventional	5·40[b]	4·37[c]
	pelvic	4·65[c]	4·39[c]
m.semimembranosus:	conventional	4·50[b]	4·11[b,c]
	pelvic	4·13[b,c]	3·78[c]
m.biceps femoris:	conventional	5·17[b]	5·14[b]
	pelvic	4·57[c]	3·90[d]

[a] Fast chilling in tunnel at −18°C after dressing, then in chill room at 2–4°C until 24 h *post mortem.* Slow chilling in chill room at 2–4°C for 24 h.
[b,c,d] Means within a muscle with different superscripts are significantly different ($P < 0{\cdot}05$).
Moller *et al.* (1987).

results of a recent study are shown in Table 20. Fast chilling toughened all three muscles investigated including *m.biceps femoris* (which does not usually toughen in beef) and the effect was partly, although not completely, overcome by pelvic hanging.

Abattoir effects on tenderness associated with different chilling regimes were found by Barton-Gade *et al.* (1987). The abattoir producing the toughest meat (abattoir 1, Table 21) used very rapid chilling but this was not obviously different from abattoir 3 which produced the most tender meat. Cold shortening occurred in abattoir 1 as shown by the shorter sarcomere lengths and this may have been due to conditions in the chilling tunnel (carcasses experienced a much greater air velocity). This report showed the importance of marbling fat and muscle pH in tenderness in all factories. The toughest pork occurred when pH_{45} exceeded 6·5 and total lipid was less than 1·3%. The most tender pork occurred in the pH range 6·1–6·5 and above 2·0% total lipid.

Conditioning/Ageing of Meat

The tenderising effect of ageing is due to the action of proteolytic enzymes, the cathepsins and calpains. These weaken myofibrils in the

Table 21
Tenderness of grilled pork chops obtained from different abbatoirs employing different chilling systems

	Abattoir			
	1	2	3	*4*
Time (temp.) of chilling	47 min (−25·5°C)	127 min (−10°C)	60 min (−24°C)	25 h (−10/+4°C)
n	38	30	37	32
pH_1[d]	6·35[b]	6·37[b]	6·45[a]	6·43[a]
Shear force value[e]	117·8[c]	87·4[b]	73·5[a]	85·6[b]
Tenderness[f]	−0·51[c]	1·86[a,b]	2·27[a]	1·59[b]
Chops with unacceptable eating quality (%)	57	12	9	11
Sarcomere length (μm)	1·75[c]	1·86[b]	1·86[b]	1·90[a]

[a,b,c]Means within a row with different superscripts are significantly different ($P < 0{\cdot}05$).
[d]In *m.longissimus*, 45 min after slaughter.
[e]See Tables 6 and 10.
[f]Taste panel scores −5 (very tough) to +5 (very tender).
Barton-Gade *et al.* (1987).

region of the Z-line and introduce breaks into the muscle structure causing a reduction in toughness (Etherington *et al.*, 1987; Mikami *et al.*, 1987). Since cathepsins are much more active at low pH it has generally been believed that ageing begins at rigor. Post-rigor ageing occurs at different rates in different species. Dransfield *et al.* (1981) concluded that 80% of ageing is normally complete in 5 days in pork, 9 days in beef and 2 days in chicken. Chicken proteases are particularly effective tenderising enzymes.

Post-rigor conditioning cannot overcome the toughening caused by severe cold shortening (Taylor *et al.*, 1972) but can be beneficial if cold shortening has had less severe effects, in lamb (Shorthose *et al.*, 1986) and pork (Brown *et al.*, 1988).

Table 22
Effects of high temperature conditioning on tenderness in lean and fat beef carcasses

	Lean carcasses		*Fat carcasses*	
	High temp.	*Low temp.*	*High temp.*	*Low temp.*
Lee & Ashmore (1985)				
Temperature regime	35°C for 3 h then 0°C to 24 h	0°C for 24 h	Same as for lean carcasses	
Fat thickness (12th rib, mm)	3·6		12·6	
Toughness[e]	8·13[a]	8·35[a]	7·83[a]	6·88[b]
Tenderness[f]	5·17[a]	4·86[a]	4·99[a]	5·56[b]
Lochner et al. *(1980)*				
Temperature regime	9°C for 7·5 h then −2°C to 24 h	−2°C for 24 h	Same as for lean carcasses	
Fat thickness (12th rib, mm)	5·0		26·0	
Toughness[g]	7·4[b]	8·8[a]	6·7[c]	6·2[c]
Tenderness[h]	2·9[b]	2·6[a]	3·8[d]	4·1[c]

[a,b,c,d]Means in a row with different superscripts are significantly different ($P < 0{\cdot}05$).
[e]Warner–Bratzler shear force/cm^2 after ageing at 4°C for 7 days.
[f]Scale 1 (very tough) to 8 (very tender).
[g]Warner–Bratzler shear force/cm^2 after ageing for 2 days.
[h]Scale 1 (very tough) to 7 (very tender).

Marsh (1981, 1986) showed that tenderising could also occur in pre-rigor muscle if carcasses were held at high temperatures (10–30°C) for a short time (4–6 h) after slaughter. This could be due to the more rapid pH fall which occurs at high temperatures (Cassens & Newbold, 1967) which could either protect against any subsequent effect of rapid chilling or release and activate lysosomal enzymes. Moeller *et al.* (1976) showed that holding beef carcasses for 4 h at 22°C post-slaughter increased the activity of the cathepsin C enzyme compared with control sides held in a 2°C cold room.

Against this evidence of beneficial effects of 'high temperature conditioning', other workers have observed toughening effects (Lee & Ashmore, 1985; Koh *et al.*, 1987; Koohmaraie *et al.*, 1988). This could be due to 'rigor shortening', muscle shortening having originally been shown to occur above a critical temperature range as well as below it (Locker & Hagyard, 1963).

A positive effect of carcass fatness during ageing could occur if high early post-mortem temperatures were indeed beneficial (Marsh, 1981). Thus fatter and bulkier carcasses would retain higher temperatures longer, encouraging a faster pH fall and enhanced proteolytic enzyme activity. However, there is no direct evidence of this effect of fatness. In two American studies (Table 22) high temperature conditioning actually produced tougher meat in fat carcasses but had the opposite effect in lean carcasses.

CONCLUSIONS: DIRECT AND INDIRECT EFFECTS OF FATNESS ON MEAT QUALITY

In general it can be said that fatter meat has higher quality in terms of presentational and particularly eating characteristics. This is true whether fatness variations are achieved by changing breeds, sexes or feeding systems on farms. Part of this effect is direct and occurs however fatness is manipulated but some is indirect because other factors, perhaps more important, are changing simultaneously with fatness.

Direct Effects of Fatness

Fat within the muscle (marbling fat) affects juiciness and tenderness. Levels of 2–3% ether-extractable lipid in loin (*m.longissimus*) steaks have been shown to be necessary for optimum tenderness and juiciness in beef and lamb. These levels are easily achievable, even in meat with low levels of separable fat. A concentration of 2–3% extractable lipid in *m. longissimus* has also been suggested as necessary in pork although it is frequently not achieved in the white skinned European breeds at fat thickness levels acceptable to most consumers. Research conducted in the UK, shows that pork with highly acceptable juiciness and tenderness can be found at 0·8–1·0% extractable lipid, levels around 1% being preferred.

Indirect Effects of Fatness

There are several examples of changes in production practices lowering quality to different degrees whilst simultaneously lowering fatness. In these cases the reduction in fatness is not responsible for the reduction in quality. For example:

Bulls are leaner than steers but may have lower tenderness due to changes in collagen.
Entire male pigs are leaner than castrates but may have poorer flavour and softer fat. Tenderness is unaffected.
Feeding on low-energy diets or at a restricted level has often reduced tenderness by an amount greater than expected from the reduction in fatness.
β-Agonists produce leaner carcasses but slightly tougher meat. Pigs homozygous (and less obviously, heterozygous) for the halothane gene are lean but produce PSE muscle which tends to be tough and dry.

A major indirect of fatness occurs during processing. Fat on the outside of the carcass (subcutaneous fat) affects the cooling rate of muscles, especially superficial muscles such as *m.longissimus,* and this is critical when rapid chilling systems are used. Toughness due to cold shortening develops when a particular combination of pH (> 6·0) and temperature (< 10°C) conditions is achieved and can be avoided.

Effects on Quality which are Independent of Fatness

Other factors, unconnected with fatness, also regulate quality:

Age has small effects through reducing collagen solubility. Preslaughter stress generally produces dark meat (high ultimate pH)

which has poor flavour but possibly high tenderness.
Pre- or post-rigor ageing/conditioning have important effects on tenderness.

Conclusions

In the commercial situation some or all of these factors will be present which explains why a clear-cut effect of fatness is frequently not observed. Overall, the positive effects of fatness are small, especially in the upper range, and are not large enough to counteract the clear public demand for meat with low levels of fat, for health and other reasons.

The involvement of so many factors in meat quality, especially in tenderness, implies that quality can be manipulated by making changes on the farm, in the abattoir and during processing. The overall level of quality can be raised and variability reduced even in very lean meat by carefully controlling inputs and events during these stages.

REFERENCES

Aberle, E. D., Reeves, E. S., Judge, M. D., Hunsley, R. E. & Perry, T. W. (1981). Palatability and muscle characteristics of cattle with controlled weight gain : time on a high energy diet. *J. Anim. Sci.,* **52,** 757–63.

Arthaud, V. H., Mandigo, R. W., Koch, R. M. & Kotula, A. W. (1977). Carcass composition, quality and palatability attributes of bulls and steers fed different energy levels and killed at four ages. *J. Anim. Sci.,* **44,** 53–64.

Ashgar, A. & Pearson, A. M. (1980). Influence of ante- and postmortem treatments upon muscle composition and meat quality. *Advances in Food Research,* **26,** 54–213.

Barton-Gade, P. A. (1984). Influence of halothane genotype on meat quality in pigs subjected to various pre-slaughter treatments. *Proceedings of the 30th European Meeting of Meat Research Workers,* pp. 8–9.

Barton-Gade, P. A. (1987). Meat and fat quality in boars, castrates and gilts. *Livestock Prod. Sci.,* **16,** 187–96.

Barton-Gade, P. A., Bejerholm, C. & Borup, U. (1987). Influence of different chilling procedures on the eating quality of pork chops. *Proceedings of the 33rd International Congress of Meat Science and Technology,* pp. 181–4.

Batcher, D. M. & Dawson, E. H. (1960). Consumer quality of selected muscles of raw and cooked pork. *Food Technol.,* **14,** 69–73.

Bejerholm, C. (1984). Experience in taste testing fresh pork at the Danish Meat Research Institute. *Proceedings of the 30th European Meeting of Meat Research Workers,* pp. 196–7.

Bejerholm, C. & Barton-Gade, P. A. (1986). Effect of intramuscular fat level on eating quality of pig meat. *Proceedings of the 32nd European Meeting of Meat Research Workers,* pp. 389–91.

Bendall, J. R., Ketteridge, C. C. & George, A. R. (1976). The electrical stimulation of beef carcasses. *J. Sci. Food Agric.,* **27,** 1123–31.

Bensadoun, A. & Reid, J. T. (1965). Effect of physical form, composition and level of intake of diet on the fatty acid composition of the sheep carcass. *J. Nutr.,* **87,** 239–44.

Blumer, T. N. (1963). Relationship of marbling to the palatability of beef. *J. Anim. Sci.,* **22,** 771–8.

Boccard, R. L. (1978). Development of connective tissue and its characteristics. In *Patterns of Growth and Development in Cattle,* eds H. de Boer & J. Martin. Martinus Nijhoff, The Hague, pp. 73–89.

Boccard, R. L., Naude, R. T., Cronje, D. E., Smit, M. C., Ventner, H. J. & Rossouw, E. J. (1979). The influence of age, sex and breed of cattle on their muscle characteristics. *Meat Sci.,* **3,** 261–80.

Bouton, P. E., Carroll, F. D., Fisher, A. L., Harris, P. V. & Shorthose, W. R. (1973). Effect of altering ultimate pH on bovine muscle tenderness. *J. Food Sci.,* **38,** 816–20.

Bowling, R. A., Smith, G. C., Carpenter, Z. L., Dutson, T. R. & Oliver, W. M. (1977). Comparison of forage-finished and grain-finished beef carcasses. *J. Anim. Sci.,* **45,** 209–15.

Broad, T. E. & Davies, A. S. (1980). Pre- and post-natal study of the carcass growth of sheep. 1. Growth of dissectible fat and its chemical components. *Anim. Prod.,* **31,** 63–71.

Broad, T. E. & Davies, A. S. (1981). Pre- and post-natal study of the carcass growth of sheep. 3. Growth of dissectible and chemical components of muscle and changes in the muscle : bone ratio. *Anim. Prod.,* **32,** 235–43.

Brown, S. N., Bevis, E. A. & Warriss, P. D. (1989). An estimate of the incidence of dark cutting beef in the UK. *Meat Sci.,* **27,** 249–58.

Brown, T., Gigiel, A. J., Swain, V. L. & Higgins, J. A. (1988). Immersion chilling of hot cut, vacuum packed pork primals. *Meat Sci.,* **22,** 173–88.

Butler-Hogg, B. W., Francombe, M. A. & Dransfield, E. (1984). Carcass and meat quality of ram and ewe lambs. *Anim. Prod.,* **39,** 107–13.

Callow, E. H. (1948). Comparative studies of meat. II. The changes in the carcass during growth and fattening and their relation to the chemical composition of the fatty and muscular tissues. *J. Agric. Sci., Camb.,* **38,** 174–99.

Callow, E. H. (1961). Comparative studies of meat. VIII. The percentage of fat in the fatty and muscular tissues of steers and the iodine number of the extracted fat, as affected by breed and level of nutrition. *J. Agric. Sci., Camb.,* **58,** 295–307.

Campbell, R. G., Steel, N. C., Caperna, T. J., McMurtry, J. P., Solomon, M. B. & Mitchell, A. D. (1988). Interrelationships between energy intake and endogenous porcine growth hormone administration on the performance, body composition and protein and energy metabolism of growing pigs weighing 25 to 55 kilograms live weight. *J. Anim. Sci.,* **66,** 1643–55.

Campion, D. R., Crouse, J. D. & Dikeman, M. E. (1975). Predictive value of USDA beef quality grade factors for cooked meat palatability. *J. Food Sci.,*

40, 1225–8.
Carpenter, Z. L. & King, G. T. (1965). Tenderness of lamb rib chops. *Food Technol.,* **19,** 1706–8.
Cassens, R. G. & Newbold, R. P. (1967). Temperature dependence of pH changes in ox muscle post-mortem. *J. Food Sci.,* **32,** 13–14.
Cole, J. W., Ramsey, C. B., Hobbs, C. S. & Temple, R. S. (1964). Effects of type and breed of British, Zebu and Dairy cattle on production, carcass composition and palatability. *J. Dairy Sci.,* **47,** 1138–55.
Crenwelge, D. D., Terrell, R. N., Dutson, T. R., Smith, G. C. & Carpenter, Z. L. (1984). Effects of time post-mortem of electrical stimulation and post-mortem chilling method on pork quality and palatability traits. *J. Food Sci.,* **49,** 294–7.
Cross, H. R., Smith, G. C. & Carpenter, Z. L. (1972). Palatability of individual muscles from ovine leg steaks as related to chemical and histological traits. *J. Food Sci.,* **37,** 282–5.
Cross, H. R., Carpenter, Z. L. & Smith, G. C. (1973). Effects of intramuscular collagen and elastin on bovine muscle tenderness. *J. Food Sci.,* **38,** 998–1003.
Cross, H. R., Berry, B. W. & Wells, L. H. (1980). Effects of fat level and source on the chemical, sensory and cooking properties of ground beef patties. *J. Food Sci.,* **45,** 791–3.
Cross, H. R., Crouse, J. D. & MacNeil, M. D. (1984). Influence of breed, sex, age and electrical stimulation on carcass and palatability traits of three bovine muscles. *J. Anim. Sci.,* **58,** 1358–65.
Crouse, J. D., Seideman, S. C. & Cross, H. R. (1983). The effects of carcass electrical stimulation and cooler temperature on the quality and palatability of bull and steer beef. *J. Anim. Sci.,* **56,** 81–90.
Davies, A. S. & Pryor, W. J. (1977). Growth changes in the distribution of dissectible and intramuscular fat in pigs. *J. Agric. Sci., Camb.,* **89,** 257–66.
DeVol, D. L., McKeith, F. K., Bechtel, P. J., Novakovski, J., Shanks, R. D. & Carr, T. R. (1988). Variation in composition and palatability traits and relationships between muscle characteristics and palatability in a random sample of pork carcasses. *J. Anim. Sci.,* **66,** 385–95.
Dikeman, M. E. (1987). Fat reduction in animals and the effects on palatability and consumer acceptance of meat products. *Proceedings of the 40th Reciprocal Meat Conference,* pp. 93–103.
Dransfield, E. (1977). Intramuscular composition and texture of beef muscles. *J. Sci. Food Agric.,* **28,** 833–42.
Dransfield, E. (1981). Eating quality of DFD beef. In *The Problem of Dark Cutting in Beef,* eds D. E. Hood & P. V. Tarrant. Martinus Nijhoff, The Hague, pp. 344–58.
Dransfield, E. & Lockyer, D. K. (1985). Cold-shortening toughness in excised pork. *Meat Sci.,* **13,** 19–32.
Dransfield, E., Nute, G. R., MacDougall, D. B. & Rhodes, D. N. (1979). Effect of sire breed on eating quality of cross-bred lambs. *J. Sci. Food Agric.,* **30,** 805–8.
Dransfield, E., Jones R. C. D. & MacFie, H. J. H. (1981). Tenderising in *m. longissimus dorsi* of beef, veal, rabbit, lamb and pork. *Meat Sci.,* **5,** 139–47.
Dransfield, E., Nute, G. R. & Francombe, M. A. (1984). Comparison of eating

quality of bull and steer beef. *Anim. Prod.*, **39,** 37–50.

Dransfield, E., Nute, G. R., Mottram, D. S., Rowan, T. G. & Lawrence, T. L. J. (1985). Pork quality from pigs fed on low glucosinate rapeseed meal: influence of level in the diet, sex and ultimate pH. *J. Sci. Food Agric.*, **36,** 546–56.

Dryden, F. D. & Marchello, J. A. (1970). Influence of total lipid and fatty acid composition upon the palatability of three bovine muscles. *J. Anim. Sci.*, **31,** 36–41.

Duniec, H., Kielanowski, J. & Osinska, Z. (1961). Heritability of chemical fat content in the loin muscle of baconers. *Anim. Prod.*, **3,** 195–8.

Edwards, S. A., Wood, J. D., Moncrieff, C. B. & Porter, S. J. (1990). Comparison of the Duroc and Large White as terminal sire breeds and their effect on pigmeat quality. *Anim. Prod.* (In press.)

Emmons, D. B., Kalab, M., Larmond, E. & Lowrie, R. J. (1980). Milk gel structure. X. Texture and microstructure in cheddar cheese made from whole milk and from homogenized low-fat milk. *J. Texture Studies,* **11,** 15–34.

Etherington, D. J., Taylor, M. A. J. & Dransfield, E. (1987). Conditioning of meat from different species. Relationship between tenderising and the levels of cathepsin B, cathepsin L, calpain 1, calpain 11 and β-glucuronidase. *Meat Sci.*, **20,** 1–18.

Field, R. A. (1971). Effect of castration on meat quality and quantity. *J. Anim. Sci.*, **32,** 849–58.

Fisher, A. V., Wood, J. D., Stevens, G. & Robelin, J. (1983). The relationships between carcass fatness and the lipid and protein content of beef. *Proceedings of the 29th European Meeting of Meat Research Workers,* pp. 48–54.

Fjelkner-Modig, S. (1985). *Sensory and biophysical properties of pork.* Swedish Meat Research Institute, Kavlinge.

Gerrard, D. E., Jones, S. J., Aberle, E. D., Lemenager, R. P., Dikeman, M. E. & Judge, M. D. (1987). Collagen stability, testosterone secretion and meat tenderness in growing bulls and steers. *J. Anim. Sci.*, **65,** 1236–42.

Gigiel, A. J. & James, S. J. (1984). Electrical stimulation and ultra-rapid chilling of pork. *Meat Sci.*, **11,** 1–12.

Griffiths, B. (1985). The role of fats in foods. *Food,* March, pp. 43–5.

Harris, P. V. (1976). Structural and other aspects of meat tenderness. *J. Texture Studies,* **7,** 49–63.

Hinch, G. N. & Thwaites, C. J. (1982). Chemical and histological parameters associated with the tenderness of hemicastrate and steer beef. *Proceedings of the New Zealand Society of Animal Production,* **42,** 131–2.

Hiner, R. L., Thornton, J. W. & Alsmeyer, R. H. (1965). Palatability and quantity of pork as influenced by breed and fatness. *J. Food Sci.*, **30,** 550–5.

Honikel, K. O. (1986). Influence of chilling on biochemical changes and quality of pork. In *Recent Advances and Developments in the Refrigeration of Meat by Chilling.* International Institute of Refrigeration, Paris, pp. 45–51.

James, S. J. & Bailey, C. (1986). Temperature changes, weight loss and product loads during beef chilling. In *Recent Advances and Developments in the Refrigeration of Meat by Chilling.* International Institute of Refrigeration, Paris, pp. 105–14.

James, S. J., Gigiel, A. J. & Hudson, W. R. (1983). The ultra rapid chilling of pork. *Meat Sci.*, **8,** 63–78.

Jensen, P., Craig, H. B. & Robison, O. W. (1967). Phenotypic and genetic associations among carcass traits of swine. *J. Anim. Sci.*, **26,** 1252–60.

Johnson, D. D., Lunt, D. K., Savell, J. W. & Smith, G. C. (1988). Factors affecting carcass characteristics and palatability of young bulls. *J. Anim. Sci.*, **66,** 2568–77.

Johnson, E. R. (1987). Marbling fat in beef. *Meat Sci.*, **20,** 267–79.

Jones, S. D. M., Schaeffer, A. L., Tong, A. K. W. & Vincent, B. C. (1988). The effects of fasting and transportation in beef cattle. 2. Body component changes, carcass composition and meat quality. *Livestock Prod. Sci.*, **20,** 25–35.

Kay, R. N. B., Sharman, G. A. M., Hamilton, W. J., Goodall, E. D., Pennie, K. & Coutts, A. G. P. (1981). Carcass characteristics of young red deer farmed on hill pasture. *J. Agric. Sci., Camb.*, **96,** 79–87.

Kemp, J. D., Mahyuddin, M., Ely, D. G., Fox, J. D. & Moody, W. G. (1981). Effect of feeding systems, slaughter weight and sex on organoleptic properties and fatty acid composition of lamb. *J. Anim. Sci.*, **51,** 321–30.

Kempster, A. J., Cook, G. L. & Grantley-Smith, M. (1986*a*). National estimates of the body composition of British cattle, sheep and pigs with special reference to trends in fatness. A review. *Meat Sci.*, **17,** 107–38.

Kempster, A. J., Dilworth, A. W., Evans, D. G. & Fisher, K. D. (1986*b*). The effects of fat thickness and sex on pig meat quality with special reference to the problems associated with overleanness. 1. Butcher and consumer panel results. *Anim. Prod.*, **43,** 517–33.

Kirkegaard, E., Moller, A. J. & Wismer-Pedersen, J. (1979). Relationship between fat content, connective tissue and objective tenderness measurement in porcine longissimus dorsi. *Proceedings of the 25th European Meeting of Meat Research Workers,* pp. 311–17.

Kirton, A. H., Winger, R. J., Dobbie, J. L. & Duganzich, D. M. (1983). Palatability of meat from electrically stimulated carcasses of yearling and older entire male and female sheep. *J. Food Technol.*, **18,** 638–49.

Koh, K. C., Bidner, T. D., McMillin, K. W. & Hill, G. M. (1987). Effects of electrical stimulation and temperature on beef quality and tenderness. *Meat Sci.*, **21,** 189–201.

Koohmaraie, M., Seideman, S. C. & Crouse, J. D. (1988). Effect of subcutaneous fat and high temperature conditioning on bovine meat tenderness. *Meat Sci.*, **23,** 99–109.

Kregel, K. K., Prusa, K. J. & Hughes, K. V. (1986). Cholesterol content and sensory analysis of ground beef as influenced by fat level, heating and storage. *J. Food Sci.*, **51,** 1162–90.

Larmond, E. & Moran, E. T. (1983). Effect of finish grade and internal basting of the breast with oil on sensory evaluation of small white toms. *Poult. Sci.*, **62,** 1110–12.

Lawrie, R. A. (1961). Studies on the muscles of meat animals. 1. Differences in composition of beef *longissimus dorsi* muscles determined by age and anatomical location. *J. Agric. Sci., Camb.*, **56,** 249–59.

Leat, W. M. F. (1975). Fatty acid composition of adipose tissue of Jersey cattle during growth and development. *J. Agric. Sci., Camb.*, **85**, 551–8.

Lee, Y. B. & Ashmore, C. R. (1985). Effect of early postmortem temperature on beef tenderness. *J. Anim. Sci.*, **60**, 1588–96.

Liboriussen, T., Andersen, B. B., Buchter, L., Kousgaard, K. & Moller, A. J. (1977). Crossbreeding experiment with beef and dual-purpose sire breeds on Danish dairy cows. IV. Physical, chemical and palatability characteristics of longissimus dorsi and semitendinosus muscles from crossbred young bulls. *Livestock Prod. Sci.*, **4**, 31–43.

Lochner, J. W., Kauffman, R. G. & Marsh, B. B. (1980). Early post-mortem cooling rate and beef tenderness. *Meat Sci.*, **4**, 227–41.

Locker, R. H. & Hagyard, C. J. (1963). A cold shortening effect in beef muscles. *J. Sci. Food Agric.*, **14**, 787–93.

Malmfors, B. & Lundstrom, K. (1983). Consumer reactions to boar meat—a review. *Livestock Prod. Sci.*, **10**, 187–96.

Marsh, B. B. (1981). Properties and behaviour of prerigor meat. *Proceedings of the 34th Reciprocal Meat Conference*, pp. 75–80.

Marsh, B. B. (1986). The tenderizing mechanisms of electrical stimulation. In *Recent Advances and Developments in the Refrigeration of Meat by Chilling*. International Institute of Refrigeration, Paris, pp. 75–80.

Marsh, B. B. & Leet, N. G. (1966). Studies in meat tenderness. III. The effects of cold shortening on tenderness. *J. Food Sci.*, **31**, 450–9.

Marsh, B. B., Woodhams, P. R. & Leet, N. G. (1968). Studies in meat tenderness. V. The effects on tenderness of carcass cooling and freezing before the completion of rigor mortis. *J. Food Sci.*, **33**, 12–18.

McGloughlin, P., Allen, P., Tarrant, P. V. & Joseph, R. L. (1988). Growth and carcass quality of crossbred pigs sired by Duroc, Landrace and Large White boars. *Livestock Prod. Sci.*, **18**, 275–88.

Meat and Livestock Commission (1983). *Very Lean Pigs. Planning and Development Team Report.* Meat and Livestock Commission, Milton Keynes.

Meat and Livestock Commision (1989). *Stotfold Pig Development Unit. Results of the First Trial*. Meat and Livestock Commission, Milton Keynes.

Meat Industry Research Institute of New Zealand (1985). Accelerated Conditioning. Bulletin No. 11. Meat Industry Research Institute of New Zealand Inc.

Medeiros, L. C., Field, R. A., Menkhaus, D. J. & Russell, W. C. (1987). Evaluation of range-grazed and concentrate-fed beef by a trained sensory panel, a household panel and a laboratory test market group. *J. Sensory Studies,* **2**, 259–72.

Mikami, M., Whiting, A. H., Taylor, M. A. J., Maciewicz, R. A. & Etherington, D. J. (1987). Degradation of myofibrils from rabbit, chicken and beef by cathepsin L and lysosomal lysates. *Meat Sci.*, **21**, 81–97.

Moeller, P. W., Fields, P. A., Dutson, T. R., Landmann, W. A. & Carpenter, Z. L. (1976). Effect of high temperature conditioning on subcellular distribution and levels of lysosomal enzymes. *J. Food Sci.*, **41**, 216–17.

Moller, A. J. & Vestergaard, T. (1986). Effects of altered carcass suspension during rigor mortis on tenderness of pork loin. *Meat Sci.*, **18**, 77–87.

Moller, A. J., Sorensen, S. E. & Larsen, M. (1981). Differentiation of myofibrillar and connective tissue strength in beef muscles by Warner–Bratzler shear parameters. *J. Texture Studies,* **12,** 71–83.

Moller, A. J., Kirkegaard, E. & Vestergaard, T. (1987). Tenderness of pork muscles as influenced by chilling rate and altered carcass suspension. *Meat Sci.,* **21,** 275–86.

Moody, W. G. & Cassens, R. G. (1968). A quantitative and morphological study of bovine *longissimus* fat cells. *J. Food Sci.,* **33,** 47–52.

Moore, V. J. & Bass, J. J. (1978). Palatability of cross-bred beef. *J. Agric. Sci., Camb.,* **90,** 93–5.

Moran, E. T. (1986). Manipulation of fat characteristics in animal products. In *Recent Advances in Animal Nutrition—1986,* eds W. Haresign & D. J. A. Cole. Butterworths, London, pp. 31–45.

Mortensen, A. B., Bejerholm, C. & Pedersen, J. K. (1986). Consumer test of meat from entire males, in relation to skatole in backfat. *Proceedings of the 32nd European Meeting of Meat Research Workers,* pp. 23–6.

Mottram, D. S. & Edwards, R. A. (1983). The role of triglycerides and phospholipids in the aroma of cooked meat. *J. Sci. Food Agric.,* **34,** 517–22.

Mottram, D. S., Wood, J. D. & Patterson, R. L. S. (1982). Comparison of boars and castrates for bacon production. 3. Composition and eating quality of bacon. *Anim. Prod.,* **35,** 75–80.

Murphy, M. O. & Carlin, A. F. (1961). Relation of marbling, cooking yield, and eating quality of pork chops to backfat thickness on hog carcasses. *Food Technol.,* **15,** 57–63.

Newbold, R. P. & Harris, P. V. (1972). The effect of pre-rigor changes on meat tenderness. A review. *J. Food Sci.,* **37,** 337–40.

Powell, V. H., Harris, P. V. & Shorthose, W. R. (1986). Beef tenderness—Australia, 1985. *Food Technol. in Australia,* **38,** 230–3.

Purchas, R. W., O'Brien, L. E. & Pendleton, C. M. (1979). Some effects of nutrition and castration on meat production from male Suffolk cross (Border Leicester—Romney Cross) Lambs. II. Meat quality. *New Zealand J. Agric. Res.,* **22,** 375–83.

Purslow, P. P. (1985). The physical basis of meat texture: observations on the fracture behaviour of cooked bovine *m.semitendinosus. Meat Sci.,* **12,** 39–60.

Rhodes, D. N. (1970). Meat quality: Influence of fatness of pigs on the eating quality of pork. *J. Sci. Food Agric.,* **21,** 572–5.

Rhodes, D. N. (1971). A comparison of the quality of meat from lambs reared intensively indoors and conventionally on grass. *J. Sci. Food Agric.,* **22,** 667–8.

Rhodes, D. N. (1972). Consumer testing of pork from boar and gilt pigs. *J. Sci. Food Agric.,* **23,** 1483–91.

Rhodes, D. N., Jones, R. C. D., Chrystall, B. B. & Harries, J. M. (1972). Meat texture. II. The relationship between subjective assessments and a compressive test on roast beef. *J. Texture Studies,* **3,** 298–309.

Riley, R. R., Savell, J. W., Murphey, C. E., Smith, G. C., Stiffler, D. M. & Cross, H. R. (1983). Effects of electrical stimulation, subcutaneous fat thickness and masculinity traits on palatability of beef from young bull carcasses. *J.*

Anim. Sci., **56,** 584–91.
Robertson, J., Ratcliff, D., Bouton, P. E., Harris, P. V. & Shorthose, W. R. (1986). A comparison of some properties of meat from young buffalo (*Bubalus bubalis*) and cattle. *J. Food Sci.*, **51,** 47–50.
Sahasrabudhe, M. R. & Smallbone, B. W. (1983). Comparative evaluation of solvent extraction methods for the determination of neutral and polar lipids in beef. *J. Amer. Oil Chem. Soc.*, **60,** 801–5.
Salmon, R. E. (1979). Slaughter losses and carcass composition of the Medium White Turkey. *Brit. Poult. Sci.*, **20,** 297–302.
Savell, J. W., Cross, H. R. & Smith, G. C. (1986). Percentage ether extractable fat and moisture content of beef *longissimus* muscle as related to USDA marbling score. *J. Food Sci.*, **51,** 838–40.
Schroeder, J. W., Cramer, D. A., Bowling, R. A. & Cook, C. W. (1980). Palatability, shelflife and chemical differences between forage- and grain-finished beef. *J. Anim. Sci.*, **50,** 852–9.
Schworer, D., Blum, J. K. & Rebsamen, A. (1986). Phenotypic and genetic parameters of intramuscular fat in pigs. *Proceedings of the 32nd European Meeting of Meat Research Workers,* pp. 433–6.
Seideman, S. C., Koohmaraie, M. & Crouse, J. D. (1987). Factors associated with tenderness in young beef. *Meat Sci.*, **20,** 281–91.
Shorthose, W. R., Powell, V. H. & Harris, P. V. (1986). Influence of electrical stimulation, cooling rates and ageing on the shear force values of chilled lamb. *J. Food Sci.*, **51,** 889–928.
Smith, G. C., Carpenter, Z. L., King G. T. & Hoke, K. E. (1970). Lamb carcass quality, II. Palatability of rib, loin and sirloin chops. *J. Anim. Sci.*, **31,** 310–17.
Smith, G. C., Dutson, T. R., Hostetler, R. L. & Carpenter, Z. L. (1976). Fatness, rate of chilling and tenderness of lamb. *J. Food Sci.*, **41,** 748–56.
Smith, G. C., Carpenter, Z. L., Cross, H. R., Murphey, C. E., Abraham, H. C., Savell, J. W., Davis, G. W., Berry, B. W. & Parrish, F. C. (1984). Relationship of USDA marbling groups to palatability of cooked beef. *J. Food Qual.*, **7,** 289–308.
Solomon, M. B., Campbell, R. G., Steele, N. C., Caperna, T. J. & McMurtry, J. P. (1988). Effect of feed intake and exogenous porcine somatotropin on *longissimus* muscle fibre characteristics of pigs weighing 55 kg live weight. *J. Anim. Sci.*, **66,** 3279–84.
Tarrant, P. V. (1987). Review: Muscle biology and biochemistry. *Proceedings of the 33rd International Congress of Meat Science and Technology,* pp. 1–5.
Tatum, J. D. (1981). Is tenderness nutritionally controlled? *Proceedings of the 34th Reciprocal Meat Conference,* pp. 65–7.
Taylor, A. A., Chrystall, B. B. & Rhodes, D. N. (1972). Toughness in lamb induced by rapid chilling. *J. Food Technol.*, **7,** 251–8.
Truscott, T. G. (1980). A Study of the Relationships Between Fat Partition and Metabolism in Hereford and Friesian steers. PhD thesis, University of Bristol.
Truscott, T. G., Wood, J. D. & MacFie, H. J. H. (1983). Fat deposition in Hereford and Friesian steers. 1. Body composition and partitioning of fat between depots. *J. Agric. Sci., Camb.*, **100,** 257–70.
Vickery, J. R. (1977). *Influence on Carcass Lipids of the Condition of Cows at Slaughter.* Division of Food Research and Technology, paper No. 42. Melbourne, Australia, CSIRO.

Wakefield, D. K., Dransfield, E., Down, N. F. & Taylor, A. A. (1989). Influence of post-mortem treatments on turkey and chicken meat texture. *Int. J. Food Sci. Technol.*, **24,** 81–92.

Warriss, P. D. & Brown, S. N. (1985). The physiological responses to fighting in pigs and the consequences for meat quality. *J. Sci. Food Agric.*, **36,** 87–92.

Warriss, P. D., Kestin, S. C., Brown, S. N. & Wilkins, L. J. (1984). The time required for recovery from mixing stress in young bulls and the prevention of dark cutting beef. *Meat Sci.*, **10,** 53–68.

Warriss, P. D., Kestin, S. C., Rolph, T. P. & Brown, S. N. (1990). The effects of the beta-adrenergic agonist salbutamol on meat quality in pigs. *J. Anim. Sci.*, **68,** 128–36.

Webb, A. J., Carden, A. E., Smith, C. & Imlah, P. (1982). Porcine stress syndrome in pig breeding. Proceedings of the 2nd World Congress on Genetics Applied to Livestock Production, vol. 5, pp. 588–608.

West, R. L. & Myer, R. O. (1987). Carcass and meat quality characteristics and backfat fatty acid composition of swine as affected by the consumption of peanuts remaining in the field after harvest. *J. Anim. Sci.*, **65,** 475–80.

Westerling, D. B. & Hedrick, H. B. (1979). Fatty acid composition of bovine lipids as influenced by diet, sex and anatomical location and relationship to sensory characteristics. *J. Anim. Sci.*, **48,** 1343–8.

Wood, J. D. (1984). Fat deposition and the quality of fat tissue in meat animals. In *Fats in Animal Nutrition*, ed. J. Wiseman. Butterworths, London, pp. 407–35.

Wood, J. D. (1989). Meat yield and carcass composition in turkeys. In *Recent Advances in Turkey Science*, eds C. Nixey & T. C. Grey. Butterworths, London, pp. 271–88.

Wood, J. D., Dransfield, E. & Rhodes, D. N. (1979). The influence of breed on the carcass and eating quality of pork. *J. Sci. Food Agric.*, **30,** 493–8.

Wood, J. D., MacFie, H. J. H., Pomeroy, R. W. & Twinn, D. J. (1980). Carcass composition in four sheep breeds: the importance of type of breed and stage of maturity. *Anim. Prod.*, **30,** 135–52.

Wood, J. D., Jones, R. C. D., Francombe, M. A. & Whelehan, O. P. (1986). The effects of fat thickness and sex on pig meat quality with special reference to the problems associated with overleanness. 2. Laboratory and trained taste panel results. *Anim. Prod.*, **43,** 535–44.

Wood, J. D., Edwards, S. A. & Bichard, M. (1988). Influence of the Duroc breed on pigmeat quality. *Proceedings of the 34th International Congress of Meat Science and Technology*, pp. 571–2.

Wood, J. D., Enser, M. B., Whittington, F. M. & Moncrieff, C. B. (1989). The effects of fat thickness and sex on the composition of backfat in pigs. *Livestock Prod. Sci.*, **22,** 351–62.

Zembayashi, M., Nabita, H. & Motutsuji, T. (1988). Effects of breed and nutritional planes on intramuscular lipid deposition of fattening steers. *Japanese J. Zootech. Sci.*, **59,** 39–48.

Chapter 9

Developments in Low-Fat Meat and Meat Products

R. GOUTEFONGEA & J. P. DUMONT
Laboratoire d'Etude des Interactions des Molécules Alimentaires, INRA, Nantes, France

INTRODUCTION

Food availability is quite uneven throughout the world. Whilst most people do not have enough to eat, a minority (mainly in Western Europe and North America) overeat and face health disorders such as obesity, cardiovascular disease, and so on. In developed countries, people are now taking an interest in the composition of their food. They tend to exercise control over their fat intake and to reduce the proportion of animal fats in the diet.

To the layman, it very often appears that meat and meat products contain much more fat than other food commodities. In France, this opinion is commonly perpetrated by the media (TV, magazines, papers) who often rely on a single set of obsolete data for making exaggerated statements about lipids in meat and meat products. No doubt the consumer's call for less fat and cholesterol in his staple food must be satisfied but it would make more sense if the claims were substantiated with actual figures.

As we believe that becoming a vegetarian is not an inevitable choice for most readers of this chapter, we will try to establish a clearer picture of the true intake of animal fats by the consumer when eating meat and meat products.

First of all, meats rather than meat must be considered. Therefore, we will try to review briefly information on the fat content of different cuts obtained from the different breeds and species in different countries.

Considering that studies should focus on the amount of ingested fat rather than on data relative to available fat, we will examine processing and cooking practices in order to trace the dishes and products that clearly deserve fat content reduction. Proposals that intend to meet the specific requirements of different meat products will be put forward and, eventually, a general scheme for lowering fat in meat and meat products will be proposed.

FAT IN MEAT AND MEAT PRODUCTS

Much Information, Little Knowledge

A considerable amount of information on the fat content of foods is available to consumers via booklets, etc., although unfortunately the information on meat is frequently wrong or misleading. For example, data relating to whole carcasses are often used to describe meat purchased at retail. Sometimes, data are borrowed from older tables based on experimental material collected forty or fifty years ago. It is well known that fat content in meat has significantly decreased in the last twenty years. For instance, the fat content of pork has dropped by 30% in Finland and by 15% in the United States (Niinivaara, 1985). According to Kauffman & Breidenstein (1983), carcass fatness has been reduced by about 6% in beef, 23% in pork and 9% in lamb during this period.

Very often, an animal species is associated with only a single value. This is quite unrealistic since, in a given species, breed, sex, age, degree of finish all introduce large differences between animals. This is quite obvious from Tables 1 and 2 which show the effects of breed, sex and age, in the bovine species, under normal French conditions of management. Data from Dikeman and Crouse (1975) indicate that under the conditions prevailing in the USA, the fat proportion in the carcass is much higher. Values as high as 38·5, 35·9, and 31·1% were measured on crossbreed steer carcasses (Hereford×Angus, Simmental×Angus and Limousin×Angus, respectively). Moreover, the large differences in fat content between individual retail cuts taken from the same carcass make single species values of fat content of little significance.

Table 1
Percentage of fat deposits in beef carcasses according to breed and age (Frebling *et al.*, 1982)

Age category	*Limousin*	*Charolais*	*Maine-Anjou*	*Hereford*
Young bull (15–18 months)	12·0	13·2	16·0	20·9
Steer (30–32 months)	12·6	15·3	16·6	—
Young cow (< 5 years)	16·9	19·9	20·3	26·5
Adult cow (> 5 years)	12·2	13·1	13·2	—

Table 2
Percentage of fat deposits in carcasses from young purebred and crossbred bulls (16 months)

Breed	*Holstein*	*Normande*	*Charolais*
Holstein	17·8	17·5	15·3
Normande		15·0	14·8
Charolais			13·2

If the intention of surveys is to obtain a reliable estimation of the consumer's fat intake, then only fat contained in the edible part of meat dishes should be taken into account. Whether meat is eaten fresh in pieces from which fat can be easily removed, or in processed meat products, data dealing solely with carcasses are not likely to supply the information needed.

Location of Fat in Meat

Based on the anatomical location in animal tissues, four different categories of fat can be considered: internal; subcutaneous; intermuscular; and, intramuscular.

Fat distribution between the four categories shows some variations related to such parameters as animal species, breed and degree of finish. Table 3 reports general data relative to fat distribution in bovine and porcine carcasses. It must be emphasized that comparisons should be made

Table 3
Distribution of fat deposits in bovine and porcine carcasses (data summarized from Henry, 1977; Robelin, 1978; Demarne, 1982; Kempster *et al.*, 1982; Renand, 1983)

Species	*% of total fat*			
	Internal fat	*Subcutaneous fat*	*Intermuscular fat*	*Intramuscular fat*
Bovine (young bull 550 kg live weight)	25–30	15–20	50–55	2–3
Porcine (castrated male 100 kg live weight)	5	70–75	20	1–2

at the same live weight or the same degree of finish in order to minimize variations not directly governed by biology. From Kempster *et al.* (1976), Robelin (1978) and Fortin *et al.* (1981) it can be concluded that meat-type breeds of British origin (Angus and Hereford) are characterized by a large extent of subcutaneous fat and little intermuscular and internal fat. Dairy breeds (Holstein, Friesian and Jersey) are characterized by high internal fat contents and an extent of subcutaneous fat intermediate between Angus and Charolais (meat-type of continental origin showing very little fattening). As shown in Fig. 1, it can be considered that, at the same total fat proportion, early maturing breeds show more subcutaneous fat than late maturing breeds, which, in turn, have more intermuscular fat.

There is a proportionality between intramuscular and total fat. On average, intramuscular fat increases by about 1% for each 5% of total fat. Figure 2 illustrates the relationship for cattle produced in two different types of management.

In the course of carcass division into primal cuts, most of the internal fat is removed. Subcutaneous fat is also removed more or less extensively depending on local custom, animal species and the required degree of finish. For instance, in France, excess subcutaneous fat is trimmed off beef carcasses at the slaughterhouse. In many countries, backfat is removed from pork carcasses and used in the formulation of different kinds of meat products but in other cases, the thickness of the fat layer remaining on the

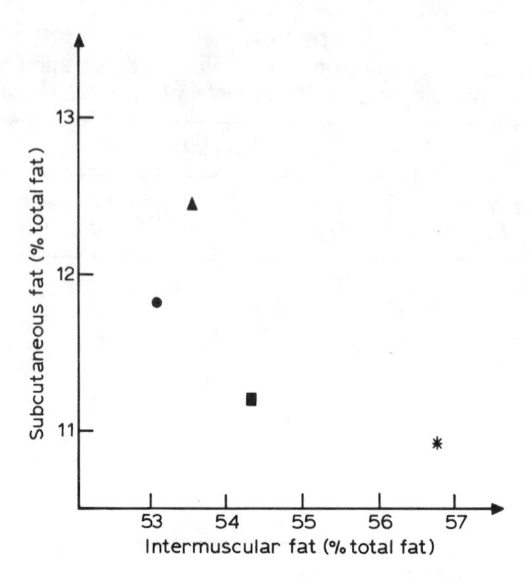

Fig. 1. Distribution of subcutaneous and intermuscular fat in young bull carcasses at the same degree of finish. Total fat = 11% carcass weight (Robelin, 1978). ▲, Limousin; ●, Friesian; ■, Salers; ★, Charolais.

carcass is adjusted according to prevailing local habits. Intermuscular fat is removed partially as retail cuts are prepared. Data on the mean fat content in different retail cuts are shown in Tables 4 and 5. For comparison, data on the lipid content in commercial meat products are reported in Tables 6 and 7.

Intramuscular fat would represent the only fat eaten in lean-only cuts if they were carefully trimmed before cooking. Information showing the broad variation in intramuscular fat is given in Tables 8–11.

SHOULD THE FAT CONTENT OF MEAT AND MEAT PRODUCTS BE LOWERED?

As the lipid fraction is involved in most if not all food properties, the different functions of fat in meat and meat products should be considered.

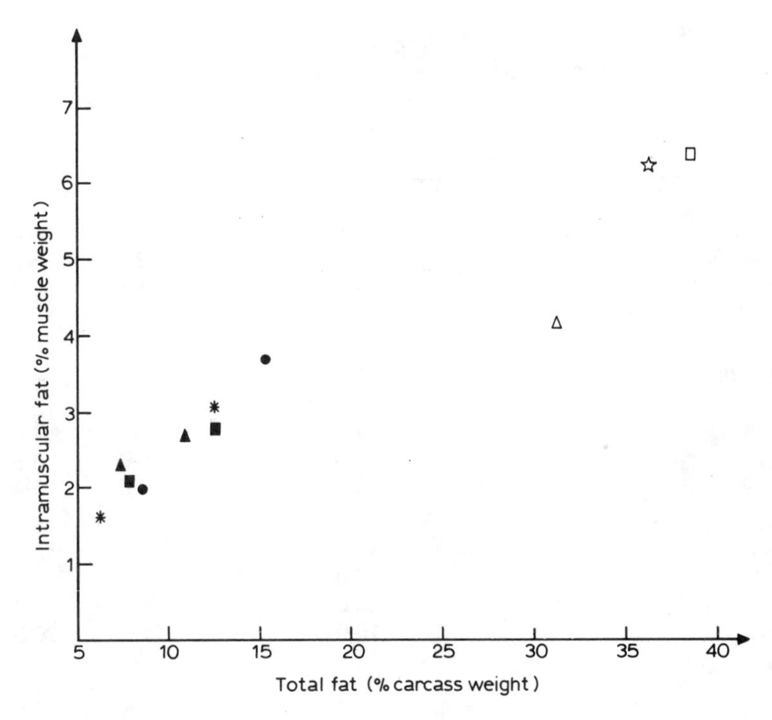

Fig. 2. Relationship between intramuscular fat and total fat. Young bulls (Robelin, 1978): ▲, Limousin, ●, Friesian; ■, Salers; ★, Charolais. Steers (Dikeman & Crouse, 1975): △, Limousin×Angus; ☆, Simmental×Angus; □, Hereford×Angus.

Effect of Fat on Nutrition

Fat has twice the energy content of carbohydrate and protein (38 vs 17 kJ/g dry matter) and has little, if any, water binding ability. The total solid content of fatty foods is commonly higher than in low-fat counterparts that are far less energy dense on a wet weight basis.

Animal fat is also a source of liposoluble substances like vitamins A, D, E and K and also of essential fatty acids although less so than fish oils. It should be remembered that fat composition is very different from one species to another: poultry and corn-fed pigs have high proportions of polyunsaturated fatty acids compared with ruminants whose tissues are higher in saturated fatty acids.

Table 4
Lipid content of retail cuts (Souci *et al.*, 1981)

	Average value % of fresh weight	*Min.*	*Max.*
Beef			
Brisket	21·0	10·0	36·0
Tenderloin	4·4	2·9	6·5
Top round	4·2	1·1	6·3
Sirloin	10·0	4·5	19·5
Veal			
Shoulder	2·9	1·0	5·0
Loin	1·5	1·0	2·0
Neck	2·5	0·8	4·6
Leg	1·6	0·6	3·6
Chop	3·0	1·6	5·4
Pork			
Belly	42·0	28·0	49·0
Loin	10·0	2·2	12·0
Ham	23·0	18·8	26·7
Chop	19·0	18·0	20·0
Lamb			
Loin	3·4	1·9	4·9
Leg	18·0	—	—
Chop	32·0	—	—
Poultry			
Boiling hen	20·0	11·0	25·0
Broiler	5·6	4·5	7·2
Duck	17·2	6·0	28·0
Goose	31·0	26·0	38·0
Turkey	6·8	5·3	8·2

—, no data available.

Effect of Fat on Sensory Properties

Flavour

Three different kinds of fat–flavour interactions can be considered.

First of all, fat is able to absorb hydrophobic flavour compounds either present in the live animal (mutton flavour or boar taint) or formed during

Table 5
Lipid content of retail cuts (Dumont, 1986)

	Average value % fresh weight (n = 14)	*Coefficient of variation*
Beef for stew	8·9	27·0
Beef shank cross cut	8·0	41·3
Pork loin rib chop	33·9	15·2
Blade pork chop	18·4	15·2
Pork belly	19·6	19·9
Veal riblets	8·2	41·5
Veal breast	24·7	50·2
Neck of lamb	20·8	26·0
Lamb rib chop	27·0	23·7
Lamb breast riblets	29·1	27·5

Table 6
Lipid content (%) of some meat products (Souci *et al.*, 1981)

	Average value	*Min.*	*Max.*
Cooked ham	12·8	5·6	23·0
Smoked country-style ham	33·3	25·0	40·0
Luncheon meat	25·4	12·8	34·5
Blockwurst	45·0	42·0	50·0
Frankfurter	24·4	—	—
Mortadella	32·8	29·7	35·8
Wiener	24·4	20·8	25·0
Knackwurst	33·7	25·5	25·0
Liver sausage	41·2	36·5	45·9

processing (roasted and smoked flavour). The binding capacity of lipids as well as the pattern of the subsequent desorption are largely dependent on the composition of the fat fraction and the organization of the solid crystalline network.

Secondly, fat is a known precursor of a large number of flavour compounds (aldehydes, ketones, lactones, volatile fatty acids, secondary alcohols) that can contribute to desired as well as undesired tastes and aromas (rancid and warmed-over flavour).

Table 7
Lipid content (%) of some French meat products (Anon., 1986*b*)

	Maximum authorized value
Sausage for broiling	40
Spreadable sausage	60
Dry sausage	45
Salami	45
Cooked sausage	40
Country-style 'pâté'	40
Liver 'pâté'	45

Thirdly, it has been reported that phospholipids are responsible for noticeable changes in the odour quality of processed meats. In some cases they interfere in the course of the Maillard reaction that develops during cooking (Mottram & Edwards, 1983). Lipids also interact with sodium nitrite during meat curing (Goutefongea *et al.*, 1977).

Colour

Fat is usually low in colour. Mixing fat and lean intimately results in a fading of the original lean meat colour. Using the CIE 1931 colorimetric system (Wyszecki & Stiles, 1982) it appears that whilst the dominant wavelength in reflected light (λ_d) is unchanged, the luminance factor Y is increased and the excitation purity (P_e) is lowered.

In meat and in coarsely ground meat products, the human eye is able to distinguish fat from lean. Visual assessment of the degree of marbling of meat or of the fat content of a meat product can normally be done unless fat has been dyed red with carmine as may happen. Unfortunately, instrumental measurement of colour (reflectance) in meat and in coarsely ground meat is sometimes severely impeded by large fat patches.

Texture

Rheological and structural characteristics must be considered. Tenderness and juiciness are pre-eminent factors in the sensory evaluation of meat texture that tend to increase with increasing meat marbling. Finely ground meat products are often high in fat content, and they usually have a soft texture that appears to be strongly dependent on the physical characteristics of the fat. Frequently, temperature is a critical parameter as illustrated by spreadable products.

Effect of Fat on Technological Properties

The most extensively used functional property of animal fats is the broad variety of their melting patterns. Very different meat products can be obtained by incorporating different fat types: for instance, dry sausage manufacture requires hard fat while meat patties are made with soft fat. One of the main drawbacks is fat sensitivity to oxidation which makes long-term storage of deep-frozen fatty foods difficult. It should also be remembered when positive thermal treatments are involved that severe heating can badly affect fat quality and lead to the formation of potentially hazardous molecules (malonyldialdehyde and benzopyrene for instance).

Maintaining the Balance Between Nutrition and Palatability

In many instances, reducing the fat content of foods will not adversely affect the overall quality of most dishes. However, fat should not be reduced to too low a level since it has been commonly observed that nutritional benefits gained through extensive defatting are more than outweighed by a correlated loss in palatability.

PROPOSALS FOR LOWERING THE FAT CONTENT OF MEAT PRODUCTS

There are two possible approaches to the lowering of fat content in meat and meat products: removal and substitution.

Trimming adipose tissue off the meat is the only action that can be taken when the intrinsic structure of the muscle is expected to be found in the finished food (e.g. steak, roast, ham).

Control of the fat content is much easier when ground meat is involved. In the first stage, meat is trimmed in order to reduce subsequent incorporation of depot fat and ground thereafter. In the latter stage, fat usually

Table 8
Intramuscular fat in beef muscles

Reference	Breed	Age (months)	Sex	Muscle	Intramuscular fat (% fresh muscle)
Link *et al.* (1970)	Angus	12	Castrated and ♀	*L.D.*	2·6
	Angus	22	Castrated and ♀	*L.D.*	13·2
Terrell *et al.* (1969)	Angus	15	Castrated and ♀	*P.M.*	11·4
				L.D.	10·0
				T.A.	9·9
				S.M.	6·2
				T.B.	6·1
				S.T.	6·0
Lawrie (1961)	Ayrshire × Red Poll	16	♂	*P.M.*	2·3
				L.D.	2·0
	Friesian	18	Castrated	*P.M.*	4·5
				L.D.	2·0
	Highland	32	Castrated	*P.M.*	9·5
				L.D.	13·8
Lawrie (1985)	Ayrshire × Red Poll	12	♂	*L.D.*	1·0
			Castrated	*L.D.*	3·0
Valin *et al.* (1975)	Normande and Friesian	18	♂	*L.D.*	1·5
				P.P.	1·2

Goutefongea & Valin (1978)	Normande and Friesian	84	♀	*L.D.*	2·4
Valin & Goutefongea (1982)	Limousin	9	♂	*L.D.*	1·3
				T.F.L.	1·7
		16	♂	*L.D.*	1·3
				T.F.L.	1·4
		24	♂	*L.D.*	1·6
				T.F.L.	1·8
Marinova *et al.* (1985)	Brown Bulgarian ×Holstein	24	♂	*T.F.L.*	1.1
				L.D.	0·9
				P.M.	1·7
				D.	4·1
			Castrated	*T.F.L.*	4·8
				L.D.	2·8
				P.M.	4·9
				D.	8·4
	Brown Bulgarian ×Murrah	24	♂	*T.F.L.*	1·4
				L.D.	1·4
				P.M.	1·8
				D.	3·6
			Castrated	*T.F.L.*	1·3
				L.D.	1·4
				P.M.	1·6
				D.	3·7

L.D., Longissimus dorsi; P.M., Psoas major; T.A., Transversus abdominis; S.M., Semimembranosus; T.B., Triceps brachii; S.T., Semitendinosus; P.P., Pectoralis profundus; T.F.L., Tensor fasciae latae; B.F., Biceps femoris; R.F., Rectus femoris; S.S., Supraspinatus; D., Diaphragma.

Table 9
Intramuscular fat in pork muscles

Reference	*Breed*	*Live weight*	*Sex*	*Muscle*	*Intramuscular fat (% fresh muscle)*
Lawrie *et al*. (1963)	Large White	70 kg	Castrated	*L.D.*	2·8
	Large White	70 kg	Castrated	*P.M.*	1·7
Lawrie *et al*. (1964)	Large White	90 kg	Castrated	*L.D.*	3·3
	Large White	90 kg	Castrated	*P.M.*	1·5
	Large White	90 kg	♀	*L.D.*	2·5
	Large White	90 kg	♀	*P.M.*	1·6
	Large White	120 kg	Castrated	*L.D.*	4·0
	Large White	120 kg	Castrated	*P.M.*	1·7
	Large White	120 kg	♀	*L.D.*	2·7
	Large White	120 kg	♀	*P.M.*	1·5
Girard *et al*. (1983)	Large White	100 kg	Castrated	*L.D.*	1·9
Gandemer *et al*. (1985*a*)	Large White	100 kg	Castrated	*L.D.*	2·1
	Large White	100 kg	Castrated	*T.A.*	2·7
	Large White	100 kg	Castrated	*B.F.*	2·5
Pinkas *et al*. (1985)	Large White	100 kg	Castrated	*L.D.*	1·4
	Large White	100 kg	Castrated	*S.M.*	1·3
	Large White	100 kg	Castrated	*R.F.*	1·1
	Large White	100 kg	Castrated	*T.B.*	2·7
	Belgian Landrace	100 kg	Castrated	*L.D.*	1·9
	Belgian Landrace	100 kg	Castrated	*S.M.*	2·0
	Belgian Landrace	100 kg	Castrated	*R.F.*	1·0
	Belgian Landrace	100 kg	Castrated	*T.B.*	2·7
	Pietrian	100 kg	Castrated	*L.D.*	2·8
	Pietrain	100 kg	Castrated	*S.M.*	2·5
	Pietrain	100 kg	Castrated	*R.F.*	1·7
	Pietrain	100 kg	Castrated	*T.B.*	2·4

L.D., Longissimus dorsi; P.M., Psoas major; T.A., Transversus abdominis; S.M., Semimembranosus; T.B., Triceps brachii; B.F., Biceps femoris; R.F., Rectus femoris.

Table 10
Intramuscular fat in lamb muscles

Reference	*Breed*	*Age (weeks)*	*Muscle*	*Intramuscular fat (% fresh muscle)*
Pinkas *et al.* (1982)	Karaktchanska	22	*L.D.*	2·5
	Karaktchanska	22	*S.S.*	2·7
	Karaktchanska	22	*R.A.*	2.8
	Karaktchanska	33	*L.D.*	2.8
	Karaktchanska	33	*S.S.*	3·1
	Karaktchanska	33	*R.A.*	3·0
Ono *et al.* (1984)	Suffolk	18	*L.D.*	4·9
	Suffolk	32	*L.D.*	4·4
Pinkas *et al.* (1983)	Ascanian	12	*L.D.*	2·4
	Ascanian	12	*S.M.*	2.0
	Ascanian	20	*L.D.*	3·2
	Ascanian	20	*S.M.*	2·3
	Ascanian	52	*L.D.*	3·9
	Ascanian	52	*S.M.*	3·7

L.D., Longissimus dorsi; S.M., Semimembranosus; T.A., Transversus abdominis; S.S., Supraspinatus; R.A., Rectus abdominis.

Table 11
Intramuscular fat in poultry

Reference	*Species*	*Breast*	*Thigh*
Bodwell & Anderson (1986)	Chicken	1·7	4·3
	Turkey	1.6	4·4
Girard *et al.* (1985)	Chicken	1·1	3·6
	Turkey	2·0	—
Marion (1965)	Chicken	1·0	2·5
Kim & Gandemer (1987)	Chicken	1·0	5·5
Touraille *et al.* (1981)	Chicken	1·0	3·6
Ricard *et al.* (1983)	Chicken	1·8	—
Wangen *et al.* (1971)	Turkey	1·0	3·7

added for functional purposes in the formulated food can be partly substituted by functional food additives (starch, texturized vegetable proteins, whey proteins, dietary fibres).

Lean-Only Cuts

Trimming

Most subcutaneous fat can be trimmed off the carcass just after dehiding. Remaining external fat is trimmed off primal cuts and, when necessary, defatting is completed on retail cuts.

Trimming carried out at the early stage may have some non-desirable consequences—removal of the subcutaneous fat soon after slaughter may result in an increased sensitivity of the carcass, deprived of its effective insulating barrier, to 'cold shortening' during fast chilling. However this problem can be overcome by means of electrical stimulation which strongly reduces or even suppresses such sensitivity.

A detrimental effect on the sensory properties of the leaner cuts might occur as the result of extensive trimming. Indeed, it is frequently reported that in the consumer's mind, abundant external fat is associated with high organoleptic qualities in meat. In a recent study, Coleman *et al.* (1988) compared tenderness, juiciness and flavour of a set of beef cuts taken on either side of the carcass at corresponding anatomical locations. Cuts from the one side were trimmed of external fat exceeding 1·3 cm thickness whilst all external fat was trimmed off the mirror images. Strip loin (*longissimus lumborum*) steaks were broiled (internal temperature 70°C), top round (*semimembranosus*) and eye of round (*semitendinosus*) roasts were roasted (internal temperature 66°C) and arm pot roasts (*triceps brachii*) and briskets (flat-half) (*pectoralis profundus*) were braised (internal temperature 85°C). The authors reported that briskets were the only cuts whose sensory properties (tenderness $P < 0{\cdot}01$; juiciness $P < 0{\cdot}05$) were adversely affected by removal of external fat prior to cooking. As Choice quality grade (US grading) carcasses were used in this experiment, the studied cuts had relatively high levels of intramuscular fat (6–10% of the net weight). It is likely that there was sufficient marbling fat to impart good sensory properties to the meat in the absence of any contribution from external fat. However, it is probably unfair to extend these results to quality grades corresponding to lower levels of intramuscular fat than Choice cuts. It should also be kept in mind that consumers' cooking and eating habits can be very different from one country to another. For instance, it has been reported by Goutefongea and Valin (1978) that, for French panellists, the more desirable sensory properties of striploin steaks broiled without external fat at a rare degree of doneness (internal temperature 40–45°C) are obtained at intramuscular fat contents close to 3·5%. Data for retail cuts commonly available on the French market are shown in Table 12.

Table 12
Lipid content of edible part of retail cuts (Dumont, 1986)

	Average value % fresh weight	*Coefficient of variation*
Ground beef	5·8	78·8
Sirloin	3·5	28·0
Chuck (for braising)	3·9	36·8
Veal cutlet	1·8	40·0
Lamb rib chop (eye only)	4·1	33·6

Although equivalent trimming effectiveness cannot be attained for all the different commercial cuts (Table 13), modern meat technology makes thorough trimming of meat much easier than in the past. Formerly, cured cooked ham was deboned without muscle separation and defatting was

Table 13
Effect of trimming on lipid content of cuts from beef forequarter (Hermant & Goussault, 1983)

	Not trimmed		*Trimmed*	
	Lipid (% fresh weight)	*Coefficient of variation*	*Lipid (% fresh weight)*	*Coefficient of variation*
Neck	10·9	33·2	7·8	25·7
Chuck steak	14·0	27·5	12·0	32·3
Chuck top blade steak	7·9	23·7	4·9	31·5
Cross cut shank	7·8	20·9	5·9	25·9
Brisket thick cut	23·5	24·4	14·9	29·1
Chuck cross rib pot roast	25·9	27·3	21·0	27·8

limited to removal of external fat. Nowadays, legs are dissected, individual muscles are trimmed extensively and reassembled by tumbling or massaging before cooking. The lipid content in the edible products that was currently about 10% (Schimann & Goutefongea, 1973) now ranges between 4–5%.

Cooking Methods

During meat cooking, part of the solid lipid network melts and some fat is released in the liquid state. Intramuscular as well as external fat may spread out from the meat or be recaptured by the fibrous protein network in the lean. Depending on which cooking method has been selected (broiling, roasting, barbecuing, braising, steaming, boiling), fat may or may not infiltrate into the lean portion.

Apart from fat, drippings and meat juice are composed of water and small amounts of protein, non-protein nitrogen and minerals. It is common practice, when comparing the nutrient content of cooked vs uncooked foodstuffs, to present data as w/w percentages. This is misleading because of the large amount of water which is lost during cooking. The absolute amount of lipid in the cooked portion is the only really useful information for the consumer although he has no access to objective measurements on the raw material except through assessment of marbling. Figures normalized on the basis of 100 g edible portions have been published by Bodwell and Anderson (1986) for raw and cooked pork: ribs contained 7·53 g of fat when raw but this had increased to 14·94 g after broiling, and 13·8 g after roasting. The fat content of ham was reported to range from 5·41 g (raw) to 11·03 g (roasted). Corresponding data relative to beef cuts show a similar trend for Choice grade (US grading) round and rib with a lower fat content in raw than in braised and in roasted meats (round 6·28; 9·96; 10·84; rib 8·52; 11·64; 14·27).

The true retention percentage (TR) was proposed by Murphy *et al.* (1975). For any nutrient it can be calculated according to the equation:

$$\text{TR}\ (\%) = (\text{NCc}\times\text{g food after cooking})/(\text{NCr}\times\text{g food before cooking})\times 100$$

where: NCc = nutrient content per g of cooked food;
NCr = nutrient content per g of raw food.

This index allows the comparison of different cooking methods and the assessment of the amount of lipid in ready-to-eat meat portions. Numerical values below 100 are normally expected for the whole piece of meat but it is obvious that fat transfer into the muscle will result in indexes higher than 100.

Moss *et al.* (1983) calculated TR indexes for several types of pork retail cut cooked by different methods: shoulder blade steaks were braised, loin chops were broiled (internal temperature 75°C) and arm picnic, blade loin roasts, sirloin roasts, butt and shank were roasted (internal temperature 75°C). TR indexes obtained for the visible lean part ranged from 111 (braising) to 137 (roasting) with broiling (129) intermediate. These results show that lean gained fat in all cases but it should be recorded that the subcutaneous fat layer (thickness from 2·3–4·4 cm) had not been trimmed off before cooking. In an experiment carried out on pork loins carefully trimmed, Gandemer *et al.* (1983) obtained very different results. The TR index calculated from their data was 80 for roasted meat (internal temperature 85°C). It can be concluded that knowledge of the extent of trimming is a prerequisite for any reliable comparison of experimental data.

Some information on the intrinsic effects of the cooking method can be found in a paper by Renk *et al.* (1985). Trimmed pork chops and roasts (*longissimus dorsi*) were broiled and roasted respectively. Samples of both types were cooked until selected internal temperatures (68 and 79°C) were attained. In addition, samples displaying either slight or moderate marbling were compared. Although the number of samples is too small to allow any statistical treatment of the data, there seems to be a trend toward higher TR indexes for broiling at the lower internal temperature and for roasting at the higher one. In a similar experiment carried out on beef meat, the opposite trend was observed by the authors. Obviously, high variability in biological samples makes any definite conclusion uncertain.

Johansson and Laser-Reuterswärd (1987) investigated the effect of different kinds of cooking (panbroiling, ovenroasting and boiling in water) on the fat content of beef and pork meats. Retail cuts corresponding to a wide range (2–35%) of fat content in raw tissue were selected. The authors concluded that basically most (above 85%) of the fat is recovered in the cooked meat regardless of total fat content, fat distribution in the raw samples or cooking method. They found only one exception regarding thin (under 15 mm) meat slices that were found to show a TR proportional to the thickness of the slice. Furthermore, considering that most fat can be trimmed off as visible fat before eating, they proposed (for Sweden) 88% as a reference TR for estimating the actual fat supply associated with meat dishes.

A TR index of 79 can be calculated from the data of Gandemer *et al.* (1985*b*) who reported that 100 g of beef (*pectoralis profundus*) containing 3·4% of fat, boiled in water for 4 h, yielded 58·3 g of cooked meat with a fat content amounting to 4·6%.

Beef ribs from different marbling classes (practically devoid, traces, slight, small, modest and moderate) and having external fat (0·6 cm thickness) removed or not were either broiled or microwave cooked (Berg *et al.*, 1985). Lower fat levels were measured in the broiled than in the microwave cooked samples. Trimming of fat was found to be detrimental to the eating quality of ribs showing little marbling but was an advantage in samples belonging to the upper two marbling classes. However, these comparisons may not be very pertinent as the heating schedules needed to reach the required internal temperature (65·5°C) were very different from one sample to another (microwave cooking: fat removed 7 min, control 15 min; broiling: fat removed 10 min per side, control 15 min per side).

Perchais (1984, pers. comm.) studied the effect of French style roasting on the lipids in beef loin: after the external fat and connective tissue layer had been thoroughly trimmed off *longissimus dorsi*, meat was cut into pieces (600 g each), and roasted in an electric oven at 220°C for 30 mins. It was reported that the internal temperature did not exceed 40°C. Three different regions were distinguished according to the degree of doneness: the external area was well done and brown in colour whilst the intermediate area was medium (pink) and the central area was rare (red).

The lipid content (4·3% in the raw meat) increased slightly in the external area (5%) with a corresponding decrease in TR index (91·5). On the other hand, the other regions displayed no measurable variation in lipid or dry matter content. As it is common practice to wrap roasts in a thin slice of pork backfat (barding) before cooking, lipids from the outer lean part of barded roasts were analysed. No transfer from pork unsaturated fatty acids was found, suggesting that fat migration from the outside to the inside, if any, had been marginal. It may well be that an aqueous barrier is formed impeding the lipid transfer that is reputed to occur between external fat and inner lean.

Ono *et al*. (1984) conducted an extensive study on the cooking of lamb meat in which matched pairs of cuts were taken from both sides of 16 carcasses. Cooking was carried out without any preliminary trimming of subcutaneous fat (unless fat thickness exceeded 0·5 inch). Fore shanks and arm chops were braised whilst blade and loin chops were broiled and rib roasts, sirloin half of leg and shank half of leg were roasted. The lean part of each retail cut was carefully dissected out and the concentrations of fat and moisture measured.

The cooking retention of lipid (percent) was respectively:

—Braising: 130 (fore shank) and 136 (arm chop).
—Broiling: 115 (blade chop) and 126 (loin chop).
—Roasting: 146 (rib roast), 156 (leg sirloin) and 137 (leg shank).

In all cases, values for the retention index were higher than 100%, suggesting that lipid absorption by the lean had occurred during cooking at the expense of subcutaneous and/or intermuscular fat. The total fat retention was higher in the roasted samples than in the broiled and braised samples. According to the authors, the recorded differences may well be linked to the fact that samples submitted to roasting were the only samples processed with the subcutaneous fat side up. In such a system, subcutaneous fat could have contributed heavily to the overall fat migration whereas in braised and broiled systems, intermuscular fat probably produced most of the lipid gained. As a consequence, broiling and braising resulted in lower fat retention than roasting.

Broad differences in the fat content of cooked poultry meat have been observed and generally related to the cooking mode used. The skin of poultry contains much fat and, nowadays, skinned breast muscle meat as well as whole carcasses are cooked. Data from Bodwell and Anderson (1986) on chicken and turkey are summarized in Table 14.

Cooking (roasting) resulted in an increase of the lipid level in the flesh,

Table 14
Effect of cooking on the main components of poultry meat (Bodwell & Anderson, 1986)

Per 100 g edible portion	*Light meat*				*Dark meat*			
	Flesh and skin		*Flesh only*		*Flesh and skin*		*Flesh only*	
	Raw	*Cooked*	*Raw*	*Cooked*	*Raw*	*Cooked*	*Raw*	*Cooked*
Chicken								
Water	68·6	60·5	74·9	64·8	65·4	58·6	76·0	63·1
Protein (N×6·25)	20·3	29·0	23·2	30·9	16·7	26·0	20·1	27·4
Lipid	11·1	10·9	1·7	4·5	18·3	15·8	4·3	9·7
Turkey								
Water	69·8	62·8	73·8	66·3	71·1	60·2	74·5	63·1
Protein (N×6·25)	21·6	28·6	23·6	29·9	18·9	27·5	20·1	28·6
Lipid	7·4	8·3	1·6	3·2	8·8	11.5	4·4	7·2

even if data are corrected for moisture losses. At the same time the overall lipid concentrations of the whole carcass showed variable changes.

According to Kim and Gandemer (1988) and Kim (1989), chicken roasting results in increases of the lipid level from 1% to 3·4% in light meat (breast) and from 5·5% to 11·2% in dark meat (drumstick). Meanwhile, the fat content of skin shows a 50% decrease. About 90% of the loss is recovered in the cooking juice and the remaining 10% is absorbed by the underlying flesh.

Comparable figures for light meat broiled without skin (turkey or chicken escalope) are much lower. According to Gandemer (1988), the lipid content increased from 1% to 1·3% during cooking.

In general then it can be concluded that the trimming of external adipose tissues (poultry skin or external fat in mammals) before cooking is the most effective way of reducing the fat in edible cuts. If, for the sake of presentation or local custom, meat must be served surrounded with fat, barding is a sensible alternative that leads to little increase of the fat level in the food. In any case, poultry skin should not be eaten by people wanting to control their fat intake.

Ground Meat Products and Restructured Meats

A large number of different meat products can be obtained simply by grinding, chopping, emulsifying, chunking, flaking, sectioning and slicing (Field, 1982). On a practical basis, processed products can be classified into two main categories.

The first category is ground and restructured products whose manufacturing involves little processing. These are made from cuts with a high connective tissue content whose cooking normally requires preliminary gelatinization of collagen. Destruction of the original structure and subsequent reforming into loaves or pieces enable the use of any conventional high temperature short time cooking mode as well as microwave heating. Organoleptic properties as close as possible to tender solid muscle cuts are often expected for such foods.

The second category is composed of a variety of very different processed meat products. Comminuted to a greater or lesser extent, often cured, either cooked or dried, sometimes smoked, fermented or not, the products display very different sensory characteristics (flavour, colour, texture, granule density, sliceability, spreadability, etc.). Foods from this category are usually higher in fat as they are commonly made with coarsely trimmed meat. Moreover, some may feel encouraged to use fat as a cheap food ingredient since consumers are rather non-discriminating as long as fat assessment is not readily possible through visual inspection. Relatively high levels of fat are sometimes required to impart specific characteristics (juiciness, bite, spreadability) to some country-style traditional foods. Some products, like blood sausages, show large solid inclusions of fat in a low-fat mixture that are of prime importance for the flavour and the texture of the food.

Reduction of the fat content in ground products is mainly the responsibility of the manufacturer since the consumer is unable to trim fat off and cooking is not expected to result in noticeable fat losses.

Products Containing Less Fat

In most instances, it is possible to incorporate less fat in formulated products. More extensive trimming of the raw material associated with a tight control on fat incorporated during processing is an effective although simple means to reach the proposed aim. One obvious drawback to production of fat-reduced or low-fat meat products is cost. Raising the lean-to-fat ratio results in extra costs for the manufacturer and subsequently in higher prices for the consumer. Ground meat products are essentially cheaper alternatives to top quality solid cuts and little commercial future can be forecast for expensive, fatless, ground products in as much that they do not show outstanding sensory qualities. In general, ground meat products are tasty and nutritionally good quality meaty foods available to low income groups.

There is no technical hindrance to the control of fat content in ground meat patties or restructured steaks since cuts employed in manufacture can be trimmed more or less extensively before grinding or flaking. Finished products containing as little fat as lean-only cuts (1–6% according to species, anatomical location, degree of finish, etc.) could be obtained but, as previously pointed out, they are not attractive commercially. It has also been reported that ground meat requires a higher fat level than solid cuts in order to obtain desirable sensory qualities. Cross *et al.* (1980) compared sensory properties of ground beef patties containing 16, 20, 24 and 28% fat. They found that tenderness and juiciness increased significantly with fat content. These results were confirmed by Berry and Leddy (1984) and by Kregel *et al.* (1986).

Acceptability is the key factor to success when such foods are to be developed. Texture, colour and flavour should not differ (too much) from those of the normal product. Eating habits are quite different from one country to another and it should be realized that there may well be no universal objective criterion to cope with the fat-in-food problem. USDA tables of food composition (Breidenstein, 1987) show data on ground beef products containing 15, 20 and 27% of fat before cooking. Ono *et al.* (1985) labelled ground beef patties containing 18·5, 21·5, and 27% of fat as 'extra lean', 'lean' and 'regular' respectively. Such a classification would have been unacceptable in France where 'regular' ground beef steaks contain 15% of fat and 'lean' no more than 5%. The reason is probably linked to the French custom of cooking meat at a much lower degree of doneness than in the USA. This cooking mode ensures a better water retention in the cooked food but makes high-fat meat rather unpalatable.

According to Wirth (1988), processing low-fat (less than 10%) or fat-reduced (40% reduction in energy content compared with the normal product) cured meat products may sometimes prove rather difficult. Technological solutions have been proposed for three of the main product types: frankfurters, liver and blood sausages and dry sausages.

Lowering fat content in a food correspondingly increases the proportion of hydrophilic components in the modified product. It is commonly found that some seasonings become less strong whilst others become stronger. This is a frustrating situation that must be tackled empirically. The distribution of flavour-active compounds between the hydrophobic and the hydrophilic phase may well be quite different in the low-fat product from that in the normal reference. Sensorily, spices perform in an unpredictable manner and tailor-made seasonings have to be developed for the new product.

Frankfurters

If considered strictly from a technological point of view, the manufacture of fat-reduced frankfurter-type sausages with suitable quality characteristics involves few problems. However, practical difficulties have arisen with binding, colour development and retention and with flavour quality.

Reduction in fat content results in a correlated enhancement of the intensity of the salty note in the flavour. Very often salt addition must be cut by one-fifth to one-quarter to restore the usual intensity. Consequently, ionic strength in the corrected mix decreases with subsequent detrimental effects on the stability of the meat emulsion. Good binding can be obtained through methods and by use of ingredients which promote water binding. Hot meat processing, the use of phosphates and blood plasma adjunct or vacuum processing in the chopper have been recommended.

Safety regulations enforce the use of curing salt consisting of a mixture of common salt with a small proportion of curing agent (sodium nitrite). In normal industrial practice, salting and curing are connected. Reducing the amount of added salt will also decrease the quantity of available nitrite in the processed product. Low-fat meat products, that contain more lean meat and consequently more muscle pigment than normal, develop a strong cured colour only if nitrite concentration is kept above a critical level. In some cases, when nitrite is reduced too much, reduced colour development and stability are experienced.

It can be helpful to use curing aids containing ascorbic acid and to control oxygen incorporation into the mix by carrying out the emulsification step under vacuum. It is also advisable to pre-redden the sausages for longer and to heat treat them in such a manner that the temperature at the core will reach 72°C at least and 75°C whenever possible.

Spreadable sausages

Spreadability is an essential property in the acceptability of liver sausages. Keeping the original texture qualities of the high-fat (40% or more) food in the low-fat type is a challenge that has been matched. Reduction of the fat content by up to 50% has no detrimental effect on the palatability of the finished product. Some technical problems arise as the fat proportion drops below 20%. Formulations in which excessive fat has been entirely substituted by meat and/or liver, as in top quality liver sausage, often results in products with poor spreadability and mouthfeel. This can be obviated by incorporating and microdispersing water into the meat mix. As a result, lowering of the fat content down to around 15% can be achieved without any noticeable change in the sensory properties of the food. The level of fat in medium and standard quality products is controlled more easily by the incorporation of rind and offals other than liver and making minor processing adjustments.

Because that lean products have higher water contents and are more susceptible to a release of water vapour during manufacture and storage, the use of low permeability casings is recommended.

In blood sausages and brawns, fat is confined to large solid inclusions of adipose tissue. There is no technological hindrance to fat reduction but the aim is to retain the typical appearance. Fat-devoid inclusions from different origins have been successfully used to overcome this problem.

Dry sausages

Processed meat products in which the fat content cannot be readily reduced include the dry sausage type. Fat granules play a positive role in the drying process of firm sausages as they loosen up the mix and regulate water mass transfer. Products with small granules, like salamis, rather than products displaying large granules, can be made from low-fat meat mixes containing no more than 15% fat.

It has been found experimentally that during the drying of low-fat brands, the concentration of fat calculated on a wet basis increased from 15% to around 20% after two weeks and to 30% after four weeks. This also indicates that fat-reduced dry sausages dry much faster than the conventional equivalents. The lowest acceptable fat content in spreadable dry sausages is in the same range as in other spreadable meat products (about 20% in the finished food).

In a number of different applications, adaptation of traditional food processing to the production of low-fat although genuine meat-like products requires minor modifications only. Despite the apparent diversity of the encountered problems, it should not be overlooked that most of them and the underlying causes are basically similar.

Fat Substitution by Non-lipid Functional Additives

Keeping a good water binding capacity in the finished food is probably the main challenge that must be met when attempting to reduce the fat content of meat products. Despite the potential usefulness of agents such as phosphates, it may be difficult to cope with the increase in the water content produced by the decrease in fat. Poor water binding has a direct negative effect on sensory quality: lack of juiciness, crumbly texture and dry consistency can be encountered. Food components, like non-meat proteins, starch and some types of dietary fibre, that are cheaper than meat proteins but display high water binding ability, can be considered as worthy alternatives.

It must be noted however, that the use of non-meat proteins and non-protein components in traditional foods is restricted by legislation in Europe as well as in the United States. Nevertheless, as particular regulations may not always exist, we consider it worth reviewing these promising unconventional ingredients.

Non-meat proteins

Papers by Goutefongea (1986) and Endres and Monagle (1987) provide guidelines on the suitability of non-meat proteins for the preparation of meat products.

Fish proteins. Known in Japan for five centuries, Surimi is a white odourless and tasteless paste made of fish muscle myofibrillar proteins obtained from fish flesh after extensive washing. Besides the various meals developed earlier by Japanese people and later by technologists, it would be profitable for the meat industry to take advantage of the intrinsic nutritional and technological properties of Surimi:

The edible product can be prepared at an acceptable cost from fish species that have little or no commercial value.
Binding and emulsifying properties make it a first grade additive.
Gelation ability and the white colour suggest that it could be advantageously substituted for large fat inclusions in products like blood sausages, mortadella, and even some kinds of dry sausages.

Milk proteins. Whole casein and caseinates are well-known emulsifiers but are totally unable to bind meat pieces (Siegel *et al.*, 1979).

Up to 1970, whey had been essentially used as a food for pigs. Development of refined separation processes at the industrial scale have made whey proteins readily available. They are now appreciated functional additives that show excellent foaming and emulsifying capacities as well as useful heat gelation characteristics. Whey proteins, associated with starch, have been used to prepare liver 'pate' containing 10% of fat instead of the 40–50% in the traditional recipe (Goutefongea, 1985). Milk proteins have no associated off-flavour nor colour defects and they have good nutritional properties.

Egg albumen. Undoubtedly the oldest functional additive as a foaming and heat coagulating agent, it is also a very effective binder for meat pieces. Having very good nutritional properties, the only drawback to its use is its relatively high cost.

Wheat proteins. Gluten is a normal component of wheat flour (10–15%) from which it can be extracted either by air classification or by washing (lixiviation). Gluten swells when hydrated and an elastic and extensible paste with very good meat binding ability is obtained.

Soy proteins. Soy proteins are commonly used in the meat industry. They are commercially available in different preparations with different protein contents. Relatively crude extracts as well as elaborated intermediate products are used in a large number of food applications.

Soy flours and grits, obtained by grinding defatted flakes, contain about 50% protein. Grits are coarsely ground products whilst flours consist of fine particles able to pass through a 100 mesh screen. Soy protein concentrates are obtained after most of the non-protein components from flours or grits have been solubilized in water or in water–alcohol mixtures. The protein content in soy concentrates is around 65%. Protein isolates (90% protein) are prepared by dispersing the concentrate in slightly alkaline aqueous solutions (pH 9), the supernatant is collected and protein subsequently recovered by isoelectric precipitation.

Whether flour, concentrate or isolate, soy products are obtained in the form of a powder. In the last fifteen years, experiments have been carried out with the aim of imparting more attractive texture properties to soy protein. For instance, thermoplastic extrusion, steam texturization and fibre spinning have been proposed. Technical details on application of texturization process to soy protein can be found in review papers by Gutcho (1973), Kinsella (1978), Rhee *et al.* (1981) and Culioli (1985).

Texturized protein is currently used as a meat extender and sometimes contributes a significant proportion of the protein content of the finished food. In such cases, care must be taken to reduce the undesirable beany flavour and remove antinutritional substances. It has been reported that nitrosyl-haemoglobin can help in imparting a desirable meat colour to the soy products (Noël *et al.*, 1984; Noel & Goutefongea, 1985). When improvement in overall food quality is required, soy protein functionality must be at its best. Then, small amounts of powdered isolate are usually preferred.

In the 1970s, many studies were made to examine the effect on sensory characteristics of incorporating soy protein into meat products. For instance, it was reported by Kotula *et al.* (1976) that, on the acceptance criteria of a consumer panel, ground beef patties containing 20% textured soy protein scored almost equally to all-beef patties.

Smith *et al.* (1976) compared texturized soy protein (TSP) from seven different commercial suppliers. They found that different brands can show noticeable variations in overall acceptance. In addition, from their results it can be concluded that TSP should not make up more than 30% of the protein fraction. Experiments by Judge *et al.* (1974) and Anderson and Lind (1975) have led to similar conclusions. However, as relatively high fat levels were involved in these latter studies, it seems that textured soy proteins should be considered merely as a meat extender and not as a fat replacer. Recently, Huffman and Cordray (1987) proposed formulas for restructured meat products that contained 10–15% fat and involved a correlated incorporation of soy protein. In France, where the regular fat level in ground beef steaks is 15%, several companies have lowered the value to 12% through the addition of 20% TSP.

Protein content as well as functional properties (water absorption and retention, emulsion formation and stabilization) are positively related to the cost of the soy protein product. There is probably no definite superiority of one kind of soy protein over the others. Selecting, for instance, concentrate rather than flour or isolate depends on the planned practical application.

The effects on nutritional value of the addition of soy protein to food have been studied extensively (Kies & Fox, 1973; Noda *et al.*, 1977; Liener, 1981). Comparisons of mixed-protein vs pure-meat reference products have shown only a slight lowering of the protein efficiency ratio, provided soy antinutritional factors had been previously removed or heat-inactivated. Another problem, dealing with the inhibition of non-haem iron absorption in man by soy protein, has been raised by Cook *et al.* (1981). Later experiments by Rizk and Clydesdale (1985) have shown a positive effect of the addition of some organic acids, like ascorbic, lactic, citric and malic acids, to the meal.

Other vegetable proteins. Protein from ground-nuts, peas and beans, sunflower seeds, rapeseed and cottonseed can also be used as an ingredient in the manufacture of meat products. Functional properties of the extracted proteins have often been studied in reference to the better known soy proteins. Data have been reported by Cherry (1983) (peanut), Huffman *et al.* (1975) (sunflower), Gillberg (1978) (rapeseed), Lusas *et al.* (1977) (cottonseed), Guegen (1980, 1983, 1986), and Guegen *et al.* (1980, 1984) (peas and beans). In France, fababean protein concentrates, prepared by air-classification, are used in cured meat products, in ready-to-eat dishes, and by the bread industry, while texturized protein is incorporated into some brands of ground meat products.

Carbohydrates

Starches. Starch is industrially extracted from tubers (potato, cassava) and cereal seeds (wheat, corn, rice) and is also obtained as the main by-product of protein extraction from proteinaceous seeds. It has long been used by the meat industry as a functional additive for its binding and gel-forming capacities. When heated under wet conditions, starch imparts firmness and stickiness to the products in which it has been incorporated.

Starch can be used as a main component to bind meat pieces in spreadable products with a heterogeneous appearance. Low-fat liver 'pate' in which most of the lipids are substituted by gelatinized starch has been produced on a pilot scale (Goutefongea, 1985). It was found that starches from different plants do not perform equally well. Fababean starch enabled the fat content of country-style 'pate' to be cut down by one-third without any adverse effect on sensory properties (Laroche, 1988, pers. comm.). Chemical or enzymic modifications are industrially applied to starch with the aim to impart new or better functional properties to the resulting modified starch. Among the various products available commercially, some have proved effective fat replacers.

Dietary fibres. Different kinds of dietary fibres are able to firmly bind large quantities of water. Water-to-fibre ratios (w/w) up to 5 are not uncommon as shown for fibres prepared from sugar-beet pectins (Michel *et al.*, 1985). Apart from their recognized positive action on intestinal transit, fibres could have another nutritional function as a fat replacer. Fibres of a similar type are already used by the bakery industry in Sweden.

Other polysaccharides. Polysaccharides can prove useful in imparting good texture qualities to low-fat meat products. Currently, they are considered more as technological aids rather than established fat-replacers, although their thickening and gelling capacity could make possible the processing of foods with highly reduced fat levels. Amongst those commonly used by the food industry (guar gum, locust bean gum, carrageenans and alginates), the latter, which form heat-stable gels in the presence of calcium ions, are certainly the most versatile and therefore the most promising.

Miscellaneous

Pseudofats or cal-o-fats. Another way to reduce the contribution of fat to overall energy intake is to prevent or reduce absorption of fat-derived calories (Hamm, 1984; Anon., 1986*a*). Large industrial concerns are working hard in this area and patents have already been granted to some products. However, uncomfortable secondary effects are still very frequently experienced by consumers. It is probably too early to come to any conclusion but such alternatives may become operative in the long term.

New Products

Up to this point in the chapter, proposals for reducing the fat content in meat products have been made. In some instances, future meat products have been predicted and the aim now is to consider what form these might take.

True new products that would not aim to replace meat and meat products, but simply add to the variety of the available high protein foods can be expected. Specially designed foods would enable people to eat fat-free meals. Although not yet available on the market, such products are not unrealistic as is suggested by recent developments in the field. Nowadays, unconventional food materials exhibiting original functionalities or properties can be obtained. More flexible than their traditional counterparts, these commodities make possible the design of more creative foods.

The scope of Surimi uses has rapidly enlarged as the food industry in the western hemisphere realized that Surimi was a versatile raw material: colourless, tasteless, odourless, it can be coloured and flavoured as required by food processors. Texturization can be achieved by different processes and very diverse shapes and aspects can be imparted to the product.

Dyed TVP can be produced through extrusion-cooking (Noel *et al.*, 1984) with no off-flavour, provided proper raw materials have been selected. These products have been used on a laboratory scale to prepare ready-to-eat meals served with traditional vegetables or pasta and with tasty and/or highly-seasoned sauces.

To the layman, 'protein' is an abstract word that is strongly associated with meat and fish. Most people are used to considering these foods as valuable, nutritious and fibrous. Therefore, it is more than likely that imparting a fibrous texture to fat-free, high-protein new foods is probably the best way they will achieve success in the market.

In the United Kingdom, a myco-protein has been marketed under the trade mark 'Quorn' since 1985. It is used as a prime ingredient in the Savoury Pie currently made and sold by one major marketing company. Basically, Quorn is obtained from *Fusarium graminearum*, a microscopic plant related to the mushroom, that forms minute filaments growing into the soil. Industrial production is achieved through continuous liquid fermentation with carbohydrates as the main nutrient (March, 1983). The finished product has textural characteristics that owe much to the fibrous nature of the constituent filaments. Texture can be modified by changing the orientation of the filaments and the overall composition of the binder (that contains egg albumen and also colours and flavours if appropriate). Desired structures are stabilized by steaming in order to heat-set albumen. Quorn has a total fat content of 4%. At present, it is claimed to be a prime ingredient in various recipes produced by different food manufacturers.

Since pioneering work by Boyer (1954), a lot of papers and patents have dealt with the spinning of proteins. Reviews by Gutcho (1977) and Culioli (1981, 1985) provide valuable information on the state of the art at regular intervals. Despite extensive fundamental knowledge, no processed food prepared from spun vegetable proteins is currently available on the market. In the past, marketing tests have been carried out in the United States but they did not result in new products.

Whatever drawbacks they may yet present, these products are bound eventually to become prime ingredients in food manufacturing. Fibres with controlled consistency and diameter can be obtained at request by simply modifying spinning parameters (spinerette type, dry matter content and pH of the dope, pH and ionic strength of the coagulation bath). Easily dyed (Culioli *et al.*, 1981) and shaped into bundles, spun protein fibres can also be readily flavoured. They yield a versatile multipurpose raw material that has many potential uses.

At our laboratory, experimental samples were prepared that displayed a wide range of consistencies varying between that of beef tongue (overcooked) and that of deer leg. Samples were dyed and flavoured with reference to the meat cuts that had similar consistencies. Dishes were prepared in the traditional manner with the appropriate vegetables and sauces and assessed by a consumer taste panel. No particular distaste was experienced and the dishes met with broad acceptance. Total lipid content in the 'meat component' of the finished food was around 2% and the estimated cost price was about half of that of an equivalent beef stew serving.

CONCLUSIONS

The fat content of meat and meat products is a technical parameter that can be varied over a very wide range. Delivering information to the public which is sound scientifically sometimes proves rather difficult, although greater care should be taken in order to avoid major flaws and misinformation. The current trend in meat production toward leaner animals has made most data banks dealing with carcass composition obsolete and there is a need to keep data on meat up-to-date. Trimming and to a lesser extent, cooking are effective means of lowering the lipid content of meats marketed as lean-only cuts. At the end-point, fat may be ultimately trimmed off meat on the plate by the consumer to an extent unknown to most nutritionists. On average, the meat consumed in Europe is not an energy-rich food.

The highest fat levels are found in meat products, especially traditional dishes prepared by local pork butchers. Consumer claims for more sophisticated and more diversified dishes give an opportunity to shift toward low-fat adaptations of the original recipes. Modern technology provides food manufacturers with the technical means to elaborate less energy-dense foods that remain tasty and attractive.

In the final analysis, the dangers associated with fat ingestion must not be overemphasized. One approach is to reduce fat in foods in which a high fat level is not justified for technical reasons. Another approach is to provide good basic information to consumers in order for it to be known that animal fats are not necessarily saturated, that cholesterol in meat does not present the dangers imagined and that, moreover, knowledge is the best defence against alarmist propaganda.

REFERENCES

Anderson, R. H. & Lind, K. B. (1975). Retention of water and fat in cooked patties of beef and of beef extended with textured vegetable protein. *Food Technol.*, **29**, 44–5.

Anon. (1986*a*). Diète lipidique et biotechnologies. *Biofutur*, **45**, 19–32.

Anon. (1986*b*). *Code de la Charcuterie, de la Salaison et des Conserves de Viande* (Réglementation et Usages), 3rd edn. Centre Technique de la Charcuterie, de la Salaison et des Conserves de Viande, Paris.

Berg, P. T., Marchello, R. J., Erickson, D. O. & Slanger, W. D. (1985). Selected nutrient content of beef *longissimus* muscle relative to marbling class, fat status, and cooking method. *J. Food Sci.*, **50**, 1029–33.

Berry, B. W. & Leddy, K. F. (1984). Effects of fat level and cooking method on sensory and textural properties of ground beef patties. *J. Food Sci.*, **49**, 870–5.

Bodwell, C. E. & Anderson, B. A. (1986). Nutritional composition and value of meat and meat products. In *Muscle as Food*, ed. P. J. Bechtel. Academic Press, Orlando, pp. 321–69.

Boyer, R. A. (1954). High protein food product and process for its preparation. US Patent, 2 682 466.

Breidenstein, B. C. (1987). Nutrient value of meat. *Food and Nutr. News*, **59**, 43–55.

Cherry, J. P. (1983). Peanut protein, properties, processes and products. In *Peanuts: Production, Processing, Products*, 3rd edn, ed. G. J. Woodroof. AVI Publishing, Westport, p. 337.

Coleman, M. E., Rhee, K. S. & Cross, H. R. (1988). Sensory and cooking properties of beef steaks and roasts cooked with and without external fat. *J. Food Sci.*, **53**, 34–61.

Colleau, J. J. (1978). Analyse et décomposition de la variabilité intergénotypique de la consommation alimentaire pour des jeunes bovins. *Ann. Genét. Sél. Anim.*, **10**, 29–45.

Cook, J. D., Morck, T. A. & Lynch, S. R. (1981). The inhibitory effect of soy products on nonheme iron absorption in man. *Am. J. Clin. Nutr.*, **34**, 2622–6.

Cross, H. R., Berry, B. W. & Wells, L. H. (1980). Effects of fat level and source on the chemical, sensory and cooking properties of ground beef patties. *J. Food Sci.*, **45**, 791–3.

Culioli, J. (1981). La texturation des protéines. In *Protéines Foliaires et Alimentation*, ed. C. Costes. Gauthiers-Villars, Paris, pp. 183–209.

Culioli, J. (1985). Les procédés de texturation des matières protéiques végétales: Aspects technologiques. In *Proteínes Végétales*, ed. B. Godon. Lavoisier, Paris, pp. 489–522.

Culioli, J., Noël, P. & Goutefongea, R. (1981). Texturation par filage de mélanges de protéines végétales et de protéines de sang. *Sci. Aliments*, **1**, 169–85.

Demarne, Y. (1982). Le développement des tissus adipeux et l'origine des graisses de réserve chez les mammifères. Variations interspécifiques et effets de l'alimentation. *Rev. Fr. Corps Gras*, **29**, 485–90.

Dikeman, M. E. & Crouse, J. D. (1975). Chemical composition of carcasses from Hereford, Limousin and Simmental crossbred cattle as related to growth and meat palatability. *J. Anim. Sci.*, **40**, 463–7.

Dumont, B. L. (1986). Implications nutritionnelles de la restructuration des viandes. In *La Restructuration des Viandes*, ed. B. L. Dumont. ERTI, Paris, pp. 206–34.

Endres, J. G. & Monagle, C. W. (1987). Nonmeat protein additives. In *Advances in Meat Research, Vol. 3, Restructured Meat and Poultry Products*, eds A. M. Pearson & T. R. Dutson. AVI, New York, pp. 331–50.

Field, R. A. (1982). New restructurated meat products: Food service and retail. In *Meat Science and Technology International Symposium Proceedings*, eds K. R. Franklin & H. R. Cross. National Live Stock and Meat Board, Chicago, pp. 285–98.

Fortin, A., Reid, J. T., Maiga, A. M., Sim, D. W. & Wellington, G. E. (1981). Effect of level of energy intake and influence of breed and sex on growth of fat tissue and distribution in the bovine carcass. *J. Anim. Sci.,* **53,** 982–91.

Frebling, J., Bonaiti, B., Gillard, P., Menissier, F. & Renand, G. (1982). Comparisons of fattening and slaughter performances between Charolais, Limousin, Maine-Anjou and Hereford breeds according to various production type. *2nd World Congress of Genetics Applied to Animal Production,* Madrid, VIII, pp. 334–9.

Gandemer, G. (1988). Personal communication.

Gandemer, G., Girard, J. P. & Denoyer, C. (1983). Influence of cooking on the lipids of pork meat. *29th Europ. Meet. Meat Res. Workers Proceedings,* Parma. 503–10.

Gandemer, G., Sharma, N. & Viau, M. (1985*a*). Etude comparative des lipides de la viande de porc suivant la localisation anatomique. *Journées Rech. Porcine en France,* **17,** 55–62.

Gandemer, G., Viau, M., Maho, C., Metro, F. & Laroche, M. (1985*b*). Modifications des lipides intra-musculaires au cours de traitement thermiques. Etude d'une cuisson de type pot-au-feu. *Sci. Aliments,* **5,** 299–306.

Gillberg, L. (1978). Influence of electrolytes on the solubility of rapeseed protein isolates. *J. Food Sci.,* **43,** 1219–28.

Girard, J. P., Denoyer, C., Desmoulin, B. & Gandemer, G. (1983). Facteurs de variation de la composition en acides gras des tissus adipeux (bardière) et musculaires (*longissimus dorsi*) de porc. *Rev. Fr. Corps Gras,* **30,** 73–9.

Girard, J. P., Bucharles, C., Gerardot, L. & Denoyer, C. (1985). *Les Lipides Animaux dans la Filière Viande*, Vol. 1. APRIA, Paris.

Goutefongea, R. (1985). Réalisation de produits tartinables à faible teneur en lipides. *Sci. Aliments,* **5,** 345.

Goutefongea, R. (1986). Intérêt et limites de l'emploi des diverses sources possibles de matières premières. In *La Restructuration des Viandes,* ed. B. L. Dumont. ERTI, Paris, pp. 80–97.

Goutefongea, R. & Valin, C. (1978). Etude de la qualité des viandes de bovin. 2: Comparaison des caractéristiques organoleptiques des viandes de taurillon et d'animal adulte. *Ann. Technol. Agric.,* **27,** 609–27.

Goutefongea, R., Cassens, R. G. & Woolford, G. (1977). Distribution of sodium nitrite in adipose tissue during curing. *J. Food Sci.,* **42,** 1637–41.

Gueguen, J. (1980). Elimination des facteurs antinutritionnels de la féverole (*Vicia faba* L.) et du pois (*Pisum sativum* L.) au cours de la préparation des isolats protéiques. *Lebens. Wiss. Technol.,* **13,** 72–7.

Gueguen, J. (1983). Legume seed protein extraction and end product characteristics. *Qual. Plant.,* **32,** 267–303.

Gueguen, J. (1986). Préparation de produits alimentaires intermédiaires à partir des protéines des grains et graines. Symposium international. *Les Produits Alimentaires Industriels Intermédiaires*. APRIA, Paris, pp. 47–61.

Gueguen, J., Valdebouze, P. & Melcion, J. P. (1980). Antinutritional factors in legume seeds and their products: Effect of physicochemical and hydrothermal treatments. *Ann. Nutr. Alim.,* **34,** 312.

Gueguen, J., Vu, A. T. & Schaeffer, F. (1984). Large scale purification and characterisation of pea globulins. *J. Sci. Food Agric.,* **35,** 1024–33.

Gutcho, M. (1973). *Textured Foods and Allied Products*. Noyes Data Corp., Park Ridge.

Gutcho, M. (1977). *Textured Protein Products*. Noyes Data Corp., Park Ridge.

Hamm, D. J. (1984). Preparation and evaluation of trialkoxytricarballylate, trialkoxycitrate, trialkoxyglycerylether, jojoba oil, and sucrose polyester as low calories replacements of edible fats and oils. *J. Food. Sci.,* **49,** 419–28.

Henry, Y. (1977). Développement morphologique et métabolique du tissu adipeux chez le porc: influence de la sélection, de l'alimentation et du mode d'élevage. *Ann. Biol. Anim. Biochim. Biophys.,* **17,** 923–52.

Hermant, J. J. & Goussault, B. (1983). Variabilité de la composition biochimique des muscles de l'avant de bovins. *Viande et Produits Carnés,* **4,** 90–1.

Huffman, V. L. & Cordray, J. C. (1987). Formulations for restructured red meat products. In *Advances in Meat Research, Vol. 3, Restructured Meat and Poultry Products,* eds A. M. Pearson & T. R. Dutson. AVI, New York, pp. 383–403.

Huffman, V. L., Lee, C. K. & Burns, E. E. (1975). Functional properties of sunflower meal. *J. Food Sci.,* **40,** 70–4.

Johansson, G. & Laser-Reutersward, A. (1987). Effect of cooking on fat content of beef and pork. *33rd International Congress of Meat Science and Technology Proceedings*, Helsinki, pp. 203–7.

Judge, M. D., Haugh, C. G., Zachariah, G. L., Parmelee, C. E. & Pyle, R. L. (1974). Soya additives in beef patties. *J. Food Sci.,* **39,** 137–9.

Kauffman, R. G. & Breidenstein, B. C. (1983). A red meat revolution: Opportunity for progress. *Food and Nutr. News,* **55,** 21–4, 28.

Kempster, A. J., Cuthbertson, A. & Harrington, G. (1976). Fat distribution in steer carcasses of different breeds and crosses: distribution between depots. *Anim. Prod.,* **23,** 25–34.

Kempster, A. J., Cuthbertson, A. & Harrington, G. (1982). *Carcass Evaluation in Livestock Breeding, Production and Marketing.* Granada, London.

Kies, C. & Fox, H. M. (1973). Effect of varying the ratio of beef and textured vegetable protein nutritive value for humans. *J. Food Sci.,* **38,** 1211–13.

Kim, E. K. (1989). Contribution à la connaissance de la fraction lipidique du muscle chez le poulet. Thèse, Université de Nantes.

Kim, E. K. & Gandemer, G. (1987). Lipids and PUFA contents of muscle and skin of chicken. Influence of anatomical location. *33rd International Congress of Meat Science and Technology Proceedings*, Helsinki, pp. 213–16.

Kim, E. K. & Gandemer, G. (1988). Influence d'une cuisson de type rôti sur la composition lipidique des muscles et de la peau chez le poulet. *Viandes et Produits Carnés,* **9,** 235.

Kinsella, J. E. (1978). Texturized proteins: Fabrication, flavouring and nutrition. *Crit. Rev. Food Sci. Nutr.,* **10,** 147–207.

Kotula, A. W. Twigg, G. G. & Young, E. P. (1976). Evaluation of beef patties containing soy protein, during 12-month frozen storage. *J. Food. Sci.,* **41,** 1142–7.

Kregel, K. K., Prusa, K. J. & Hugues, K. V. (1986). Cholesterol content and sensory analysis of ground beef as influenced by fat level, heating and storage. *J. Food Sci.,* **51,** 1162–5.

Laroche, M. (1988). Personal communication.

Lawrie, R. A. (1961). Systematic analytical differences between *psoas major* and *longissimus dorsi* muscles of cattle. *Brit. J. Nutr.*, **15,** 453–6.

Lawrie, R. A. (1985). *Meat Science,* 4th edn. Pergamon Press, Oxford.

Lawrie, R. A., Pomeroy, R. W. & Cuthbertson, A. (1963). Studies on the muscles of meat animals. 3: Comparative composition of various muscles in pigs of three weight groups. *J. Agric. Sci.*, **60,** 195–209.

Lawrie, R. A., Pomeroy, R. W. & Cuthbertson, A. (1964). Studies on the muscles of meat animals. 6: Comparative composition of various muscles in boars of two weight groups in relation to hogs. *J. Agric. Sci.*, **63,** 385–6.

Liener, I. E. (1981). Factors affecting the nutritional quality of soya products. *JAOCS,* **58,** 406–15.

Link, B. A., Bray, R. W., Cassens, R. G. & Kauffman, R. G. (1970). Lipid deposition in bovine skeletal muscle during growth. *J. Anim. Sci.*, **30,** 6–9.

Lusas, E. W., Lawhon, J. T., Clark, S. P., Matlock, S. W., Meinke, W. W., Mulson, D. W., Rhee, K. C. & Wan, P. J. (1977). Glandless cottonseed (flour): Its significance, status, and prospects. *Proc. Glandless Cottonseed Conf.* Dallas. USDA Agricultural Research Service, p. 31.

March, R. A., cited by Giddey, C. (1983). Phenomena involved in the 'texturization' of vegetable proteins and various technological processes used. In *Plant Proteins for Human Food*, eds C. E. Bodwell & L. Petit. Martinus Nijhoff/Dr.Junk, Publishers, The Hague, pp. 425–37.

Marinova, P., Renerre, M., Pinkas, A., Polikhronov, O. & Lacourt, A. (1985). Comparaison de la composition des carcasses et de quelques qualités de viande de buffles et de bovins. *31st Europ. Meet. of Meat Res. Workers Proceedings*, Albena, pp. 278–82.

Marion, J. E. (1965). Effect of age, dietary fat on the lipids of the chicken muscle. *J. Nutr.*, **85,** 38–44.

Michel, F., Thibault, J. F. & Pruvost, G. (1985). Procédé de préparation de fibres alimentaires et fibres obtenues. Brevet Français No. 8516748.

Moss, M., Holden, J. M., Ono, K., Cross, R., Slover, H., Berry, B., Lanza, E., Thompson, R., Wolf, W., Vanderslice, J. & Stewart, K. (1983). Nutrient composition of fresh retail pork, *J. Food Sci.*, **48,** 1767–71.

Mottram, D. S. & Edwards, R. A. (1983). The role of triglycerides and phospholipids in the aroma of cooked beef. *J. Sci. Food Agric.*, **34,** 517–22.

Murphy, E. W., Criner, P. E. & Gray, B. C. (1975). Comparisons of methods for calculating retentions of nutrients in cooked foods. *J. Agric. Food Chem.*, **23,** 1153–7.

Niinivaara, F. (1985). Round-table held at the 31st Europ. Meet of Meat Res. Workers. *Fleischwirtsch.*, **65,** 1480.

Noda, I., Sofos, J. N. & Allen, C. E. (1977). Nutritional evaluation of all-meat and meat-soy wieners. *J. Food Sci.*, **42,** 567–9.

Noël, P. & Goutefongea, R. (1985). Saucisses du type francfort à base de protéines végétales: Stabilité et coloration de l'émulsion. *31st Europ. Meet. of Meat Res. Workers Proceedings,* Albena, pp. 786–9.

Noël, P., Culioi, J., Melcion, J. P., Goutefongea, R. & Coquillet, R. (1984). Coloration par la nitrosylhémoglobine de protéines végétales texturées par cuisson-extrusion. *Lebens. Wiss. Technol.*, **17,** 305–10.

Ono, K., Berry, B. W., Johnson, H. K., Russek, E., Parker, C. F., Cahill, V. R. & Althouse, P. G. (1984). Nutrient composition of lamb of two age groups. *J. Food Sci.,* **49,** 1233–9.

Ono, K., Berry, B. W. & Paroczay, E. (1985). Contents and retention of nutrients in extra lean, lean and regular ground beef. *J. Food Sci.,* **50,** 701–7.

Perchais, C. (1984). Personal communication.

Pinkas, A., Marinova, P., Tomov, I. & Monin, G. (1982). Influence of age at slaughter, rearing technique and preslaughter treatment on some quality traits of lamb meat. *Meat Sci.,* **6,** 245–55.

Pinkas, A., Marinova, P. & Monin, G. (1983). Influence of age on growth dynamics of muscle, fibres, their metabolic type and meat quality in sheep. *29th Europ. Meet. of Meat Res. Workers Proceedings,* Parma, pp. 140–5.

Pinkas, A., Marinova, P., Stoykov, A. & Monin, G. (1985). Influence of breed and halothane sensitivity on some traits of pig muscle. *31st Europ. Meet. of Meat Res. Workers Proceedings,* Albena, 274–8.

Renand, G. (1983). Effets de la sélection sur l'adiposité chez les bovins. *Rev. Fr. Corps Gras,* **30,** 7–12.

Renk, B. Z., Kauffman , R. G. & Schaefer, D. M. (1985). Effect of temperature and method of cookery on the retention of intramuscular lipid in beef and pork. *J. Anim. Sci.,* **61,** 876–81.

Rhee, K. C., Kho, C. K. & Lusas, E. W. (1981). Texturization. In *Protein Functionality in Foods,* ACS Symp. Ser. 147. American Chemical Society, Washington, pp. 51–8.

Ricard, F., Leclercq, B. & Touraille, C. (1983). Selecting broilers for low and high abdominal fat: distribution of carcass fat and quality meat. *Brit. Poultry Sci.,* **24,** 511–16.

Rizk, S. W. & Clydesdale, F. M. (1985). Effect of organic acids in the in vitro solubilization of iron from a soy-extended meat patty. *J. Food Sci.,* **50,** 577–81.

Robelin, J. (1978). Répartition des dépôts adipeux chez les bovins selon l'état d'engraissement, le sexe et la race. *Bull. Tech. CRZV Theix-INRA,* **34,** 31–4.

Schimann, C. & Goutefongea, R. (1973). Variations de la composition du jambon de Paris en fonction du lieu de prélèvement. *19th Europ. Meet. of Meat Res. Workers Proceedings,* Paris, pp. 487–500.

Siegel, D. G., Church, K. E. & Schmidt, G. R. (1979). Gel structure of nonmeat proteins as related to their ability to bind meat pieces. *J. Food Sci.,* **44,** 1276–9.

Smith, G. C., Marshall, W. H. & Carpenter, Z. L. (1976). Textured soy proteins for use in blended ground beef patties. *J. Food Sci.,* **41,** 1148–52.

Souci, S. W., Fachmann, W. & Kraut, H. (1981). *Food Composition and Nutrition Tables,* 2nd edn. Wissenschaftliche Verlagsgesellschaft mbH, Stuttgart.

Terrell, R. N., Suess, G. G. & Bray, R. W. (1969). Influence of sex, liveweight and anatomical location on bovine lipids. 2: Lipid components and subjective scores of six muscles. *J. Anim. Sci.,* **28,** 454–8.

Touraille, C., Ricard, F. H., Kopp, J., Valin, C. & Leclercq, B. (1981). Qualité du poulet. 2: Evolution en fonction de l'âge des caractéristiques physicochimiques et organoleptiques des viandes. *Arch. Geflügelk.,* **45,** 97–104.

Valin, C. & Goutefongea, R. (1982). Qualité des viandes de taurillons: la com-

position des muscles. *Bull. Tech. CRZV Theix-INRA,* **48,** 30–4.
Valin, C., Palanska, O. & Goutefongea, R. (1975). Etude de la qualité des viandes de bovin. 1: Etude biochimique de la maturation des viandes de taurillon. *Ann. Techn. Agric.*, **24,** 47–64.
Wangen, R. M., Marion, W. W. & Hotchkiss, D. K. (1971). Influence of age on total lipids and phospholipids of turkey muscle. *J. Food Sci.*, **36,** 560–2.
Wirth, F. (1988). Technologies for the manufacture of fat-reduced meat products. *Fleischwirtsch.*, **68,** 160–4.
Wyszecki, G. & Stiles, W. S. (1982). *Color Science: Concepts and Methods, Quantitative Data and Formulae,* 2nd edn. John Wiley & Sons, New York.

Chapter 10

Marketing Procedures to Change Carcass Composition

A. J. KEMPSTER
Meat and Livestock Commission,
Milton Keynes, UK

INTRODUCTION

The object of this chapter is to examine how marketing procedures can improve carcass composition, and the problems involved in making the necessary changes. I will focus on the EEC situation as an example, referring to the situation in other countries as appropriate.

The success of any marketing procedure for improving carcass composition depends ultimately on three factors:

(1) The accuracy with which carcass composition can be measured under commercial circumstances (carcass classification);
(2) The size of the price differentials applied to different types of carcasses (grading); and
(3) The interaction of the types of carcass and price differentials with other factors which affect the overall profitability of meat production.

Carcass classification and grading tend to be species-specific and will be considered separately for the pig, beef and sheep industries. The interactions with producer profitability will be considered in general terms because the principles are similar for the three species.

The third factor can be very important because many things besides price differentials affect profitability and producer decisions. In some cases these can be directly antagonistic to the production of leaner carcasses (for example, the need to maximise unit carcass weight to cover breeding herd costs in the beef suckler industry); in others they can accelerate the trend towards leaner carcasses (as, for example, does the positive association between leanness and the efficiency with which feed is converted into meat in the pig industry).

CARCASS CLASSIFICATION AND GRADING

The Rôles of Classification and Grading

The principle of classification is that carcasses are described, using a common language which is understood by everyone trading in the market. The language relates to characteristics of commercial importance to traders but does not normally impute value, the essence being that the individual trader is able to identify the relative value of different classes for his particular business and specify price premia (grade) accordingly. This encourages the production of carcasses more suited to market requirements. (The principles and definitions involved in classification and grading were discussed by De Boer, 1984.)

Classification followed by individual trader grading is a relatively new development. Grading schemes developed early in the major meat exporting countries of the world, providing carcass description to facilitate trade at a distance. The foremost examples are Danish bacon and New Zealand lamb, for both of which the United Kingdom was the dominant market, although the same principles applied to a lesser extent to the export of beef from Australia and Argentina. They were normally the responsibility of statutory bodies set up to control or influence trading. The schemes quickly took on a promotional significance. In the USA carcass grading arose from the recognition in the 1920s that effective price reports were dependent on a uniform grading system. Price was highly dependent, especially for beef, on the association with particular breeds, fed in particular ways. There was strong pressure to base the grades on criteria closely related to eating quality, particularly marbling and maturity and to structure them in such a way that only beef considered of high quality would achieve the higher grades to which names attractive to the consuming public could be applied. These developments were controversial and still cause fierce debate today with the increasing conflict between leanness and 'quality'. The historical development of grading and classification schemes was reviewed by Kempster *et al.* (1982).

With the exception of the trade in bacon from Denmark, the same incentives to create grading systems have not, until recently, existed in Europe. In Britain, for example, the operation of market support schemes with a quality standard has added another dimension: 'grading' came to be thought of as a process carried out by government employees or their agents as an integral part of the market support system, rather than as a device to facilitate trading or provide market information to producers. Grading has not been seen to have a rôle in promotion or for consumer information.

Developments in carcass classification and grading have been slow in many countries. Despite the laudable objectives, that improvements in production and distribution efficiency depend on producers being given strong signals about the type of carcass in demand by the wholesale trade, the subject has been a controversial one. To many academics and economists, it seems self-evident that the efficiency of the distribution system must be open to improvement if the basic produce being traded—the carcass—is described in a uniform manner understood by all parties. The fact that such a description can never describe the infinite variations of carcass and meat qualities, or even the factors which would be brought into his assessment by any individual buyer, is not seen as negating the basic principle. Benefits are seen to accrue even when the quality of the information falls short of perfection, provided the net benefits in improved efficiency exceed any costs added to the distributive process by the introduction of grading. (Cost–benefit analysis has been attempted by several people—for example, Griffith (1976).)

While some traders can see benefits to their own business in such carcass description, many still remain sceptical. They suspect that the short-term added costs will fall on them, while the benefits, if any, are long term and realised elsewhere. Perhaps, most important of all, they are concerned that their individual skills in sorting, allocating and the associated bargaining will be devalued and even, in due course, superseded. More realistically, economic circumstances will dictate that the traditional skills in disposing of variable supplies to customers with varying demands and varying degrees of market power will have to be replaced gradually by new skills of minimising variability and assuring the retail buyer and his customers that tighter control of the product itself has been achieved. In these circumstances, effective methods of carcass description will prove a valuable component but not the 'be-all and end-all' of the process.

The European aspect of this subject is now dominated by the Common

Agricultural Policy of the EEC. The systems of supporting the market for beef, pig meat and sheep involve standardised methods of reporting national prices to the Commission headquarters, so that support measures can be triggered. A common pig grading scheme has been used for price reporting since the early 1970s and a common beef scheme was introduced in 1981. While it has been the somewhat idealistic intention of EEC officials that these schemes should also be used as a common basis of trading throughout the Community, such a development is still some years away even for pigs.

Pig Classification and Grading

Grading and classification schemes for pig carcasses throughout the world now tend to concentrate on grouping carcasses by their estimated lean content. This is because the pig has historically been a fat animal and the need to increase leanness has been a dominating factor. Meat quality factors, such as colour and wetness of the lean, and softness and oiliness of fat, are sometimes taken into account, but have not had the same importance relative to compositional variations as they have for beef, where fatness has not been such an important issue. This concentration on leanness and the application of substantial price differentials against fatness have contributed considerably to the success of pig industries of many countries in meeting consumer demand for leaner meat. Where success has been less evident, for example in the USA, other factors, such as the predominance of live weight selling and a sophisticated packing industry able to use raw material effectively whatever the composition, have mitigated against reductions in fatness. In contrast to the USA, Canada introduced an indexed pricing system based on estimated meat yield in the late 1960s and major improvements in carcass leanness have been achieved (Fredeen, 1976).

The EEC Grading Scheme has evolved greatly since the accession of Denmark, Eire and UK to the Community in 1971 and is now a model international system. The latest revision to the Scheme was implemented in all member states in January 1989. The revised scheme aims to improve the accuracy and consistency of application of commercial value across member states. The scheme is compulsory (except for abattoirs slaughtering fewer than 200 pigs per week). It embodies the following principles: (a) a standard carcass description, (b) individual carcass identification and (c) the use of objective measurements to predict the lean meat content of the carcass (which determines the grade). The standard carcass description is an important feature of any classification scheme because minor differences in the way carcasses are dressed can have an important effect on carcass value and override price differentials due to quality. The lean meat content (or the equivalent grade) has to be recorded either on the carcass or the documentation to provide feedback to producers and information to retailers. The objective methods for predicting carcass composition must comply with a minimum level of precision before they can be approved for use. Different methods are being employed in different countries but they must all relate to a common definition of carcass lean meat content. (Details of the methods employed and the problems in standardisation between countries were discussed recently by Kempster & Cook (1989).)

At present, the arrangements above have no statutory influence on the price paid for carcasses; this is determined by agreement between buyer and seller. Consequently the decisions on the measurements used for payment purposes will continue to depend on local considerations and domestic agreement within each country. But in most countries there are strong price differentials against fatness: these are typically 1·0–1·5 p/kg carcass weight for each additional one percent estimated lean in the carcass (the price relationships in the principal pigmeat-producing countries were compared by the Meat and Livestock Commission (1987)). This has resulted in significant increases in carcass lean content. In the UK, for example, the lipid content of the average pig carcass has been reduced by an estimated 66 g/kg over the past ten years (see Table 1).

As pigs have become progressively leaner, there has been increasing concern about a possible deterioration in meat quality. For many years pale, soft, exudative muscle (PSE), associated with porcine stress syndrome, has been the most important meat quality problem because of its obvious effect on commercial value due to reduced water holding capacity, although in the UK, and to a lesser extent in other countries, interest is now shifting towards presentational and eating quality problems associated with overleanness (Kempster & Wood, 1987).

There is a strong interest, therefore, in the measurement of meat quality (PSE; dark, firm, dry muscle (DFD); soft fat, etc.) routinely in the abattoir and several instruments are now available for doing this. Major developments are also taking place in the field of sensor technology which are likely to provide a more accurate estimation of these characteristics (Schaertel & Firstenberg-Eden, 1989). It is difficult to predict how these developments will influence grading but it is likely that carcasses with

Table 1
Estimates of the national average composition of British cattle, sheep and pig carcasses[a] in 1975/77 and 1987

	Cattle[b]		*Sheep*		*Pigs*	
	1975	*1987*	*1977*	*1987*	*1975*	*1987*
Carcass weight (kg)	249	278	17·5	17·2	61·3	62·5
Lean in carcass (g/kg)	602	607	559	562	476	564
Separable fat in carcass (g/kg)	258	254	259	254	274	192
Lipid in carcass (g/kg)	202	196	235	232	223	157

[a] Definitions of carcass as in normal commercial practice. Pig carcass includes head.
[b] Including perinephric and retroperitoneal fat.
Source: Kempster, *et al.* (1986), updated using 1987 classification results.

deficiencies will be culled from higher grades and rejected from quality assurance schemes. Some of the issues involved here and requirements for future research were discussed by Kempster (1989).

The development of techniques for measuring the concentrations of androstenone and skatole in pig fat, which are responsible for unpleasant odour (boar taint), requires special mention in this context. Entire male pigs produce lean meat about 20% more efficiently than castrated males and their carcasses typically contain 3 percentage units less fat, yet castration is practised in many countries because of fears that the meat may be tainted. (The extent to which entire males are used in European countries is shown in Table 2.) On-line methods are available for measuring skatole and will almost certainly be introduced in Denmark in 1990 and possibly some other countries over the next five years. Similar equipment for measuring androstenone is being developed. The ability to screen out tainted carcasses will lead to the greater use of entire males and a dramatic increase in carcass lean content. (The pros and cons of using entire males were discussed by Kempster (1988).)

Beef Classification and Grading

Methods of measuring beef carcasses in classification and grading have followed one of two pathways. The USA and Canada have focused on the measurement of the eye muscle and overlying fat on the exposed surface

Table 2
A review of the situation on the slaughter of entire males in some European countries, 1989

	Pigs slaughtered (million)	*Average live weight (kg)*	*Percentage of male pigs left entire*	*Industry interest in increasing the use of entire males*
Denmark	15	100	1	High
UK	14·5	85	90	
Eire	1·5	90–95	> 95	
Netherlands	21	105	2	Moderate[a]
Norway	1	100	< 1	None
Spain	16	95	90	
Sweden	4	110	< 1	None
West Germany	38	110	< 1	None
Yugoslavia	5	100–110	0	Low
Greece	2	90	100	

[a] Interest if the percentage of tainted carcasses could be reduced.
Original source: Lundstrom, *et al.* (1985), updated at the EAAP Working Group Meeting on entire males (Gerona, Spain, 1989).

of the cold ribbed carcass, whereas European countries have tended to use visual assessments of external fat cover and conformation of the hot carcass.

No carcass grading scheme has had more influence on the industry than that for beef in the United States of America. The country has a massive beef industry and *per capita* consumption of beef is high, while the increasing expenditure on beef has been seen by many as an index of afluence and, indeed, of national prosperity. This has ensured that the grading scheme itself has been in the forefront of industry politics. The scheme has been widely researched and comprises both yield grades (carcass meat yield) and quality grades (predicted eating quality). The higher quality grades are usually associated with high levels of fatness and commonly embrace beef cattle from high-energy feeding systems.

Yield grades are estimated using a prediction equation involving fat thickness over the eye muscle (*m.longissimus*), eye muscle area, the percentage of kidney knob and channel fat, and carcass weight. There is an element of visual assessment in this because fat thickness is adjusted by eye for unusual fat distribution across the carcass, and there have been criticisms of the use of visual assessment. However, the equation has stood up well to examination over the years (Abraham *et al.*, 1980) and there is now support internationally for the view that the particular fat thickness used (12th rib at a position 75% of the eye muscle width) has special value. A comparable fat measurement is used in the Canadian beef-grading system.

Over the years there have been certain changes in the United States Department of Agriculture (USDA) beef quality standards, with increased pressure from cattlemen to lower the fatness requirements of the top quality grades because of the feed costs of producing fat cattle. Most of the proposed changes over the years have been aimed at reducing the importance of the major component, i.e. requiring less marbling (intramuscular fat) at a given level of maturity and diminishing, and eventually eliminating, the importance of conformation. Meat traders have generally tried to resist these changes. They have argued that any deterioration in eating quality will have a signficant effect on beef consumption, although at the kind of levels of fatness at which cattle are slaughtered in the States there is little research evidence to support such views.

A major review of dietary issues in the US, carried out by the National Research Council (1988) recommended that government play a more constructive rôle in encouraging consumers to reduce fat and cholesterol consumption. It proposed the uncoupling of the yield and quality grades to encourage the marketing of leaner cattle, and that excess carcass fat should be trimmed in the processing sequence before weight determination and grading (to penalise the producer of over-fat cattle). Changes to the nomenclature of grades have already been introduced: the new 'Select' designation established by USDA in November 1988 will allow consumers a greater range of low-fat beef under a more appealing name, while allowing producers to market lean beef more aggressively.

The USDA grading scheme, the last bastion of the 'fatness means quality' philosophy, has now been under siege for some time and a major breakdown of the system to one encouraging leaner beef production is now taking place.

Another reason for the possible breakdown of the USDA scheme is the need to accommodate major technological changes at the abattoir, such as hot boning. There is a good deal of interest in this technique, plus accumulating evidence that it could make a substantial contribution to increasing efficiency in the abattoir. But in the USA the major obstacle to its wider adoption is seen as the grading system requiring the measurement of cold ribbed carcasses, a clear example of the servant being the master. Nevertheless, there is increasing interest in the possible use of probes for measuring fat thickness on the hot carcass, stimulated particularly by research work in Canada. Recent developments in Canada were reviewed by Jones (1989). A major study has recently been carried out to compare the Canadian and US Beef Grading Systems in an attempt to understand them and to eventually enhance trade between the two countries (Talbot & Campbell, 1989).

Visual assessments of external fat cover and conformation have played a more important rôle in the development of grading systems in Europe. A group of technical experts from various countries agreed a standard method of assessment which was published in 1974 by De Boer *et al.* The group concentrated on achieving comparability in judgements, with particular reference to the distinction between muscularity (thickness of muscle in relation to skeletal size); fleshiness (thickness of muscle plus intermuscular fat in relation to skeletal size); and conformation (thickness of muscle plus fat in relation to skeletal size). This was done by reference to photographic standards for 15-point scales for both fat covering and fleshiness (5 basic classes, subdivided with −, 0 and +).

Given this background work by scientists, the development of a common classification scheme for the market management purposes of the EEC's Common Agricultural Policy should have been facilitated. Unfortunately, however, by the time the EEC project began (1979), several member states had well-established systems of their own. No-one likes to change a working method, and the EEC Classification Scheme was set up in such a way as to integrate existing systems (Commission Regulation 1208/81). It is based on separate classifications of visual conformation and fat cover, in both cases in five basic classes (E, U, R, O and P for conformation; 1 to 5 for fat cover). For domestic purposes, member states may subdivide the basic classes into three. The main challenges for the EEC Scheme in the 1990s are still to achieve standardised application across member states and replace the visual assessment with more accurate objective measurement. Kempster *et al.* (1984) have argued that greater flexibility will be required to encourage and facilitate the uptake of new objective measurement techniques, and that in particular this requires a base line in terms of carcass lean content rather like the EEC Pig Scheme. As far as the author is aware, no progress has yet been made on this issue.

The significance of the EEC Scheme in Britain exemplifies the situation. Its use remains low, with less than 45% of slaughterings being classified. This figure reflects the fact that some 50% of all cattle are sold in the live auction market. The perceived advantages of the auction market system include clear evidence of price competition, avoidance of contentious 'grading' at a distant plant and immediate payment. The method of operating the Special Beef Premium scheme, now modified, has also encouraged live-weight marketing.

But live-weight marketing blurs the price differentials for quality from the meat trade because assessment of composition is less accurate on the live animal than the carcass, and other factors such as killing-out percentage confuse the issue. In view of the continuing importance of live auction market selling, there is now strong interest in the development of a live classification scheme for cattle using nomenclature similar to that in the EEC carcass grading scheme.

But even when cattle are sold on a dead-weight basis, strong price differentials against fatness are uncommon because many traders still see fatness as being important for eating quality (Table 3 shows typical price differentials for the EEC Scheme as applied in Britain and differences in meat yield between the classes). The lack of strong price differentials has contributed to the fact that there has been little or no reduction in the

Table 3

Saleable meat in carcass (%) for carcasses in different fatness×conformation classes the EEC Beef Classification Scheme, and observed and estimated price differentia between classes in Great Britain

		Fat class				*At fat class 4L*	
		3	*4L*	*4H*		*Observed differential*[a]	*Estimated differential*[b]
Conformation class	U+	73·2	72·3	70·9	U+ 4L		
	U	73·3	71·6	70·7			
	R	72·4	71·1	69·8		6p	7p
	0+	71·9	70·4	69·0	0 4L		
	0−	71·2	69·4			5p	4p
		R3		R4H	0− 4L		

Observed differential[a] at R = 4p.
Estimated differential[b] at R = 12p.

[a] Observed differentials for steers from the deadweight price reporting sample (198 figures for steers in the 250–299·5 kg carcass weight band).
[b] Differentials estimated from the value differences in saleable meat percentage (b taking no account of differences in meat quality).

average fatness of beef carcasses over the past ten years (Table 1).

By contrast the price differentials for conformation are more in line with estimated meat yield, carcasses with better conformation having a lower bone content and hence a higher meat yield. In these circumstances conformation is effectively identifying breed differences. Historically, cattle breeding has been dominated by size and shape. For centuries, visual assessments (and to some extent handling methods) were the only means used to identify differences between animals, and indeed they still play a major role. Shape has been considered very important, its influence stemming from the aesthetic appeal of certain types of stock. In the last century, the traditional British beef breeds—Hereford, Aberdeen–Angus and Beef Shorthorn—were all altered in shape to fit man's idea of what a 'beef' animal should look like. In making this change to blockier animals (and, at the same time, smaller and fatter animals), breeders, producers and meat traders have come to believe that these shapes are associated with more meat and better quality meat. Any movement away from this ideal has been seen by many in the beef industries of the English-speaking world as a step backwards which should be avoided. However, the traditional views have now been challenged by consumer demand for lean meat at an acceptable price and by the requirements of meat manufacturers for large, lean carcasses which are efficient to process. The conflict of views is seen in the arguments which flare up regularly in the farming and meat-trade press. The traditional view of shape and the demand for size and leanness are finding common ground in the muscular Continental breeds—in particular Charolais, Limousin and Simmental—which combine good conformation, size and leanness. Crosses by these breeds are now well suited to the buying schedules of most forward-looking meat wholesalers and retailers demanding high meat yield, and result in appropriate price differentials.

Sheep Classification and Grading

The international development of sheep carcass classification and grading schemes shows some parallels with beef. Visual fat and conformation assessments have dominated, except in New Zealand where fat thickness measurements have been used for a number of years. However, less importance has been attached to fatness from an eating quality viewpoint and more attention to conformation reflecting higher meat yield and more attractive cuts. In the same way that it has for beef, good conformation has sometimes been confused with fatness and reduced the pressure that could be applied to producing leaner carcasses (Kirton & Johnson, 1979; Kempster, *et al.*, 1981).

There is as yet no EEC dimension to sheep carcass classification, although discussions have been taking place in Brussels. The British classification scheme, operated by the Meat and Livestock Commission, was introduced about 10 years ago and is fairly typical of those operating in other countries. The scheme is based on visual fatness and conformation assessments (details given by MLC, 1988). It now covers approximately 32% of the national kill.

The British sheep industry suffers from the same marketing problems as the beef industry, with a high dependence on auction markets. Approximately 70% of lambs are sold by live weight and surveys by MLC in auction markets showed little price differential between fat classes. As in several other countries, production in Britain has been slow to respond to the demand for leaner carcasses. Table 1 shows that there has been little change over the past ten years. While there has been some reduction in the number of very fat carcasses (fat classes 4 and 5 in the MLC Sheep Carcass Classification Scheme), there has been little alteration in the balance of carcasses in the mid to lower fat classes where the majority lies.

However, there are indications that those wholesalers now using the classification scheme are applying realistic differentials, and a wide range of grade schedules and price differentials are beginning to emerge, stimulated by the development of quality assurance schemes. Premiums for conformation differ substantially between traders. They are particularly high for those exporting lambs to Germany, France and Belgium where good conformation carcasses are in especially high demand.

Alternative methods of estimating carcass composition are now being explored. Results to date suggest that the scheme could be improved by using a combination of visual scores and fat measurements taken by probe (Fisher, 1990).

WHAT SUITS THE PRODUCTION SYSTEM OR WHAT THE MARKET WANTS

The Interaction of Price Differentials with Other Factors Affecting Producers' Marketing Decisions

The meat industries in the Western world are characterised by increasing consumer emphasis. We are in a period of market reorientation and there is probably more discussion about marketing strategies today than at any time in the past 20 years. With everyone talking about the need for better marketing, it is easy to overstate the importance of quality price premiums in overall producer profitability. Price differentials, particularly for cattle and sheep, are generally small and marketing usually ranks low in comparison with production factors in terms of their contribution to profitability.

If every livestock producer sold his finished cattle or sheep as retail beef or lamb through his farm shop he would be in no doubt about the economic significance of carcass quality and meat yield, and he would have a powerful reason for doing something about it, namely his own financial incentive. However, practice is very different to this. Often, current market systems in many countries, based largely on bid price per kg live weight or negotiated price per kg dead weight, fall seriously short in getting a clear message about consumer requirements through to the producer. Changing consumer demand does exert some influence but the market signal is weakened or confused by many other factors before it reaches him. In Britain for example, the average number of cattle sold for slaughter by individual beef producers per annum is remarkably small (about 20), reflecting the subsidiary nature of the beef enterprise for a majority of producers. With such numbers, it is not surprising that the selling of cattle is of an opportunist nature for most. Table 1 shows the marked contrast between cattle and pigs in changes in carcass composition over the past ten years. However, in the case of cattle, there has been an increase of 20 kg in carcass weight. This suggests that improvements in breed and production systems (in particular the importation of the large, late maturing Continental breeds) has created the potential for leaner carcasses but that beef producers have chosen to exploit this by increasing output. In other words, carcass weight is a more significant factor in profitability than carcass composition.

This emphasises the importance of optimum slaughter point in any discussions about the production of leaner cattle. It is not difficult to effect a major reduction in fatness by restricted feeding and earlier slaughtering at lighter weights. But this is a nonsense if it is achieved at the expense of the efficiency of lean meat production. Herein lies a major constraint to the production of leaner meat.

There is an optimum slaughter point for current genotypes with existing production technology at which the overall cost of producing and distributing lean meat to the point of consumption is minimised. The optimum slaughter point is dynamic, changing with genetic progress and technological improvement. It is this change which dictates the pace of change in the production of leaner carcasses and not the demand for leanness *per se*. Consumer demand is only one of the factors in the overall balance of costs and returns. Indeed at present it is often more efficient to meet consumer demand by producing fatter carcasses and trimming the excess fat. For example, current estimates of the carcass weight at which lean pig meat is produced most efficiently is in the region of 65–75 kg carcass weight. Lighter carcasses are leaner but the lean has a higher production and distribution cost.

Retailer Requirements and Consumer Demand

Consumer demand for joints differing in fatness can be accomodated to some extent by the traditional butcher preparing individual cuts on demand. However, the gradual replacement of traditional meat retailing by self-service supermarkets in many countries and by specialist multiple retailers and independent butchers preparing large quantities of meat for display, gives more emphasis to the need for leaner meat. The experience of self-service stores in which pre-packs have to be severely and uniformly trimmed of fat, to avoid the risk of the pack being left at the end of the day, is now also common to their modern counter service competitors. They are, thus, becoming increasingly aware of the economic advantages of carcasses with high saleable meat percentages, are becoming more and more demanding when selecting carcasses and are increasingly unwilling to pay the same price for lean and fat carcasses. This will lead to greater demand for classification and grading and tighter specifications. Any significant deterioration in carcass and and meat quality will cause buyers to become more discriminating—the larger European multiple retailers have the power to tighten requirements and expect wholesalers to comply.

In the context of an overall strategy for changing carcass fatness, there are four key lean to fat ratios:

(1) lean to fat ratio when animals are slaughtered at the optimum slaughter point defined in terms of the overall efficiency of meat production (applies where no price differentials for quality exist);
(2) lean to fat ratio perceived by meat traders to provide best consumer satisfaction (often a lower ratio than the consumer actually wants because the meat trade overestimates the fatness required to ensure good eating quality);

(3) lean to fat ratio demanded by consumers;
(4) lean to fat ratio necessary to ensure good eating quality.

Each is obviously a distribution of ratios depending on the range of production systems involved and the range of meat trader and consumer judgements, but for convenience each can be considered as an average ratio.

The important thing is that the industry focuses on the crucial ratios (1) and (3) and puts (2) and (4) into perspective. The long-term objective then becomes clear: it is to match (1) and (3) by identifying breed stock and production systems which produce the ratios demanded by consumers at the optimum slaughter weight.

Market forces will tend to bring (1) and (3) together and production is likely to follow demands for: increasing leanness; increasing size; greater muscularity; youthfulness, with carcasses the most extreme in these respects being increasingly in demand to satisfy the lean meat market. The demand for fatter traditional beef will continue but this will almost certainly be a strictly limited market.

On this basis, it is possible to define a target area in the EEC beef classification grid where market demand will tend to be concentrated. The target area is illustrated in Table 4. Different wholesalers will use the scheme in a way which best meets their requirements, yet there is an overlapping of the top grades in the target area. For example, few British wholesalers would regard fat class 4L as too lean; even the high fat 'quality' specification (c) has a demand for some target area carcasses.

The extent to which producers should try to gear their production to the precise and tighter specifications and link in directly with the requirements of individual multiple retailers is a subject of debate. Beef production, being what it is, must inevitably produce a range of weight, fatness and conformation, and the industry will be fully stretched to increase significantly the proportion of carcasses which beef producers can get into a broad target area.

Taking the results of an MLC examination of classification records of 12 beef producers using Friesian steers in a very similar 18-month production system, only 58% of the cattle fell into the carcass weight range that embraces most multiple preferences (about 50 kg wide). Although they were tightly bunched for fatness, the variation in conformation was more than most multiples would wish. In terms of two individual specifications the proportions 'to spec' would have been 56% and 24%. So, although planned production systems and careful selection for slaughter will

Table 4

Target area and the preferences of different buyers

TARGET AREA

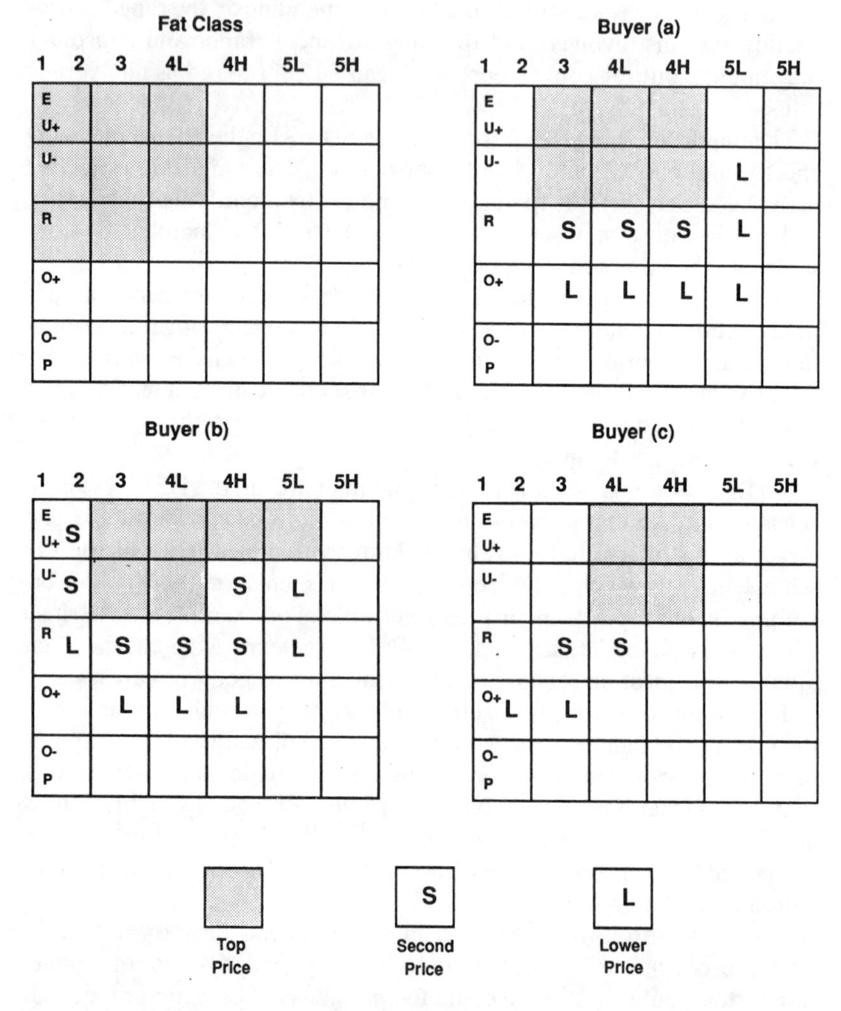

narrow the variation, it is still unlikely that more than two-thirds of the output of a group of large planned units will fall within the range of most multiples' specifications—let alone any one detailed specification.

Even pigs, which produce fairly uniform carcasses, are marketed to broad sector requirements rather than to the detailed requirements of individual retailers. There is a national bacon contract rather than a British Home Stores or a Marks and Spencer contract; pork dead weight schedules vary from buyer to buyer in detail but are not directly related to retailer specifications.

Provision of beef carcasses or cuts to a tight specification is primarily achieved by selection—at the farm, the market, the abattoir and in the cutting room—and is wholly dependent on others willing and able to utilise efficiently that which is not to specification.

EXPLOITATION OF INFORMATION TECHNOLOGY

The exploitation of information technology is likely to have a major impact on carcass classification and the feedback of carcass information in the future and to improve market transparency. Animal identification is particularly important so that carcass classification and grading results can be reported to the primary producer even when the livestock are sold through the live auction market.

The unique and tamper-proof identification of livestock from birth through to slaughter is becoming more important for reasons other than grading feedback. It is necessary in breed improvement schemes to follow progeny groups through the marketing chain (as attention continues to shift from growth performance towards carcass and meat quality characteristics), for disease control programmes and for quality assurance schemes guaranteeing that the animals have been produced under specific conditions (for example, without the use of growth promoters).

Major technological developments have taken place in the last ten years in electronic identification devices, and transponder tags are now being used widely in other industries: they are also being used in agriculture for identifying dairy cattle in dairy parlour automation schemes. Recent developments in microelectronics have led to very much smaller devices which are suitable for implantation in animals. Although their use in the livestock industry is feasible, the most suitable implantation sites are not established: the questions of animal welfare, the method of implantation and whether implantation could only be carried out by a veterinary surgeon also remain to be answered. Foolproof methods of ensuring that implanted devices are recovered before meat passes to the consumer are also required for obvious reasons. The issues involved in electronic animal identification were reviewed by Moncaster (1984).

Up until now transponders have been prohibitively expensive for livestock marketing functions, but the indications are that prices are beginning to fall because of competition between manufacturers. Several countries are now exploring the feasibility of electronic systems for national livestock identification, mainly for cattle, and there is a strong case for such an EEC initiative. International co-ordination is important because of the need to standardise equipment and the wider ramifications for classification and grading schemes and statutory market support schemes.

Another important development is electronic marketing. This is not a new concept but it has taken on new forms as telecommunications and data processing systems continue to improve. Several systems are up and running in Australia and North America. The most advanced systems involve computer-assisted remote bidding (CARB) in which livestock are bought unseen by buyers bidding via computer terminals in their offices, the sale being co-ordinated by a central computer (see for example, Bell *et al.*, 1983; Whan, 1988). In some circumstances the buyer may see a video film of the animals on offer or a declaration of their quality by independent assessors.

A key element of CARB systems is the link with classification information and the emergence of price differentials based on carcass classification. Bids can be based on assumed average quality with a retrospective adjustment determined by actual carcass classifications obtained after slaughter, in which case there must be rapid feedback of data from the abattoir to the organisation running the auction. Alternatively, a moving average system predicting the quality of a producer's pigs from his previous track record of classifications can be used; in this case, bid and price received would be based on estimated carcass classifications. Either CARB system can strengthen the use of carcass classification, relating it to competitive spot prices.

Important developments are now taking place in the area of automatic measurement and data processing on the slaughterline which are likely to make carcass classification and grading more effective in the future. Commercial applications are most advanced in pigs, with the use of reflectance-based recording probes for measuring fat and muscle thickness; and increasingly sophisticated process and yield control procedures (Clausen, 1984). Recent developments in on-line methods for measuring quality were reviewed recently by Sorensen *et al.* (1989).

Visual assessments continue to be an essential element of 'the state of the art' in beef and sheep classification and it is clear that they will be with us for some time to come. Developments in video image analysis and ultrasonics show considerable promise (Newman & Wood, 1989) but the difficulties involved in designing sophisticated measuring equipment for routine operation in the harsh abattoir environment should not be underestimated. The experience with automatic probes, which are technologically much simpler, indicates that the lead time between research prototypes and large-scale commercial application is substantial. To encourage and facilitate the use of new technology, greater flexibility will be required in the methodology of classification and grading schemes with classes or grades defined in terms of carcass composition. The main challenge is to make international classification schemes such as the EEC schemes receptive to innovation and the special needs of individual countries because of their market requirements or their particular livestock populations. If there is evidence of important differences between national livestock populations in the relationships between carcass composition and the measurements taken in classification schemes, specific prediction equations would be needed for these populations. The use of more sophisticated objective measurement techniques will facilitate such developments.

Developments in information technology are also leading to improved and speedier methods of data analysis and decision making and communication of requirements to the producer (for review, see Kempster, 1986). Although the abattoir sector in most countries has been slow to embrace this computer revolution, there are signs that attitudes are changing. There are several computer systems available which can provide highly effective data handling in abattoirs. Used in association with the new measuring equipment and electronic tags, which provide the means of linking the identification of live animals, carcasses and cuts at various points in the slaughtering/processing chain, the facility will exist to trace quality problems back to their origins efficiently and quickly.

Further integration of production and slaughter/wholesaling in the beef, pig and sheep industries, following the poultry industry's example, would also have major implications and, in particular, influence the strategies adopted in national classification schemes. But there are problems of producer independence. Producers may continue with their traditional independence, exercising their freedom to move their products to buyers according to short-term changes in the prices offered. If so they must accept that grading will be quick and cheap and, therefore, somewhat inaccurate. Alternatively, producers may become much more closely integrated with slaughterers and processors so that the buyer comes to know a great deal about each producers' stock as a result of detailed carcass and meat quality assessments. Information would then flow much more freely in both directions; each side could then appreciate the problems and the cost consequences of his actions. This implies the vetting of producers and their production practices in some detail in advance of purchase of their livestock. It also raises questions about the future of classification and grading as we know them today. Assessments made on the hot carcass at the classification/weigh point on the slaughter-line, or on the quartered carcass as in the USDA Grading Scheme, may be replaced by detailed measurements taken on cuts by quality assurance inspectors at the wholesale/retail interface. Electronic identification of cuts would relate them to the carcass from which they came and thence to the producer.

With the increasing emphasis on quality, there may be a reduced requirement for low fatness, and other more detailed factors affecting the value of carcasses to the processing industry and the retailer will become more important. The speed with which the emphasis changes will depend on the development and commercial application of new measurement techniques. Given the measurement capability, wholesalers are likely to react by introducing more demanding specifications, this trend being emphasised by the demands of the large multiples for consistency of product and facilitated by the data capture and processing revolution. This means that breeders and producers will become increasingly concerned with carcass and meat quality characteristics beyond weight and fat thickness.

The other factor which will have an increasingly important effect on the consumer/quality emphasis is the creation of the Single European Market. This will have a dynamic and self-perpetuating effect on the meat trade in Europe with trends towards fewer and bigger food organisations, a higher degree of market transparency, changes in the traditional channels of distribution, concentration of resources for development and research, and internationalisation of companies.

But an important question is how greater consumer awareness and more demanding schedules will interact with the overall efficiency of meat production in the medium term. It would not be in the best interests of the meat industry for the advances in technology to constrain producers to such an extent that there is a serious increase in cost per unit of output. Rigid specifications for fatness, muscle colour and texture will need to be tempered with an understanding of the importance of flexible production methods which allow optimum production efficiency. Buyers must also realise the effect they are likely to have on the overall distribution of carcass fatness and the impossibility of compressing variation below the inherent biological variation.

REFERENCES

Abraham, H. C., Murphey, C. E., Cross, H. R., Smith, G. C. & Franks, Jr, W. J. (1980). Factors affecting beef carcase cutability: an evaluation of the yield grade for beef. *J. Anim. Sci.*, **50,** 841–51.

Bell, J. B., Hendersen, D. R., Holder, D. L., Purcell, W. D., Russell, J. R., Sporleder, T. L. & Ward, C. E. (1983). *Electronic marketing—what, why, how?* A regional publication of Ohio State University, Oklahoma State University, Texas A. and M. University, Virginia Tech. and the Extension Service/USDA (Publication 448-004).

Clausen, V. (1984). Danish experience of on-line data capture and data processing. *Proceedings of the MLC Seminar 'Data Handling in Abattoirs'* held at the University of York, 1984.

De Boer, H. (1984). Classification and grading—principles, definitions and implications. In *Carcass Evaluation in Beef and Pork: Opportunities and Constraints.* ed. P. Walstra. Satellite Symposium EAAP, The Hague, Netherlands, pp. 9–20.

De Boer, H., Dumont, B. L., Pomeroy, R. W. & Weniger, J. H. (1974). Manual on EAAP reference methods for the assessment of carcase characteristics in cattle. *Livest. Prod. Sci.*, **1,** 151–64.

Fisher, A. V. (1990). New approaches to measuring fat in the carcasses of meat animals. In *Reducing Fat in Meat Animals.* eds J. D. Wood & A. V. Fisher. Elsevier Science Publishers, London, pp. 255–343.

Fredeen, H. T. (1976). Recent trends in carcass performance of the commercial hog population in Canada. *J. Anim. Sci.*, **42,** 342–51.

Griffith, G. R. (1976). The benefits of a national pig carcase measurement and information service. *Proceedings of the Carcase Classification Symposium,* Adelaide, May 1976.

Jones, S. D. M. (1989). Future trends in red meat processing—an overview. *Can. J. Anim. Sci.*, **69,** 1–5.

Kempster, A. J. (1986). A longer term view of technical developments. Paper presented at the MLC Conference 'Computer systems for meat plants; why and how?', Harrogate, October, (mimeo).

Kempster, A. J. (1988). Entire males for meat: facts and prejudices. *Pig Veterinary Society Proceedings,* **21,** 140–1.

Kempster, A. J. (1989). Carcass and meat quality research to meet market needs. *Anim. Prod.*, **48,** 483–96.

Kempster, A. J. & Cook, G. L. (1989). Errors in carcase lean prediction with special reference to the EC Grading Scheme. In *New Techniques in Pig Carcass Evaluation.* EAAP Publication No. 41. ed. J. F. O'Grady. Centre for Agricultural Publication and Documentation, Pudoc, Wageningen, pp. 28–36.

Kempster, A. J. & Wood, J. D. (1987). A national programme on factors affecting pig meat quality. In *Evaluation and Control of Meat Quality in Pigs,* eds P. V. Tarrant, G. Eikelenboom & G. Monin. Martinus Nijhoff, Dordrecht, pp. 359–69.

Kempster, A. J., Croston, D. & Jones, D. W. (1981). Value of conformation as an indicator of sheep carcass composition within and between breeds. *Anim. Prod.,* **33,** 39–49.

Kempster, A. J., Cuthbertson, A. & Harrington, G. (1982). *Carcase Evaluation in Livestock Breeding, Production and Marketing.* Granada, London.

Kempster, A. J., Cuthbertson, A. & Harrington, G. (1984). Beef carcass classification and grading—methods, developments and perspectives. In *Carcass Evaluation in Beef and Pork: Opportunities and Constraints,* ed. P. Walstra. Satellite Symposium EAAP, The Hague, Netherlands. pp. 21–9.

Kempster, A. J., Cook, G. L. & Grantley-Smith, M. (1986). National estimates of the body composition of British cattle, sheep and pigs with special reference to trends in fatness. A review. *Meat Sci.,* **17,** 107–38.

Kirton, A. H. & Johnson, D. L. (1979). Inter-relationships between GR and other lamb carcase fatness measurements. *Proc. NZ Soc. Anim. Prod.* **39,** 194–201.

Lundstrom, K., Malmfors, B., Vahlun, Sv., Kempster, A. J., Andresen, O. & Hagelso, M. (1985). Recent research on the use of boars for meat production—report from the EAAP Working Group in Denmark. *36th Annual Meeting of the European Association for Animal Production,* Halkidiki.

Meat and Livestock Commission (1987). *Pig Yearbook, 1987.* MLC, Milton Keynes, Bucks.

Meat and Livestock Commission (1988). *Sheep Yearbook, 1988.* MLC, Milton Keynes, Bucks.

Moncaster, M. E. (1984). Developments in live animal identification. *Proceedings of the MLC Seminar 'Data Handling in Abattoirs'* held at the University of York, January. MLC, Milton Keynes, Bucks.

National Research Council (1988). *Designing Foods: Animal Production Options in the Marketplace.* National Academy Press, Washington, US.

Newman, P. B. & Wood, J. D. (1989). New techniques for assessment of pig carcasses—video and ultrasonic systems. In *New Techniques in Pig Carcass Evaluation.* EAAP Publication No. 41 ed. J. F. O'Grady. Centre for Agricultural Publication and Documentation, Pudoc, Wageningen, pp. 37–51.

Schaertel, B. J. & Firstenberg-Eden, R. (1989). Biosensors in the food industry: present and future. *J. Food Protection,* **51,** 811–20.

Sorensen, S. E., Stoier, S., Andersen, J. R., Nielsen, T. & Olsen, A. E. (1989). Utilization of on-line methods for measuring quality parameters in carcasses—future possibilities and challenges. *Proceedings of the 35th International Congress of Meat Science and Technology.* pp. 16–29.

Talbot, S. & Campbell, C. (1989). *Characterization of Canadian and Beef Carcase Grading Systems.* Meat and Poultry Division, Agriculture, Canada.

Whan, I. F. (1988). Reform of the Australian beef cattlemarket. *Proceedings of the 34th International Congress of Meat Science and Technology, Industry Day.* Brisbane, Australia, pp. 50–4.

Index